Springer Praxis Books

Popular Astronomy

This book series presents the whole spectrum of Earth Sciences, Astronautics and Space Exploration. Practitioners will find exact science and complex engineering solutions explained scientifically correct but easy to understand.Various subseries help to differentiate between the scientific areas of Springer Praxis books and to make selected professional information accessible for you.
The Springer Praxis Popular Astronomy series welcomes anybody with a passion for the night sky. Requiring no formal background in the sciences and including very little to no mathematics, these enriching reads will appeal to general readers and seasoned astronomy enthusiasts alike.

Many of the books in this series are well illustrated, with lavish figures, photographs, and maps. They are written in a highly accessible and engaging style that readers of popular science magazines can easily grasp, breaking down denser aspects of astronomy and its related fields to a digestible level.

From ancient cosmology to the latest astronomical discoveries, these books will enlighten, educate, and expand your interests far beyond the telescope.

Marion Dolan

Decoding Astronomy in Maya Art and Architecture

Springer

Marion Dolan
Deerfield Beach, FL, USA

Springer Praxis Books
ISSN 2626-8760 ISSN 2626-8779 (electronic)
Popular Astronomy
ISBN 978-3-031-61091-2 ISBN 978-3-031-61092-9 (eBook)
https://doi.org/10.1007/978-3-031-61092-9

This Springer imprint is published by the registered company Springer Nature Switzerland AG
The registered company address is: Gewerbestrasse 11, 6330 Cham, Switzerland

Preface

Our understanding of the enormous accomplishments of the ancient Maya Civilization continually expands as archaeologists, anthropologists, and astronomers analyze the vast quantity and quality of their architectural, artistic and literary remains. The complex, brilliant cultures of Mesoamerica excelled in architecture, artworks, astronomy, mathematics, calendrics and literature. The Maya created the most sophisticated and highly developed writing system, which set them above all other cultures in Pre-Columbian America.

When writing an earlier book, *Decoding Astronomy in Art and Architecture*, the main focus was aimed at the astronomy of a variety of advanced cultures around the world. At that time I hoped to include a chapter discussing the astronomical knowledge that the ancient Maya encoded into their art and architecture. Upon reviewing my notes, books, and many scholarly journal articles on the ancient Maya, I realized that it was impossible to encapsulate the enormous accomplishments in astronomy achieved by the Maya to merely a single chapter. Their knowledge of astronomy was encoded not only into their art and architecture but also into every part of their daily life. Astronomy filled their mythology, their religion, their rituals, their kingship traditions and their cosmology.

This book presents an overview of established data as well as exciting new findings in Maya studies that reach across their wide geographic area of habitation and throughout their long historical period of urban development. This publication can only investigate a small sampling of the most famous and most visited Mesoamerican sites that have been carefully excavated by archaeologists and thoroughly studied by astronomers. These well-researched Maya cities and established beliefs discussed here cannot be dismissed as speculative for textual inscriptions verify the Maya's intentions. This work will examine the various ways that the Maya protected their vital astronomical knowledge by encoding it into their architecture, myths, rituals, histories and artworks. These different means of retention enabled their understanding of heavenly phenomena to be passed down to assist later generations. This survey is not intended for experts in these fields but instead for those readers interested in the history of astronomy and the history of art and architecture, for those who want to know how they are interrelated.

Observations of the night sky were enormously important to Maya societies. Their awareness of the apparent motions of the Sun, Moon, planets and stars is demonstrated in the abundant astronomical data found written in their codices, inscribed on their monuments and attested in their advanced calendric system. Their public and ceremonial buildings were usually oriented astronomically which helped to celebrate their ritual observances at appropriate times in the celestial cycles. The study of astronomical knowledge that was integrated into a society's buildings, temples, tombs and even whole city designs is called archaeoastronomy.

The science of archaeoastronomy is a relatively new science that was gradually accepted by a very dubious academic community. When studies of archaeoastronomical orientations in early, less developed societies were first published, the academic community was skeptical and slow in accepting that these more primitive cultures, even back to prehistoric eras, had the ability to understand the motions of the heavens. Originally the few archaeoastronomers studying this field focused mainly on megalithic structures and ancient architecture, such as Stonehenge, Avebury and Newgrange. Now this study of encoded astronomy has expanded widely, finding astronomy in ceramics, paintings, poetry, weaving, legends and myths.

Currently the field of archaeoastronomy has advanced across the divides of the past and has become an interdisciplinary field including contributions from a wide range of academic areas.

The overwhelming amount of scholarly literature focused on the Maya is now generated not only by archaeologists but also by astronomers, anthropologists, epigraphers, linguists and art historians. Maya researchers delve into more than the general history of the Maya; they have studied and written on their philosophy, mythology, cosmology, religion, mathematics and science. Experts in different fields now confer and work together in holistic ways to improve and combine their research for clearer insights into the ancient cultures. Scholars have rediscovered the crucial role astronomy played in the daily lives of the Maya. Anthropologists and ethnologists work with contemporary Maya people to gain understanding into their long-held traditions revealing a clearer idea of their values, traditions and belief systems. Some scholars believe that the contemporary Maya people hold the keys to a truer interpretation of the ancient Maya cosmology.

Because archaeoastronomy today delves into so many areas beyond their architectural ruins, most scholars refer to the study as cultural astronomy and sometimes ethno-astronomy, more comprehensive titles. These shared studies in cultural astronomy have contributed valuable data to each related area. A result of these interconnected studies has enabled scientific studies to accumulate slowly but accurately to enrich their common goal of providing a better understanding of human history through time.

Recent translations of Mayan glyphic texts offer a more precise understanding and provide extensive examples of their use of astronomical symbolism. The long, tedious efforts to decode the Maya hieroglyphs proved to be the prime factor that encouraged acceptance for the new field of archaeoastronomy. Once astronomers and epigraphers could accurately read the dates, the names and the historic events recorded on pyramids, tombs, temple walls and monolithic, no one could deny that the Maya had an accurate understanding of the movements of the heavens. The few Mayan books that survive are filled with observed data and astronomical charts. The ancient sky-watchers not only followed the nightly progression of the heavenly bodies, they included that celestial information into almost every aspect of their society. These ancient records together with archaeological evidence help explain the relationship between functional and symbolic astronomical knowledge.

This book is not meant to teach the technical aspects of archaeoastronomy, nor is it a manual for undertaking serious field studies. This text will discuss the fundamental information needed to grasp the abilities and celestial alignments of the Maya skywatchers. It will help readers understand the purpose of the monumental architecture created by the architects and artisans and explain the astronomical data that they incorporated into their designs. As a university professor teaching introduction to world art history, I would ask my students what they had learned about the Maya civilization in high school. Their answers varied from nothing, to almost nothing, to one paragraph. Students should be introduced to the ancient Maya's history, which is as essential as Greek and Roman history. The Maya should be admired and acknowledged for their 2000 years of outstanding achievements. This study's central concern is directed to the history and function of the astronomical content that has been found encoded into temples, tombs, paintings, engraved stelae and ceramics.

The second chapter of this monograph describes the grueling adventures and amazing discoveries of two very early visitors to the Maya ruins, John Lloyd Stephens (American) and Frederick Catherwood (British). They were the first English speakers to explore the lands of the ancient Maya and to discover many of their architectural ruins. The writings and illustrations of their incredible findings introduced the ancient Maya and also the contemporary Maya people to the rest of the world. Their two exploratory journeys in 1839 and again in 1841 were spent struggling through hundreds of miles of dense jungle overgrowth and fighting diseases, insects and nature, but they managed to locate and document more than 40 unknown Maya cities. Most settlements they uncovered were from the Classic Period, cities that thrived in the third to the tenth centuries. The explorers recorded and illustrated many inscribed monuments but these puzzling texts could not tell their ancient stories until the mysterious hieroglyphic inscriptions were at last deciphered, a long and tedious process. Their discoveries inspired many future expeditions into the tropical jungles which archaeologists still struggle with today. In spite of more than 100 years of investigation into the vast Maya lands, many unknown cities, pyramids and monumental sculptures still lie buried, swallowed whole by the impenetrable forest canopy.

The second chapter introduces the basic information necessary to gain a fundamental understanding of the ancient Maya Civilization, such as the relevant geographical areas within which they settled and a chronology of the culture's development and change through time. Their calendar, religious traditions, scientific knowledge and trade networks were not completely new undertakings since much had been inherited from centuries of contact with other Mesoamerican cultures, especially the Zapotecs of Monte Alban and the Olmecs who settled the Gulf Coast lowlands. Important influences from earlier societies will be discussed. The Maya were not the first society in Mesoamerica to create lofty pyramids and decorated temples, but their pyramids rival those of ancient Egypt in size, workmanship and in their accurate astronomical orientations. When first investigated, the architecture of the ancient Maya impressed archaeologists with their towering height, astonishing beauty and enormous size. Their buildings were all colorfully painted and decorated both inside and outside with symbolic images, meaningful numbers, historic dates and honored gods.

The third chapter explains the fundamentals of archaeoastronomy and cultural astronomy, introducing the reader to the type of information that has been gathered by archaeologists and archaeoastronomers and why it is important. The Sun's rising and setting positions on the horizon were an important factor in determining the desired alignment of Maya architecture. The key astronomical events to which buildings were oriented were the equinoxes, the solstices and the zenith passages, especially the Sun, Moon and Venus. The Maya realized that the five planets were distinct from the fixed stars and moved in different recurring cycles, which they carefully tracked. By far the most important planet for Maya astronomers was certainly Venus; they applied numerous names to the brightest planet according to its position as the Morning or Evening Star and its location in its 584-day cycle. The other planets were of lesser concern and are not mentioned nearly as often as Venus. Jupiter was more involved in Maya myths than the other three planets. Certain star patterns, such as the Big Dipper, Pleaides, Gemini, Orion, were recognized and watched as well, but they had their own names and numbers which they associated with these asterisms. The research of archaeoastronomers today is greatly assisted by the ability to accurately reproduce the night sky for any date in history; this can be determined with little effort. Computer programs now calculate any stellar or solar position on any past date and at any geographical location in an instant.

Chapter 4 focuses on the types of architecture that the Maya included in their cities along with the city's basic design. The meaning behind the structural form tells much about the Maya interests and values. Experts have determined that the Maya architectural and urban planning complied to a complex set of rules that incorporated astronomical factors. The functions of their architectural concepts were embedded in their cosmological understanding. Orientation to the Sun's position on the horizon was common and was intended to indicate scheduling of events and agricultural activities. Many Maya buildings were specifically aligned to record Venus' extreme positions in its celestial cycle. Some important buildings were oriented to particular features of the natural landscape, such as mountains, hills, cenotes and standing monuments. Architectural alignments corresponding to the major and minor extremes of the Moon (also called standstill positions) have been discovered in some cities. Structures along the northeast coast of the Yucatan peninsula, where a lunar cult was active, have shown significant lunar orientations, especially full Moon extremes.

Chapter 5 demonstrates that the mythology and cosmology of the Maya are rich in the personification of the forces of nature, the Sun, Moon, stars and planets, minor deities and heroes. These forces play the chief roles in their creation stories. The key figures of their mythological tales have been reconstructed chiefly by studying their artworks, personifications and texts. The myths define the ritual relationship between humans and the environment, both natural and supernatural. Their myths provide explanations for the origin of the heavenly bodies as well as human and animal life. Important spiritual forces in their myths are the sacred mountain, clouds, rain, thunder and lightning. Also important is their creation myth that is acted out in the Popol Vuh in a sequence of four efforts at creation. First were animals, next wet clay, wood and then last, the creation of the first ancestors from maize dough. These hero myths were performed and also colorfully painted on their various ceramic vessels.

The Maya created their own variations of the pantheon of gods and goddesses; the deities and heroes had many different names and various appearances. Their stories were repeated with changing names, shifting shapes and diverse depictions. Scenes in nature along with human and animal figures changed with confusing speed. But beneath this seeming confusion, a sense of an orderly, structured world and proper behavior toward the gods played an essential role in maintaining harmony and balance within the living community.

The more precise translations of Mayan script on monuments and in the few books that survive provides evidence of their consistent use of astronomical symbolism, especially in their myths and religious beliefs. These ancient records together with archaeological evidence help clarify the relationship between functional and symbolic astronomical knowledge. Both are encoded within their cosmological understanding. The glyphic texts offer graphic evidence that structures served as calendric markers and as sacred stages for seasonally timed rituals established through their reliance on cosmic connections.

The cyclical movements of Venus were thoroughly tracked and played a role in their myths. The symbolic relationship of Venus with rain and maize is manifested in many areas of Mesoamerica and in different time periods. The Venus-rain-maize connection was thought crucial to their agricultural life. This connection was associated mainly with the evening appearances of the bright planet. Honoring the god/planet with the proper ritual observances, performed by the king was vital to the production of abundant maize, necessary for survival of the Maya people. The mythical association of Venus-rain-maize was often pictured in iconographic sculptures applied to sacred temples and public monuments.

Chapter 6 discusses Maya literature especially in reference to astronomical almanacs and star charts. The Maya civilization was the only Mesoamerican culture that created a fully developed written language; it was produced in total isolation from Europe and from the rest of the world. Maya literature was not produced in a single language since 27 different languages were spoken within the wide-spread Maya territories. Many separate Mayan language variations are still spoken today but they are not connected with the Classic Mayan language which was a literary language used only by the elite, just as Latin was used by scholars and the Church for many centuries.

The Maya produced innumerable screen-fold books written and illustrated on bark paper. Unfortunately for historians, only four Maya codices have survived the passage of time, nor the humid, soggy climate of the tropics, nor the fervor of the Spanish clergy. None of these surviving books were written during the Classic period, the pinnacle of Maya culture. Scholars realize that ancient Maya books covered many more subjects than are represented in the four examples that survive. Not only are the survivors few in number, they all have numerous missing pages.

When considering the Classic period, the best-preserved samples of Maya writing are not found in books but were carved on their megalithic monuments, in particular around doorways and on their great rising stairways. Regrettably, the inscriptions on monuments were almost eliminated in the Classic cities during the ninth century. The textual inscriptions that more readily survive were those painted or carved on durable surfaces, mainly stone. Ceramic vessels also provide an important source of text since they often include images that help to explain what was written. Maya script has also been found on jade, shells, wood and bones.

The few Mayan books that survive often include important astronomical information. The Dresden Codex contains almanacs that track the sidereal motions of the Moon and the celestial patterns of Venus, the Great Star called Chak Ek'. This codex also includes eight pages listing the solar and lunar eclipses which were dreaded events, bad luck days. Astronomical data was also inscribed on monuments; for example, the inscriptions embedded on the tablets commissioned by Sun-Eyed Snake Jaguar at Palenque contain precise information about the periodicity of Jupiter, signifying that Jupiter was among the planets tracked in their almanacs.

In Chap. 7 the Preclassic period is described as well as the astronomical knowledge held by several Preclassic cities of the Maya. The dates of the original Maya occupations in Mesoamerica continue to be pushed back to earlier and earlier eras. The first known settlements now date to about 2600 BCE in Belize at which time they were already planting crops. By 1800 BCE settlers in Maya areas were producing pottery and fired clay figurines. By 750 BCE evidence of cities has been found, and by 400 BCE a developed hieroglyphic script was being inscribed on monuments and ceramics. Soon after that early date large standing stelae were being carved and raised. The earliest well known lowland city was Nakbe in the Peten, Guatemala. At the same time larger, more advanced cities were developing in the highlands as well.

El Mirador in Guatemala, north of Tikal, grew to be the largest Preclassic Maya site; the city flourished for over 500 years, from 300 BCE to CE 250. The complex city of El Mirador was founded in the center of a series of radiating causeways, reaching more than a mile from east to west, even larger than Tikal in size. The significant building complexes, pyramidal temples and individual structures of El Mirador were even greater than cities built in later periods of Maya history. The level of societal and political complexity of El Mirador was as great as cities such as the central Mexican city of Teotihuacan. Archaeoastronomical studies have indicated that many key structures of the Preclassic period were already oriented in specific relationships to the heavens, mainly to sunrises and sunsets on significant dates. The solar alignments of El Mirador represent the earliest of these observational alignments in the Maya area to date. These building orientations signify the importance of astronomical and calendrical factors in architectural design and even in city planning. Archeologists still hotly debate the different factors that caused the decline and abandonment of the Preclassic cities; they suggest it was connected to the rise of greater emerging powers, in particular the colossal city of Teotihuacan.

Chapter 8 examines the key cities of the Classic Period and surveys the astronomical information encoded into their artworks and architecture. During the Classic period (CE 250–900) the Maya civilization reached its peak, producing large-scale buildings and pyramids rising to the skies, inscribing enormous monuments and decorating their architecture with complicated iconographic artworks. The three major population centers of the Maya world during the Classic Period were Tikal in Guatemala, Copan in Honduras and Palenque in southern Mexico. These cities of the Classic Period grew dramatically in population density, reaching a remarkable size with 50,000 to 120,000 inhabitants. The important Classic cities formed communal networks of alliances for power, trade, expansion and warfare.

Tikal with its uniquely styled pyramids reigns as the largest and most extensively explored archaeological site in Guatemala. It has been recently shown that the ancient city was at least four times bigger than previously thought. As with most cities of the Maya, only a small portion of the ancient site has been uncovered. The city of Tikal was initially associated with the powerful city of Teotihuacan which held enormous influence over the Early Classic cities of the Maya. In CE 378 the ruler of Teotihuacan deposed the ruler of Tikal and other nearby cities and instead placed a new king on the throne, establishing a new dynasty. The leader of the takeover was Sivai K'ak' (Born of Fire); the political takeover is thought to have been a violent coup since the former king of Tikal died on that same day. A year later another new king was installed by Sivai K'ak'; this new king created an era of political dominance and a very successful reign. Tikal's leaders constructed a monumental city with numerous astronomically aligned buildings.

Copan too was an extremely important Classic city and was associated with Tikal in various ways. Copan grew in importance because of its trading network; many of its architectural structures are astronomically oriented and adorned with celestial symbolism. Admired for its abundance of ornate artworks, Copan structures portray the important myths of the ancestors and the cosmological beliefs of the Maya. Like many important Classic cities its decline began in the ninth century. After thriving for over 400 years, Copan began to lose its powerful position in the Maya economic world. One by one the long standing royal dynasties that had reigned in the larger Classic cities began to weaken and experience losses in their wealth and political status.

The great Maya city of Palenque was ruled by mighty kings; the most famous were Lord Pakal and his son, Chan Balum. They created the most impressive, astronomically aligned architecture, the Temple of Inscriptions, the Temple of the Sun, the Temple of the Cross and the Temple of the Foliated Cross. King Pakal's tomb revealed great riches and his symbolic tomb sculpture is considered one of the most impressive and important artworks.

During the failing Terminal Classic period, no new structures were undertaken and the great cities were gradually being abandoned. Subsequently, the cities no longer inscribed dates on monuments; the last known inscribed date is in 909. There is still no totally accepted explanation for the surprising collapse after such lengthy and successful regimes. Scholars lay the blame for the decline on multiple factors including weather shifts, drought, internecine warfare, overpopulation and the traditional system of rulership.

Chapter 9 considers the subsequent rise of the northern lowland cities in the Yucatan Peninsula, especially Chichen Itza which reigned supreme during the Terminal Classic period. The Postclassic era saw many momentous changes within Maya societies. As the major Classic cities were deserted, the surviving populations moved on to settle around permanent water sources. The Yucatan Peninsula has the largest extent of underground rivers and cenotes in the world, a reliable, plentiful source of water. Even while the older political systems in the southern lowlands were breaking down, the well-watered, major cities in the northern Yucatan Peninsula were still able to thrive. Activities, populations, wealth and trade moved north and settled in the northern lowlands of the Yucatan Peninsula and the Maya highlands.

The Postclassic Maya city of Uxmal stood as an important political and economic center and is most impressive because of its well preserved buildings. Its archaeological site is one of the largest and most visited of the impressive Maya cities. Uxmal reached its peak during the Late and Terminal Classic period. Numerous Uxmal structures are also astronomically oriented. The Governor's Palace is aligned to the extreme horizon positions of the planet Venus. The decorative sculpture on the palace shows a strong iconographic relationship with Venus, linking it with the southernmost rising point when Venus rises as the Morning Star. Additional symbolic features suggest it also refers to the northerly extremes when Venus is the Evening Star. The Governor's Palace holds convincing evidence due to the appearance of numbers and hieroglyphic symbols identical to those found in the Venus table in the Dresden Codex. Additional astronomical data was found recently when scholars deciphered the iconographic program sculpted on the palace; the sculpture they decoded was a zodiacal frieze.

Chichen Itza dominated the northern Yucatan as one of the grandest Maya cities with a vast number of inhabitants. The Terminal Classic period city was settled by a diverse population and therefore developed a variety of remarkable architectural styles. The oddly shaped astronomical observatory, El Caracol, is a round building set upon a huge platform; its significant height allows sky watchers an excellent 360-degree view of the skies high above the tangled jungle of scrub and trees that surround the site. The doors and three surviving windows of the Caracol are oriented to the phases of the Moon and the positions of Venus. Scholars believe that Maya astronomers used the Caracol to gather information concerning the movements of the Sun and other heavenly bodies to predict the seasons, the weather patterns for their agriculture, and for ritual observances.

By the eleventh century the city of Chichen Itza and the Puuc population began to decline in numbers. The elite activities of monumental building and sculpture ceased but the city probably still had a large number of residents. Unlike the Classic cities, the great buildings and monuments of Chichen Itza were never totally lost or buried in foliage, but large regions of the city have not been investigated or uncovered. The area around Chichen Itza was still settled with a small native population, who were still producing agriculture, when the Spanish arrived.

Chapter 10 considers the future direction of Maya studies and explores the most recent technological advances that have allowed archaeologists to uncover exciting, fresh discoveries including hundreds of unknown sites developed by the ancient Maya. The latest research tends to topple long-accepted theories and revise the "known" facts. The application of the ground-breaking Light Detection and Ranging, called LiDAR, enables scientists to apply laser pulses that are connected to a GPS system, toward new searches in impenetrable areas. These new techniques have the ability to pierce deep into the tropical forests of Mesoamerica with unbelievable results. The laser beams have the capacity to cut through the thick jungle growth that has obstructed investigators on the ground. By flying over and across the miles of forested lands, these scientific surveys created accurate plans of territories formerly occupied by the Maya. Their surprising discoveries produced a three-dimensional map of the 30,000-square-mile area between Vera Cruz and Campeche. Fortress cities have been discovered with huge defensive walls, moats, watchtowers indicating large scale warfare. Long-accepted theories are constantly being revised because of new understandings and from glyphic texts that are being decoded and better understood.

The goal of *Decoding Astronomy in Maya Art and Architecture* is to make some of the huge amount of new evidence more available for interested readers. Many of the important findings of scholars that circulate in scientific journals rarely reach the general public. Scholars write for other scholars and their specialized articles are not easy to follow by average readers. As with explorations of outer space, the amount of data now available relating to cultural astronomy is constantly expanding. This work demonstrates how wide-spread and crucial astronomical knowledge was among early Maya societies. The study examines the numerous ways that vital astronomical knowledge was protected, encoded into myths, rituals, histories and artworks, and then passed on accurately to assist later generations. Research into the long history of the Maya reveals the diversity and richly complex relationship that existed between the terrestrial world of humans and the celestial world of the stars. Hopefully this study will provide a clearer understanding of the importance and depth of astronomical knowledge possessed by the ancient Maya.

Deerfield Beach, FL, USA Marion Dolan

References and Further Reading

Aldana, G. (2016). Discovering discovery: Chich'en Itza, the Dresden Codex Venus Table and 10th century Mayan astronomical innovation. *Journal of Astronomy in Culture, 1*(1), 57–76.

Aveni, A. F. (1997). *Stairways to the stars: Skywatching in three great ancient cultures*. Wiley.

Aveni, A. F. (2001). *Skywatchers*. University of Texas.

Braswell, G. E., & May, N. P. (2014). The shadow of the pyramid. In *The ancient Maya of Mexico: Reinterpreting the past of the Northern Maya lowlands*. Routledge.

Grofe, M. J. (2014). Astronomy in Mesoamerica. In *Encyclopaedia of the history of science, technology, and medicine in non-western cultures*. Springer Science+Business Media.

Miller, M. E. (1986). *The art of Mesoamerica from Olmec to Aztec*. Thames & Hudson.

Miller, M. E., & O'Neil, M. E. (2014). *Maya art and architecture*. Thames & Hudson.

Sprajc, I. (2020). *Lost Maya cities*. Texas A&M University Press.

Sprajc, I., & Nava, P. F. S. (2016). Astronomy and architecture in the Maya lowlands. *Journal of Systems Architecture, 2*(2), 189–215.

Tedlock, D. (2010). *2000 years of Mayan literature*. University of California Press.

Contents

List of Figures

Part I
Introduction to the Maya Civilization

Chapter 1
Uncovering a Lost Civilization

Introduction

The astounding journeys of John Lloyd Stephens and Frederick Catherwood to Central America in 1839 and from 1841 to 1842 introduced the ancient, unrecognized Maya civilization to the rest of the world. Their life-threatening adventures saw them fighting their way through hundreds of miles of thick jungle vegetation, suffering from the sweltering heat, suffocating humidity, and soul-soaking rains. They were obstructed by the civil war of General Santa Anna; overwhelmingly frustrated and ultimately jailed by threatening politicians exercising their illusionary power. Despite their enormous struggles, they managed to locate and explore numerous ruined cities that flourished during the Classic period of Maya history, from the third to the tenth centuries. They were caught in the midst of a violent revolution, troubled with bats, snakes, and biting insects; both men suffered the effects of malaria, flesh-eating parasites, and days without food or water. The adventurers were the first English speakers to visit these areas of Central America and first to record many of the ancient monuments. Along their hazardous journey they met and documented many descendants of the ancient Maya, some who still spoke the same language. Many of the natives they met were anxious to assist the explorers in their explorations.

A detailed description of their treacherous journey, along with their astounding discoveries, appeared in their joint two-volume book, *Incidents of Travel in Central America, Chiapas, and Yucatan,* published in 1843. Their travel book became an instant bestseller with twelve editions printed in only 3 months which was unheard of at that time. Stephen's first-person account and lively text revealed the scope and complexity of the Maya cities. Catherwood's colorful artworks illustrated the variety of buildings, the precise details of their carvings and the intricacy of hieroglyphic writings. These ruins were indeed created by a totally unknown culture that lived in an unknown world. Stephens and Catherwood uncovered not merely new

M. Dolan, *Decoding Astronomy in Maya Art and Architecture*,
Springer Praxis Books, https://doi.org/10.1007/978-3-031-61092-9_1

architectural wonders and buried cities but an entire unknown civilization, buried in the jungles for over 1000 years.

John Lloyd Stephens (1805–1852) was American, living in New York, where he became a lawyer, but much preferred travel and searching for adventure in uncharted territories more than practicing law. In 1834 Stephens set off for a grand tour of the usual European cities but decided to continue even further in order to visit out of the way sites throughout Greece, Turkey, Jordan and Egypt. In Cairo, he disguised himself as a native; looking inconspicuous he was able to travel freely. He hired a *falookha* with a captain and crew to sail the length of the Nile and wandered unquestioned through the ancient Egyptian temples, tombs and into massive pyramids. He managed to gain access into the hidden rose-red city of Petra and walked the Biblical pathways through the Holy City of Jerusalem. So enthralled with his experience of these exotic cities, Stephens wrote his first highly praised travel book describing his far-off ventures, *Incidents of Travel in Egypt, Arabia Petraea and the Holy Land.* The lawyer-turned-adventurer proved to be a skillful writer; reviews of his popular book praised his escapades and asked to hear more of his ventures into little known areas. Profits from his book provided sufficient funds for future expeditions into enticing, unexplored terrain.

Six years older than Stephens, Frederick Catherwood (1799–1854) was born in England and became an accomplished artist, architect, businessman and explorer. Between 1824 and 1832 Catherwood too had made many trips to historic cities around the Mediterranean and had sailed the Nile to explore the ancient temples and beautifully decorated tombs of Egypt. Throughout his travels he used his artistic talent to produce watercolor drawings of the ruined temples, monuments and pharaonic statues with great accuracy, including the Egyptian pyramids, the Dome of the Rock, the Holy City of Jerusalem and numerous ruins in the Middle East. His meticulously detailed drawings of the sites brought these ancient civilizations to life for armchair travelers.

In 1836, Stephens and Catherwood happened to meet in London; they quickly discovered that they enjoyed similar interests and adventurous characteristics. Though quite different in temperament, Stephens was outgoing and enthusiastic; Catherwood was rather introverted and quiet; they soon became close friends. Their visits to intriguing foreign locales inspired them to travel to the lost ruins in Central America, particularly the city of Copan in western Honduras. Copan was one of the few known sites of the ancient Maya but it had seen few visitors. The two men decided they must investigate the area and write a definitive description of those intriguing, lost cities. Their extensive, life-changing journey to explore Maya remains in the New World set off from New York on a ship to Belize. Stephens was single but Catherwood left behind his wife. Stephen's describes their departure in his own words:

> On Wednesday, the 3rd of October 1839, we embarked at New York on board the British brig Mary Ann, Hampton, master, for the Bay of Honduras. The brig was lying in the North River, with her anchor apeak and sails loose, and in a few minutes, in company with a large whaling-ship bound for the Pacific, we were underway. It was seven o'clock in the morning:

the streets and wharfs were still, the Battery was desolate, and, at the moment of leaving it on a voyage of uncertain duration, New York seemed more beautiful than I had ever known it before.

The First Expedition to Mesoamerica

They were not the first outsiders to visit and write about the ruins of Central American. Both Stephens and Catherwood had read the story by the renowned anthropologist Alexander von Humboldt who had carried out an expedition to Central America in the early 1800s. Another pioneer Juan Galindo had explored Palenque and Copan 5 years earlier; they had been enticed by these earlier surveys of the two ancient cities. Those thrilling accounts of an unknown culture had whetted their appetites for undertaking their own exploration of these lost ruins. Stephen's adventuresome spirit and unquenchable thirst for the mysterious urged him forward. He recognized that Catherwood's training in architecture and artistic abilities would bring his writings to life; together they planned their journey into the unknown.

To enable easier passage through areas or dangers they might face, Stephens asked President Martin Van Buren to appoint him as special ambassador to the Federal Republic of Central America, comprising Honduras, Guatemala, El Salvador, Nicaragua and Costa Rica. The title was merely a formality to establish easier passage within the largely unexplored countries. He had no desire to conduct diplomatic duties, only to search for archaeological remains. Stephens was rigorous in his preparations to ease their difficult journey but could not possibly foresee the ensuing problems they would encounter in the dense tropical jungles. The dangers were unseen and unimaginable.

Stephens and Catherwood arrived in Belize, a town of 6000, equipped with basic surveying gear and machetes, along with drawing supplies, paper and paints. Upon entering the "untidy, shabby, tropical port on the Belize River, they sank to their boot tops in the muddy streets". Belize was a small village that sat along the length of the seacoast; its inner border faced "the chaos of disorder" that was Central America. They hired a local guide and the team embarked from the oldest settlements of the New World to begin their extensive explorations that eventually stretched from Copan in the south to Chichen Itza in the northern lowlands.

At that time only a few Maya ruins had already been visited but briefly, only Copan, Palenque and Tikal, (called Topoxte or Tayasal) which was buried somewhere deep within the tropical jungle. The ancient Maya civilization was barely understood, just a foggy presence on history's pages. Most early cities and tall pyramids had been totally overgrown and appeared as merely undistinguishable forested hills. The newcomers had no precise maps to guide them; they had no proper roads to follow. They had no experienced assistance, just a local guide and a few men to carry their valuable equipment. Often there was little food available; they had no protection from poisonous snakes, spiders, mosquitoes, and other bothersome insects. As they traveled, they questioned local inhabitants for information to guide them to a source for food, shelter or the next area of ruins.

The territory they struggled through was covered with a thick tangle of thorny undergrowth and vined trees; they contended with long-armed black moneys, sodden rain forests and murky, diseased swamps. Using a sketchy map from Galindo's writings, they traced the same trail he had followed in 1834 through the untamed areas, hacking through thick undergrowth with machetes. As they neared their initial goal, Copan, "instead of jungle, they now had enormous candelabra cacti, mile upon mile of strange, dully green cacti holding spiny arms aloft". With minimal assistance from Galindo's adventure, they finally arrived at the modern town of Copan. Upon their arrival, they learned that the town had no inn, no restaurant, and very little food.

Ruins of Copan

When asking about the ancient ruins, the local officials and the owner of the property would not allow them access to the site. After an ensuing conflict, they were both arrested and imprisoned by armed soldiers; but by midnight they were released. Although not sure of the reasons for either their imprisonment or release, they were free to continue to Copan. On the rugged road into their first archaeological site; they labored along a primitive pathway, sometimes on rented mules, always slogging through deep patches of mud and sludge (Fig. 1.1).

The beautiful city of Copan sits in a picturesque valley near the Guatemalan border; the city flourished between 426 and 820 CE. Although the area's history was unknown to the explorers, the earliest settlement of the Copan Valley dated back to around 1500 BCE. The first settlement of Maya people from the Guatemalan Highlands to Copan occurred around 100 CE. Upon finally entering the archaeological site, Stephens and Catherwood were thrilled with their first view of the mysterious Maya civilization that they long dreamed of visiting. They marveled at the stately remains of a once vibrant Maya city. At that time it was a country with few stone buildings; the people that they observed lived in small villages in wooden huts with grass roofs. The monumental stone structures they observed were breathtaking.

After the grueling struggle to reach their destination, Stephens and Catherwood felt enthusiastic when finally viewing the great mass of archaeological remains. They were amazed at the sight of elegant pyramids, huge standing stelae, huge public plazas, temples and ball courts, all decorated with strange carvings. Copan's houses, buildings and sacred temples had all been abandoned by the end of the ninth century. The local residents had no knowledge of who had created these wonderful monuments nor when they had been built. They had no explanation for the ruined city and fallen stones that surrounded them.

The structures and decorations appeared entirely different from anything they had observed in Egypt or the Middle East. Immediately the two amateur archaeologists began the daunting task of uncovering, cleaning, measuring, drawing and studying the remains of ancient Copan. Over the 1000 years of its decline, the site had been swallowed and strangled by the roots of ficus trees, beneath a thick maze

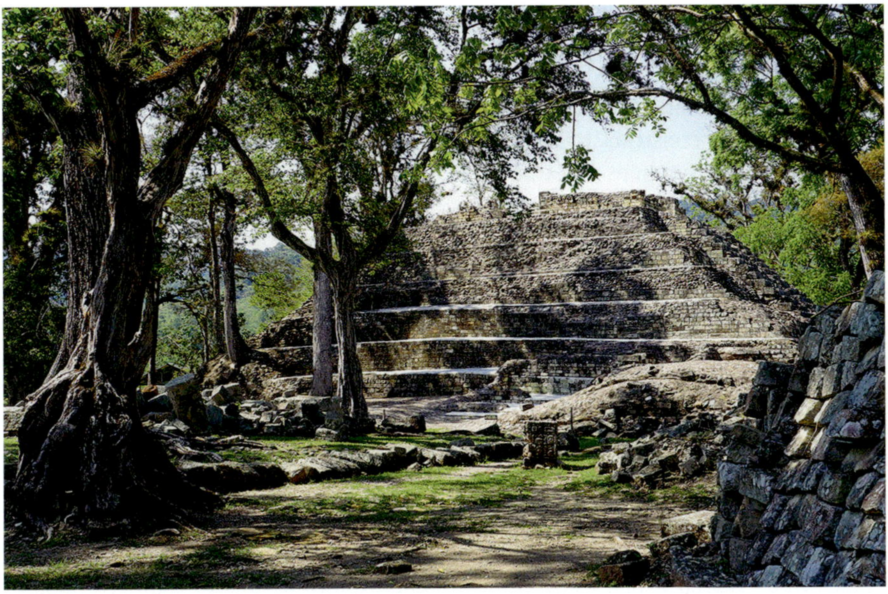

Fig. 1.1 Ruins of Copan visited by Stephens and Catherwood (Credit Inspired by Maps)

of diverse undergrowth. Searching the ancient grounds they encountered stone jaguars, cyclopean stairways, fierce-looking stone heads, carved skulls, and elaborately costumed figures wearing enormous headdresses. The pair were greatly impressed by the Hieroglyphic Stairway, Copan's most outstanding structure, adorned with 1800 individual Mayan glyphs. They were puzzled by their strange form of writing filled with odd shapes, tiny faces, dots and lines.

Once a few monuments were sufficiently cleared of foliage, Catherwood anxiously began his work. To assist in producing sketches, Catherwood had brought along an early nineteenth century invention called a *camera lucida* that could project a highly detailed image onto paper, enabling more accurate representations. This new type of camera was essential to his enormous task of capturing the intricate carvings of the Maya artists and the hieroglyphic writing of the scribes. He tossed out all of his first attempts to record these strange new art covered objects. The gigantic Copan sculptures were so complicated, so utterly different from anything he had ever seen, Catherwood had difficulty capturing their likeness at first.

After their initial visits to the ruins, they tried to hire local workers to build a small cabin at the site in order to camp close to their work. The landowner refused and threatened to have them removed by the local military. Finally Stephens offered to purchase the entire site for fifty dollars; after much argument back and forth the owner eventually agreed on the sum. The residents thought Stephens quite mad for wasting so much on that useless land filled with huge, worthless pagan idols.

After just a brief time exploring the ruins, Stephens became convinced that those who thought the Maya cities were constructed by the Lost Tribes of Israel, by

Egyptians or by Atlantean survivors were very wrong. Some had even suggested that a more advanced, sophisticated people of Greek or Roman origin must have crossed the Atlantic, created the ancient cities; then left them behind in the jungles. That the predecessors of the native population of Mesoamerica had the knowledge and expertise to construct hundred-foot high pyramids was unthinkable to most authorities. Both Stephens and Catherwood were certain that the city of Copan had been created by a totally different, native society who immortalized their culture, their leaders and their way of life in the buildings, artworks and writings. The two explorers were mystified over their findings but certain that the early residents of the Copan were the builders and artists. Who were these missing people? When was their city built? What happened to the native population and their advanced civilization? Why did they desert their cities? Where did they go?

Ruins of Palenque

After completing their work of mapping and recording at Copan, Stephens and Catherwood visited a nearby Maya site called Quiriquá at the southern edge of Maya territory. Quiriquá is a medium-sized city situated along the Motagua River but it sat at the crossroads of several important trade routes. There they found a large assembly of carved stelae and noted the tallest monumental sculpture of their entire exploration. Following a short stopover at Quiriquá, the explorers struggled through the picturesque but treacherous landscape of Guatemala heading for the border of the Mexican state of Chiapas. Upon arrival the team discovered that General Santa Anna, president of Mexico, was allowing no one to enter the city or visit the ruins. Determined to visit Palenque, they elected to ignore the travel ban and continued on their planned itinerary to the architectural site. Despite the official orders, the group set up their camp amidst the ruins. They choose the most protected area they could find, inside one of the Maya buildings, later called the Palace, but first had to strip the thick vegetation that hid its beauty (Fig. 1.2).

Despite being eaten by swarms of mosquitoes, Catherwood immediately began measuring in order to create his meticulous drawings of the impressive bas-reliefs around the main courtyard of Palenque. As in Copan, he drew accurate representations of the sculptural works. Catherwood used his *camera lucida* and re-drew the images he captured. He covered page after page with Maya carvings, rich in detail, reproducing the incomprehensible hieroglyphics, buildings and sculpted monuments. His replicas were so exact that later Mayanists who had finally deciphered the script could read them. He continued on with his recording of the engravings found on the Temple of the Foliated Cross, the Temple of the Sun and the impressive Temple of the Inscriptions where they would eventually learn that the great Lord Pakal and his treasures had lain hidden for over a thousand years. Stephens believed that the complex hieroglyphic writing they found on the temples and stelae recorded the culture's historic events but had no idea of what ancient secrets the stones held.

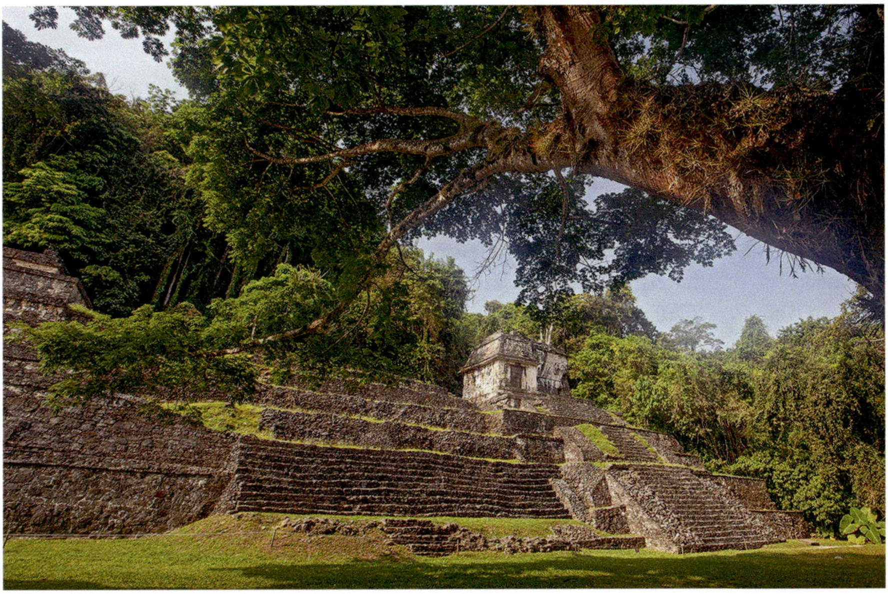

Fig. 1.2 Ruins of Palenque visited by Stephens and Catherwood (Credit Alexandra Lande)

Stephens was again thwarted by local politicians trying to block his access to Palenque. Again, he offered to buy the ancient ruins but found that owning property there was not permitted unless you were married to a Mexican. Stephens considered the prospects but quickly changed his mind. Soon after their arrival at the site, Catherwood became ill with malaria but refused to rest and insisted on working. In spite of his illness, he continued his battle with mosquitoes and with the mysterious, difficult designs. After concluding their work on the monuments of Palenque, the explorers moved on to find the nearest ruins.

They arrived at the large city of Uxmal on June 24, 1840 and recorded such buildings as the massive Temple of the Magicians. They journeyed on to visited numerous ruined cities of the Maya; in this flatter lowland area many of the sites were not so isolated nor buried in foliage. But some of the ancient ruins that they explored were so densely hidden that they were unknown even to the indigenous people of the area. Stephen's adventure tale describes their constant battle with stinging flies, ticks, bats and mice. Unfortunately Catherwood could not fully recover from malaria and fell seriously ill; eventually Stephens contracted the disease as well. They finally decided it was necessary to finalize their expedition and return to New York to regain their health. Their adventures through the amazing Maya lands had proven that the unknown people who had settled and constructed the grand cities of Mesoamerica were not savages but talented and accomplished people with a deep-seated sense of their history and a love of beauty.

Return Home

Stephens and Catherwood arrived back in New York, on July 31, 1840, almost 10 months after their departure. Upon their return, Stephens set to work on a detailed account of their adventures for the general public recreating their treacherous experiences and fantastic discoveries in Mesoamerica. The book engravings were created from the on-site drawings of Catherwood and he oversaw the process most diligently. The book was published the next year, 1841. The following year Catherwood published an additional book, his *Views of Ancient Monuments in Central America, Chiapas, and Yucatan* showing twenty-five color lithographs from watercolors he painted at the ruins. His work included ground plans of the cities, elevations of the palaces and temples, sculpted stelae, stucco bas-reliefs, clay figures and interior details.

Stephen's writings on their findings were detailed and informative, including their meeting of local characters; he turned their hazardous journey into a classic of travel literature. But it was Catherwood's magical sketches that attracted most interest to Mesoamerica and inspired future investigations by archaeological expeditions such as the Carnegie Institution of Washington, DC. Catherwood's drawings have been criticized by some as romanticized but his lithographs are highly exact and have been used by contemporary Maya scholars to aid in restoring the crumbling temples and overgrown pyramids and reading weathered glyphs. His architectural measurements are precise but he sometimes uses artistic license when placing objects or figures. His skillful sketches show details that have now eroded away.

Second Journey to Mesoamerica

After allowing time for recovering their health and gathering resources, the two adventurers planned another trek through Mesoamerica to discover and record more of the Maya ruins. Stephens proved to be a meticulous organizer; he prepared for his journeys in minute detail. Because he was fascinated with the great abundance of exotic, colorful birds they had observed on their first trip, he invited Dr. Samuel Cabot, an ornithologist, to join the second trip to fight the jungles of Mesoamerica.

In October 1841, Stephens and Catherwood set off on their second expedition. They traveled this time to the Yucatan, where the area's flatlands were much less hazardous and less difficult to traverse than the mountainous tropical jungles. Their second journey explored the cities of Kabah, Chichen Itza, Labna, Tulum and many others. They revisited a few of the earlier sites as well, especially the splendid city of Uxmal, where their time had been cut short by illness. They found that the ruins of Uxmal were in the best condition they had seen in their all their travels. They recorded a total of forty more lost Maya ruins, swallowed whole by the dense, tropical jungles. Their second journey resulted in another book by Stephens called, I*ncidents of Travel in Yucatan,* which was enhanced with 120 engravings by Catherwood. Published in New York in 1843, it too was in two volumes, retelling the incredible events, the natural wonders and their incredible findings. Stephen's second travel book was even more popular than his first. He described not only their fascinating archaeological discoveries; he told of the country grappling with war and instability. He expounded on the various people they met, their customs, festivals, music and generosity in sharing what little they had with strangers.

Conclusion

After returning safely from the second excursion to the torturous land of the ancient Maya, both Stephens and Catherwood decided they had enough adventurous exploring; nothing could exceed what they had already accomplished. Their incredible struggle through the dangerous tropical jungles, its threatening animals, snakes, insects, diseases would have deterred most explorers. The pair continued on in spite of torrential rains, extremes of heat and humidity, and oppressive natives. Their long search for fabulous cities completely engulfed by the dense, tropical jungles was entirely successful. They were the first to systematically uncover, explore and document almost fifty ancient cities of the Maya, most were completely unknown to anyone. For the remainder of their lives they were tormented by the side effects of malaria and other diseases they contracted. Sadly they both died young.

Stephens was later employed by the Ocean Steam Navigation Company and then the Panama Railroad Company. He traveled to Panama and was thought to have contracted malaria again that damaged his health even further. He died in New York when only 46. His adventuresome partner Catherwood moved to San Francisco where he opened a supply store for miners during the Gold Rush of 1849. In September 1854 Catherwood was sailing aboard the steamship Artic, crossing from Liverpool to New York when the ship collided with a French steamer called Vesta. Sadly the Artic sank; more than 350 passengers including Catherwood perished. He was only 55 years old.

By accurately recording the condition of the ruins they witnessed in the 1840s, their work has provided huge assistance to Maya scholars in the renovation and reconstruction of particular monuments and temples in the ancient cities. The length and breadth of the 3000-year-old Maya culture has been very slowly, bit by bit, released from its thousand-year sleep beneath thick, impenetrable vegetation. Even today innumerable Maya ruins still lie buried, hidden from human view, waiting for the next intrepid archaeologists to defy the grasp of the impermeable jungles.

References and Further Reading

Sprajc, I. (2020). *Lost Maya cities*. Texas A&M University Press.

Von Hagen, V. W. (1990). *Maya explorer: John Lloyd Stephens and the lost cities of Central America and Yucatan*. Chronicle Books.

Chapter 2
Introduction to the Ancient Maya

Early History

The name "Maya" is applied to a diverse assortment of peoples who varied as much as the environment in which they settled. Experts estimate that nearly 4,665,000 people lived in the Maya lowlands alone. They had many different government structures and varied cultural and kinship practices. The Maya were never a unified empire but rather a communal society with friendly relations, intermarriages, trade groups, competition, aggressive power struggles and bloody warfare between rival city-states.

Knowledge and understanding of the Maya civilization is constantly growing and changing through scholarly research. The groundbreaking work explaining the cultural and economic traits published in the 1970's and 1980's is now considered obsolete. The translation of Maya texts has become more comprehensive which indicates that initial translations were not quite adequate and had a superficial interpretation. Many words have layers of meaning not quite understood. Seemingly, translations are becoming closer to the depth of the original meaning. The latest translations have totally altered our understanding of the Maya civilization .

When first revealed to the general public by Stephens and Catherwood in the nineteenth century, the Maya were a *tabula rosa,* completely unknown. Many theories were presented concerning their civilization and just as many were later discarded. The peace-loving society of astronomer-priests contemplating the heavens, writing scholarly books and living in total harmony with the natural world was discarded long ago.

The Maya area includes central Mexico, Guatemala, El Salvador, Honduras, Nicaragua and northern Costa Rica. The Maya all shared common languages, religion and conventional beliefs; but each city-state was independent and often at war with each other. The cities developed alliances with rival cities by creating bonds through marriages. The important aspects of Maya culture were in place by

M. Dolan, *Decoding Astronomy in Maya Art and Architecture*,
Springer Praxis Books, https://doi.org/10.1007/978-3-031-61092-9_2

1500 BCE, including an accurate calendar, a base-20 numerical system with the use of zero, hieroglyphic text, illustrated codices, local maps, mythology, ritual ballgames, domesticated crops, distinctive architecture, bloodletting practices and an advanced knowledge of astronomy. The fundamental features of the Maya cities also appeared about that time. These prime components were ceremonial centers, a stratified social system, elite rulers, carved stelae, and monumental stonework. By 300 BCE the Maya had grand cities with huge plazas lined with pyramids and temples decorated with carved scenes with royal and religious iconography and free-standing stelae honoring their kings.

The Maya period of development is typically divided into three wide time periods, the Preclassic (2000 BCE–250 CE), the Classic (250–900 CE) and the Postclassic (900–1539 CE). All three of these ages were subdivided into more distinct periods. The broad time periods were classified in this manner initially because the Preclassic was formerly thought to be formative, not yet a developed civilization. The Classic period was believed to be the highest realization of the Maya culture, while the Postclassic was considered as a time of decline from Classic era peak. This was a great simplification of the actual facts now learned and is no longer the belief of scholars but the terms have stuck.

Experts have revised the classifications to make them more relevant. The more precise time periods are the Archaic period 6000 to 2000 BCE. The next phase is the Early Preclassic 2000 to 1000 BCE, the Formative 1500 BCE and the Middle Preclassic 1000 to 300 BCE. The next period is the Middle Formative begins in 800 BCE and then the Late Formative begins in 300 BCE, the Early Classic CE 250, the Late Classic CE 600, and the Postclassic CE 900 to 1539. However, all areas did not develop at the same pace and do not fit the general time periods.

Maya Early Preclassic Period

Little is known about the earliest Maya population; the tropical climate has allowed few architectural remains to survive from very early periods, but the Maya were soon able to control the steamy jungles. The Maya Preclassic period stretched for over 2000 years and saw the advance from the primitive huts of hunter-gatherers to impressive cities with monumental architecture and organized agriculture. This period includes the earliest start of settlements and villages to a complex, effective society with social stratification. Most early settlements were built along the coasts in Belize and the Yucatan Peninsula; ancient evidence is obscure but by 1000 BCE archaeological remains include ceramics and household architecture. Contacts with the developing Olmec culture would have had strong influencing effects on the developing villages and towns. The Olmec culture is considered the first advanced society in the New World; but there were multiple cradles of civilization in Maya world not just one. The Olmec developed astronomy, mathematics, a written language and accurate calendars. They also created symbolic artworks particularly superb sculptures.

The settlers of the first communities, ancestors of the later inhabitants, created the essential foundation for what was to become the Maya civilization.

Maya Middle Preclassic Period

The Middle Preclassic includes the era from 1000 BCE to 300 BCE when some Maya settlements were still small but significant and numerous enough to be recognized by archaeologists throughout most areas. Settlers began to move from the coastal regions, up river valleys and deeper into interior sections. As the populations grew in larger settlements, traces of public architecture can be found. The population increase required better management of natural resources for increased crop production. The Maya expanded in territory and complexity with chiefdoms, which were based on kinship, social status and control of the economic resources. The installation of rulership and the presence of Maya kings developed at this time. Villages began to create earth mounds and central plazas; irrigation and canals helped agriculture grew more productive.

Scholars previously thought that most of the societal advances occurred during the Classic period, but now know that the Maya were practicing writing, mathematics and calendrics by the Late Preclassic. The dates of advanced civilization in the Maya world are continually pushed earlier and earlier. Monumental public projects included major pyramids on raised platforms, ball courts, standing stelae and stone lined causeways. Buildings were painted with murals and artisans were making painted ceramics and decorative jewelry. A developed script was being used by the third century BCE. The Preclassic cities of Nakbe and El Mirador were growing to immense sizes. Tikal too was already becoming an important city by 350 BCE. Trade networks had developed with the exchange of exotic goods such as obsidian mirrors and blades, jade mosaics and pendants, feathers and pelts. Stelae appear with carvings of rulers and gods but without inscribed text. The beginnings of complex societies was underway.

Maya Late Preclassic Period

During the Late Preclassic period, 250 BCE to CE 250, the early Maya civilization developed its ultimate form. Evidence shows richer crop production, growth of the economy, and religious practices that encouraged a complex political and social order. The cities' populations were growing with more cultural interactions and roads indicating exchanges with neighboring areas.

Freestanding stelae began to appear inscribed with information on the history and events in the life of the leaders. In the lowlands massive pyramids were built decorated with elaborate stucco masks. Societies formed with increasing specialization of workers and stratification of an elite, privileged class. During the Late

Preclassic, innovative features were often applied to the forms and dimensions of their artworks and architectural programs. Maya architecture started to take its typical shape, creating massive platforms, enormous pyramids, and tiered monuments. Pyramids grew more massive and more complex, symbolizing sacred mountains, visible from far distances to broadcast the city's power. The sculpture and architectural décor depicting deities and cosmological themes were created. These same themes were applied again to Classical period monuments.

During the Preclassic period the evolution of architecture and sculpture occurred throughout the entire Maya area. Older buildings were renovated but not torn down; they constructed new buildings over top of the older ones. Archaeologists compare them to an onion, as the peel away layer after layer to expose the core structure. The available construction material influenced architecture as well. The highlands consist mainly of basaltic stone which is hard and dense which is difficult to quarry and to transport. Therefore, Preclassic buildings in the highlands were rarely over 10 m (33 ft) tall. Whereas, the lowlands had vast quantities of limestone, which is much easier to quarry and transport; as a result the lowland buildings grew to unimaginable heights.

Limestone was excellent for making lime as well; ground to powder, it provided the raw material for stucco and mortar. Maya builders covered their structures inside and out with thick layers of stucco and modeled large figures and masks for the facades. The design of cities and placement of buildings and plazas were consolidated in the Preclassic as well, preferring orientations to specific cardinal points, mainly east to west. Stone corbel vaulting was thought to have been developed during the Classic period but was already in use during the first century BCE. The corbel-vaulted palaces allowed for new designs with rooms connected by interior doors, often with slotted openings to allow sunlight and fresh air to enter. Enormous amounts of experimentation, innovation and improvements during the Preclassic led to a more widespread and effective use of architecture during the Classic period. The Maya performed terminal rituals every 52 years in which they offered fire, smoke and incense; they often made changes to their architecture at that time.

The Late Preclassic saw the Maya's testing new concepts, leading to their unification and expansion at the start of the Classic period. Their artistic expression reached new levels developing cosmological and mythical themes and symbolism. The Maya rules used ceremonial and symbolic religious practices to connect with the supernatural powers of the sky, Earth and Underworld. The Late Preclassic saw the establishment of a culture that would continue to develop knowledge and power during the next centuries.

Maya Classic Period

The Classic period (CE 200—900) was the most advanced era of the Maya civilization; evidence survives of a new type of society. During the seven centuries of the Classic period, the Maya made tremendous advances stimulated by the continual increase in population. Diverse elements created in separate places seemed to come together at the start of the Classic period to form a unified agenda, leading to their great success. Dates marking the Classic period are defined by the creation of historic monuments that used the Long Count calendar to record accurate times of important events in their history. Maya society was well-ordered with organized cities, monumental public architecture, an accurate calendar, a complex writing system, advanced astronomy and mathematics, a complete array of gods and successful agricultural production capable of feeding the enormous number of inhabitants (Fig. 2.1).

Each Maya city-state was governed by a mighty "divine king" with a royal family and royal court; an elite ruling class had also been established. Maya priests studied the movements of the Sun, Moon and planets which were honored as gods who ruled their daily lives. Through their unfailing cyclic movements, the celestial bodies announced when to plant and to reap and when to celebrate. The Classic period saw the peak of large-scale architectural programs. Maya city centers consisted of ceremonial and administrative complexes and massive pyramid-temples surrounded by elite palaces and sprawling residential areas. All buildings were covered with stucco and colorfully painted. Temple and palace walls were covered with bright mural scenes of pageantry, magnificent stone carvings, and large decorative masks. They produced painted ceramics, extravagant jewelry and finely carved jade pendants.

Enormous stelae were erected and engraved on all four sides with lengthy, detailed inscriptions, and imposing portraits of their kings. Their writing system was the most advanced in Mesoamerica; it consisted of pictorial symbols or glyphs that represent either entire words or syllables. Maya glyphs are the only Mesoamerican writing system that has been largely deciphered. The societies witnessed expert artistic development with numerous carved altars and lengthy staircases that recorded births, marriages, investiture of kings and the political, religious and social history of the Maya (Fig. 2.2).

Many of the monumental buildings and multi-storied complexes were carefully oriented for astronomical observations. With very limited instruments, Maya astronomers could determine and predict the movements of the heavenly bodies; they recorded that information in great detail. Astronomers maintained astronomical

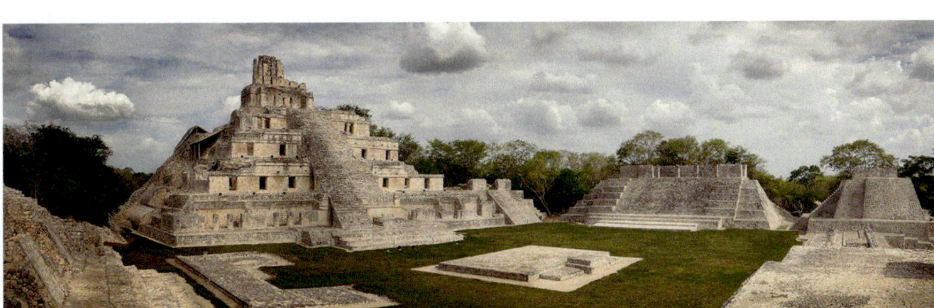

Fig. 2.1 Edzna, Campeche, The Five Story Temple, Classic Maya city (Credit Soft Light)

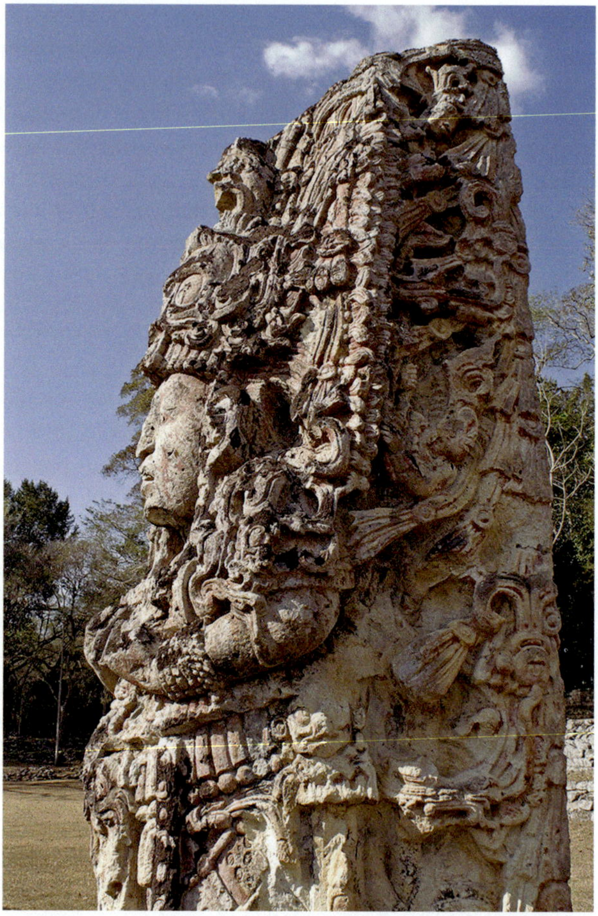

Fig. 2.2 Royal Stelae, King Waxakllajuun Ub'aah K'awiil, partial view, Copan, early eighth century (Credit Christian Wilkinson)

data on the positions of the celestial bodies, in particular the cycles of Venus. They calculated the phases of the Moon and the times of eclipses and passed down their star charts written on screen-fold codices; the four surviving codices are filled with the cycles of Venus, the phases of the Moon, prediction of eclipses and other celestial events. Artworks flourished as well; the Classic period saw the creation of large painted murals depicting the impressive pageantry and elaborate costumes of the warriors and elites. Complex networks for trade transactions and for alliances among the city-states were created. Successful cities often had smaller subordinate settlements under their power. Some enormous cities swelled to over 100,000 inhabitants.

As the centuries passed, the important cities were neither static nor constantly growing; there were periods of progress and of stagnation. Change was constant; some cities gained prestige and some cities suffered a hiatus in monumental construction. A great rivalry existed between the two largest city-states, Tikal and Calakmul; each city developed extensive networks of allies and vassals. The leaders

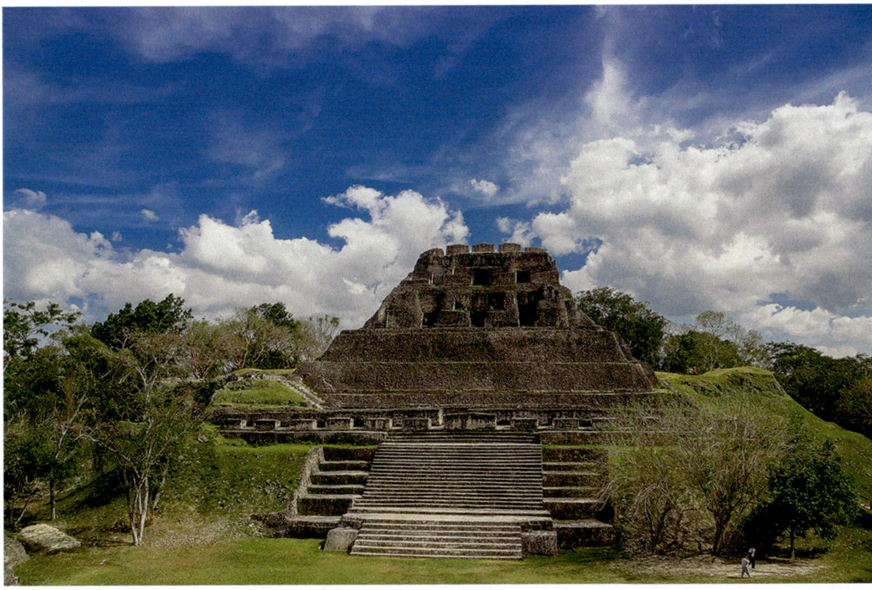

Fig. 2.3 El Castillo at Xunantunich, pyramid with large masks at Preclassic site (Credit Fotos 593)

continued to wage brutal wars to gain captives to keep as slaves or to sacrifice as offerings to the gods. By the end of the ninth century, drastic changes were occurring throughout the extensive Maya lands (Fig. 2.3).

Late Classic Period

The Late Classic period (c.CE 600 until 909) was an era of competition and construction expanded in some areas, while other areas saw a break up of dominant cities, in particular Tikal. As Tikal saw the reduction of political and economic power, other sites rose to prominence and asserted more influence. The Late Classic ended with general depletion of the populations of the massive cities, leading to the Maya's total collapse. Like all other aspects of the Maya culture, the appearance of text on monuments rapidly declined in the ninth century. The government of cities changed from a divine king to joint rule by councils with one person appointed as supreme ruler while other members served as advisors.

Maya Collapse

In the first decades of the ninth century, almost all Maya cities were impacted by a political and social crisis. By the end of the Classic period, Maya workers had stopped building palaces and temples; sculptors had quit carving monuments and artists had stopped inscribing stelae. The Maya cities were slowly being abandoned; the Maya collapse had begun, but each city had complicated factors and its own final ending. The unexpected decline spread from city to city and lasted for more than 100 years. The last century of the Early Postclassic, CE1100-1200, was very much a dark age. One by one the formerly successful cities were totally abandoned. After the collapse of Chichen Itza, very few substantial settlements survived in inland areas. The tropical jungle with its lush plants and grasping tendrils began to consume the palaces and temples; only the tall pyramids proclaimed their existence, rising high above the tree canopy to announce their presence.

Overpopulation in Maya cities, which reached its peak at the beginning of the ninth century was one of several common factors leading to collapse, estimates reach ten million people. Agricultural production would have struggled to support the huge populations. Recent climate studies suggest that a prolonged drought occurred in some areas of the Maya lowlands in the ninth century, causing crop losses and localized famines. Some scholars suggest soil exhaustion or deforestation but the Maya used non-destructive methods in their agriculture. Others proposed hurricanes but they did not build on the coasts.

Perhaps people left searching for more fertile lands, or they may have rebelled against their rulers. Warfare was another problem that undoubtedly contributed to the Maya collapse; fighting would interrupt trade routes and block roads, so that needed goods could not be transported efficiently. These difficulties would lead to shortages, economic problems and mass population migrations. During the ninth century the number of texts alluding to warfare between city-states increased and carved panels show more rulers with captured enemies. Perhaps they were competing for arable lands and dwindling resources.

Even more perplexing is the fact that 150 years later, Maya cities in the northern Yucatan Peninsula experienced a similar breakdown of their cities that seemed to imitate the ninth century collapse. The Maya civilization continued to populate traditional areas but never recovered the amazing heights it had once known. However, the descendants of the Maya nobles, priests, warriors and farmers inhabit the same lands today and continue their culture in languages and in religious rituals. Although the Maya civilization collapsed the Maya culture continues to thrive in the traditional Maya lands.

Maya Postclassic Period

The Postclassic began about CE 900, varying at different areas, a significant Maya presence remained but the major Classic cities had been deserted. Abandoned lands were not resettled quickly; surviving populations collected close to permanent water sources. The cities of the northern lowlands and the Maya highlands became more active and grew to more sophisticated levels. Near the Gulf and Caribbean coasts, new cities appeared. Some cities moved to locations that were easier to defend, which implies a more warlike conditions. Eventually the surviving cities also suffered political, social and environmental complications. Abandonment of the newer Maya cities was repeated along with a period of warfare, disease and natural disasters in the Yucatan Peninsula, which had ended just before the Spanish Conquest.

Innumerable Maya cities survived for hundreds of years but this book will cover a limited number of cities, Belize and the major cities of the highlands: El Mirador, Nakbe, Tikal, Calakmul, Copan, Palenque, Bonampak, Tula, and principal cities of the Yucatan Peninsula: Uxmal, Chichen Itza, Mayapan and the Island of Cozumel.

References and Further Reading

Biro, P. (2012). *Classic Maya Western region: A history*. BAR International Series.
Sharer, R. J. (1994). *The ancient Maya* (5th ed.). Stanford.

Chapter 3
Maya Society

Maya Rulership

The institution of kingship became fully developed during the Early Classic period when the process of sanctifying an elite ruler was established. The Maya king or *ajaw* of the Classic period was the life blood of the city-state; he acted as a channel of supernatural power, a strong, living connection between the community and the gods. The king stood alone, at the highest level of the symbolic exchanges with the heavens and declared himself to be the force that maintains the cosmic order. He became a symbol, a major force of nature expressed even in the architecture.

The Maya kings were highly revered in their royal cities; their power was displayed through elaborate public ceremonies, usually performed atop impressive pyramids or next to sacred temples. Many of these ceremonies were held in alignment with the cycles of the heavens. The relationship between kings and their gods was duplicated by the relationship between the people and their king. He could appease, interact with, and even sway the godly realm in behalf of the community; therefore, his power was equivalent to that of the gods. The ruler had control over the elements as well; it was his responsibility to keep the cosmos in its permanent order through proper ritual performances. He used religious symbols, belief systems and astronomical associations as a way to legitimize his power, gain the approval of the gods, and validate his right to rule.

By performing the necessary ritual ceremonies, sacrifices or functions, the king could maintain the order of the cosmos. In the four-part cosmos, the city stood as the fifth cardinal direction; the central plaza or acropolis was considered the center of power with the king symbolizing the living axis mundi for his subjects. The leaders of the city created sacred spaces where huge groups could watch and take part in the sacred ceremonies. The public events overseen by the king and the priestly class worked together to confirm the sense of unity within the community. Public displays of power and control were a means of social cohesion. Classic Maya rulers

M. Dolan, *Decoding Astronomy in Maya Art and Architecture*,
Springer Praxis Books, https://doi.org/10.1007/978-3-031-61092-9_3

relied upon mythology to reaffirm their regal position and assure the success of divine rule. A central element of their religious practices was ancestor worship, common among both elites and commoners.

The king wore distinctive costumes for each role he performed, as priest, war chief or civic leader. Proclaiming their high rank, Maya kings dressed in spectacular costumes; they also affiliated themselves with the spiritual deities to affirm their authority. They wore impressive headdress supported by racks on their back that towered over their heads, sometimes rising a full meter. The rack frame was decorated with meaningful objects, including shells, carved jade, feathers and textiles. Special headdresses for special occasions held the long bright green plumes of the quetzal bird and the mask of the Rain God Chaac or the Sun God Ahau Kin. The king would have jewelry and ornaments entwined in his long hair. He wore decorative ornaments in his ear lobes; his teeth displayed inlays of jade. His soft shoes were made of jaguar skins. When fully costumed with huge headdress, back rack, jewelry, ornaments, belts and weapons, the effort extended to climb the extremely steep steps of a tall pyramid (some 22 stories high) must have been enormous.

The splendor of the royal ceremonies were designed to impress the people with the crowned king arrayed in magnificent robing and seated in an impressive throne covered with jaguar skins and eagle feathers. A king was always carried on an elaborate liter. The ruler directed construction projects in order to design an ideal city that was laid out in agreement with cosmological principles, an earthly replica of the celestial order. Astronomy was an essential part of the concept of divine kingship; the orientation of the urban architecture brings down the sky to legitimate the king's right to rule. The constancy of the heavens confirmed the strength, the regularity and reliability of rulership. Powerful leaders maintained their authority even after death; the living continue to recognize the authority of the deceased. After death many kings were elevated to the status of a god, achieving their apotheosis (Fig. 3.1).

To honor Maya kings, enormous stelae were erected and engraved on all four sides with lengthy, detailed inscriptions, and imposing royal portraits. Societies witnessed expert artistic advancement with numerous carved altars and lengthy staircases that recorded births, marriages, and investiture of kings and the political, religious and social history of the Maya.

Ceramic vessels often depict Maya kings or elites dancing, which was a major part of spiritual experiences and an important ritual for specific occasions. The dancers wore symbolic masks and special costumes, sometimes dressing as birds, jaguars or even monsters, thinking they could transfer into that form. They also applied jewelry to their costumes, especially jade pendants; some would wear as much as twenty-five pounds of jade ornaments. The weight of the jewelry would help to tire them as an additional offering to appease the gods. The Maya felt that they could communicate with the gods by leaping and spinning to elevate themselves to a trance-like status that would transport them into the spiritual world. Dances took place in association with various ceremonies, such as deity

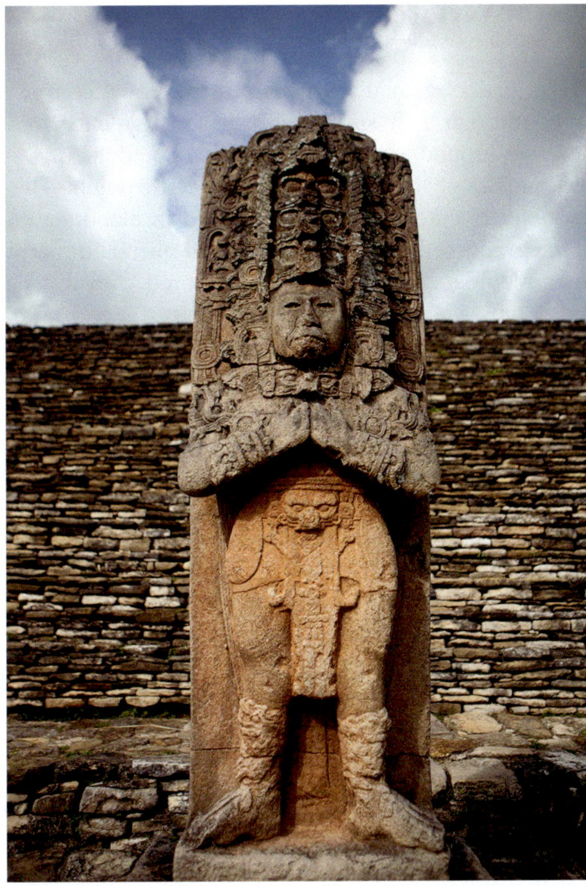

Fig. 3.1 Royal Stela, King Zots Choj Muan, Tonina, Chiapas (Credit Alpkhan Photography)

impersonation rituals, sacrificial rites, warfare, royal visits, and other dynastic events. Dancing was thought to strengthen the connection between the rulers and their gods. Particular ritual dance performances were a chief responsibility of the rulers; cities were distinguished by their distinctive dancing.

The king stood alone at the highest point of the community and proclaimed that he is the force that maintains the cosmic order. His symbolic interactions with the heavens, demonstrated through public displays of power and control, were necessary for the continuation of a successful city. It was believed that the celestial gods established kingship as the natural order. The king became a major force of nature, expressed through his construction of monumental architecture in tune with the heavens. This concept became fully developed during the Early Classic period and for the next thousand years the institution of kingship reigned within the Maya world.

Maya Warfare

Maya leaders gained respect from their own communities and from neighboring cities through their success in warfare. Evidence for warfare is found in fortified defenses around Maya cities, artistic depictions of war, and the presence of weapons, such as obsidian blades and projectile points. The high-ranking Maya warriors and leaders entered into battle attired in elaborate costumes. They wore wooden helmets adorned with brightly colored quetzal or parrot feathers that draped across their shoulders. Their faces were colored with war paint. Around their necks and wrists they wore impressive jade jewelry.

Cities conducted warfare for various reasons, for political control of people and resources. The goals and motivations for warfare are not thoroughly understood but Mayanists develop different conjectures. It appears that warfare was not conducted to increase a city's territory but the capture of prisoners for sacrificial offerings was a key motive for waging war. Sacrificing victims of defeat enhanced the leader's position in his own community and among competing cities. Another motive for warfare was the control of necessary resources; population growth increased the competition among cities, resulting in higher levels of violence.

Maya Bloodletting Ritual

Similar to all early human cultures, the ancient Maya performed rituals and ceremonies to interact with the spiritual or supernatural forces that presided over their daily lives. The primary purpose of these rituals was to assure the flourishing continuation of life, health and nourishment, as well as the continuity of the cycles of the cosmos. Royal, religious and communal ceremonies were enacted to maintain the cultural and political structure of the city-state and to legitimize the regal position of the ruling elite. To the Maya, the offering of blood was the highest level of sacrifice they could present to the deities. The blood offerings included animal and human sacrifices and especially, the solemn act of bloodletting. Pictorial evidence for these highly ritualized scenes is found on carved monuments and painted vessels that show dramatic examples of bloodletting.

Kings, nobles and priests of Maya settlements believed that they implored and pleased the gods through the ritual act of bloodletting from their own bodies. Previously, Mayanists thought that only men performed bloodletting rituals, but pictorial and textual evidence proved that royal women enacted bloodletting ceremonies as well. Bloodletting consists of the ritualized self-cutting or piercing of the person's body to shed blood which is collected on strips of bark-paper and then burned as a sacred offering. Copal incense was add to the burning to provoke abundant smoke and generate pungent odors. The sacrificial blood is thus transformed into sacred smoke rising toward the heavens (Fig. 3.2).

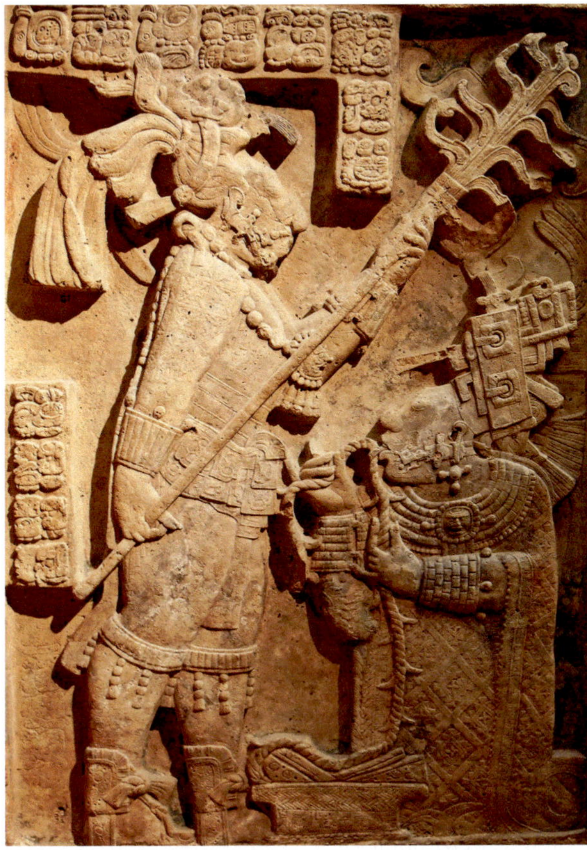

Fig. 3.2 Lady K'abal Xook kneeling, drawing blood from her tongue, Yaxchilan, Chiapas (Credit Fedor Selivanov)

Before the hallowed bloodletting ceremony, the ruler or priest/shaman was required to undergo preparatory purification acts such as fasting, tobacco smoking and certain abstinences or observances that would produce a trance-like or altered state of consciousness. They believed that the long ritual process would result in supernatural visions and communication with dynastic ancestors or with the gods of the Underworld. By entering the induced trance, the ruler or king was able to petition the sacred ancestors and the gods for rain, productive crops, and successful warfare along with other needs. The ceremony of bloodletting was thought to be essential for the well-being of the society and to maintain harmony and balance in life and with the heavenly deities.

The bloodletting ritual was sometimes conducted in private rooms inside sacred temples atop pyramids; but was more often performed on the summits of pyramids or on elevated platforms. These staged ceremonies were enacted in public spaces where great crowds would gather and witness the leader's painful sacrifice. This visual display of sacrifice worked to strengthen divine favor and to function as a means to maintain political power and to reinforce their social, political and ideological position. Their royal or sacred blood that had passed down from their honored ancestors provided them with god-like status in the earthly sphere.

The sharp instrument that the Maya used for bloodletting was usually the spondylus shell, stingray spine, or obsidian blade. Some descriptive images carved on stelae and building facades depict the ruler holding implements that have been identified as bloodletters that were used to draw blood from various soft body parts, such as the tongue, ear, cheek, lip or genitals. The Madrid Codex depicts an image of a man and a woman drawing blood from their ears. On some occasions a rope with attached thorns or obsidian flakes would be pulled through the tongue, most painful to even think of. Obsidian blades are often found in ritual or ceremonial settings such as caches, burials and caves; it is assumed that they were also used as bloodletters. At times the blood would be spread on their bodies or on the statues of the gods.

Stingray spines were dangerous tools that had to be carefully cleaned and dried before use because uncleaned they contain venom. Using them to pierce the body when unclean would have caused even greater pain and possibly effects like secondary infection or even death. Some participants may have risked using the uncleaned spines for special acts of piety or in ceremonies where the risk of death was an additional sacrificial offering. Bloodletting of this type was deemed a privilege and was willingly practiced by the nobles and priests.

Bloodletting ceremonies were usually timed to take place on significant dates and at scheduled community events noted in the Maya ritual calendar. The most important dates for bloodletting were the beginning and end of a calendar cycle, when a king ascended to the throne, and at building dedications. Events also celebrated with bloodletting rituals were life stages of the king and queen, such as births, deaths, marriages and the beginning and end of warfare When enacting the ceremony, the ruler often impersonated or assumed the identity of one the supernatural powers, often the very important Maize God or the Rain God.

The ritual act of bloodletting was based on two elementary beliefs of Maya cosmology. The first is that the gods gave life to humans by sacrificing parts of their own bodies. The second is their emphasis on blood, the core element of life. They believed that human blood was partially composed of the blood of the gods, who sacrificed their own divine blood when creating life for humanity. Therefore, to maintain the order of the cosmos, the Maya were required to offer their own blood back to the gods. The demi-god rulers gave their special blood passed down by the sacred ancestors to empower the gods in thanksgiving for bestowing them with life and divine protection.

The Maya considered bloodletting a mystical act, a process to enter an ecstatic state. A possible result of the bloodletting ritual was a manifestation of the Vision Serpent, a very prominent mythological being in Maya mythology. This serpent was a social and religious symbol thought to be the means by which the Sun, Moon and stars crossed the heavens. Maya artists depicted the Vision Serpent as having two heads with the spirit of a god or ancestor emerging from its jaws. During bloodletting rituals, the huge serpent would appear and act as a pathway to the spirit world. The gods or ancestors who were being contacted would manifest themselves through the Vision Spirit. Scholars contend that the vision may have actually been a result of massive blood loss that causes the brain to release natural endorphins which are chemically related to opiates. If a body goes into shock after blood loss, a hallucinatory vision can occur. The swirling smoke from the blood offering would provide the setting for the mystical Vision Serpent to materialize and become visible. The earliest dated stela to visually portray the Vision Serpent dates to CE 199.

The king and his family would enact the bloodletting ceremony in honor of his sacred ancestors, in behalf of himself, the city-state and his people. Religious sacrificial ritual was not intended to bring the divine from another place into the physical world. The universe was filled with the spiritual presence of the many gods who were everywhere. By sacrificing something most valuable in return for the gift of life, humans acted to assure the cyclic, immutable order of the universe. The sacred offering allowed humans to be active participants in the eternal flow of life. The sacrificing victim offers their blood to ensure the renewal of the Earth and the continuity of the cosmic cycles.

Maya Calendar

As early as 1500 BCE, Mesoamericans were monitoring the motions of the heavens; their careful observations led to the creation of a unique calendric system. For 3000 years the astronomers of Mesoamerica maintained their accurate calendar through constant sky watching and precise adjustments. Organizing and systematizing the motions of the heavenly bodies provided order and confidence that the continued reliability of the sky could be applied to order their lives and their communities. Through the interlocking cycles of their calendric system, the astronomer priests could watch for planetary influences and predict events that might influence their politics, agriculture or military affairs. It was firmly believed that the Sun, Moon and planets had compelling influences on people and on the natural world. Although average Maya people were familiar with the calendrical system, those that maintained and corrected the complicated calendrical systems were astronomer priests and those elite who preserved the esoteric knowledge and astronomical skills necessary for the tedious task.

Early Olmec art and the alignment of ancient Olmec ruins offer the earliest evidence of sky observations in Mesoamerica. The Olmec are credited with establishing the foundations of the Mesoamerican calendric system. Both the 260-day and 365-day calendars were also used by the Zapotec people in San Jose Mogote, Oaxaca beginning around 600 BCE. The Maya based their calendar on the system inherited from the Olmec civilization which had been used throughout the Mesoamerican territories. The mythological deity Itzamna was credited with teaching the knowledge of the calendrical system to the Maya ancestors; he was also thought to have introduced writing and medicine.

One of the main reasons for the Maya's fascination with the heavens was to establish an orderly system for timekeeping to manage the processes necessary for a successful existence. Their writings tell of their desire for a precise calendar that became an obsession. Through their advanced knowledge of astronomy and mathematics, the Maya created one of the most accurate calendar systems in human history. The Maya calendar is quite complicated but it fulfills their practical and ceremonial purposes. Their calendar measures time periods of varying lengths; they were determined through solar, lunar, planetary and human cycles. Deep knowledge of the complex calendar system was another source of power retained by the Maya ruler.

The Maya calendar involved two parts that worked together to form a 52-year cycle, called the Sacred and Solar rounds, and the Calendar Round of 52 years. The Sacred cycle lasted 260 days and the Solar round, called *Haab*, had 365 days. Each day of the year has a unique name. The Sacred calendar of 260 days, called the *Tzolkin,* which means "count of days", is not divided into months. The Maya considered the Solar year as consisting of 18 months of 20-day duration plus an added short month of 5 days, called *wayeb*. Those 5 days of Wayeb' were considered as a threatening period. To avoid any impending troubles, the Maya practiced rituals during Wayeb'. The Maya name for a single day was *kin*, the same name also referred to the Sun. Twenty *kins* are called a *winal;* eighteen *winal*s are considered one *tun*; twenty *tuns* are considered a *k'atun* and twenty *k'atuns* are a *bak'tun*. The Maya calendar has not just one but several day counts of different lengths.

The *Haab'* is believed to have 'begun around 550 BCE at the winter solstice. The Maya knew that the 365-day *Haab'* was different from the tropical year by about 0.25 days per year. Different intervals appear on Maya monuments which can be utilized to estimate the tropical year. The most exact of these is that the tropical year exceeds the length of the 365-day *Haab*' by 1 day every 1508 days. The date of the specific solstice in the *Haab'* will repeat after the passing of 1508 *Haab'* years of 365-days. Surprisingly, two 52-year cycles, or 104 vague solar years combine perfectly with the 584-day Venus year, equal to the actual day for 65 Venus years. This amazing connection is probably one of the reasons that the Maya did not add leap days to the 365-day year.

The Calendar Round consists of the interweaving of the *Tzolkin* and *Haab* calendars. Any combination of a *Tzolkin* day with a *Haab* day will not repeat itself for 365 days. The two rounds together were used to keep chronologies and king lists, to indicate historical events, important dates and signify the beginning of the world. The essential dates were carved onto stone stelae, painted on walls, etched on altars and building facades and written on bark paper codices.

Together the two calendars together formed a synchronized cycle lasting for 52 *Haabs'*, which they called Calendar Round, and that is still used today in rituals carried out by Maya groups in the Guatemala highlands. The Calendar Round will repeat after 52 *Haab'* years or 18,980 days, called a calendar completion. The *tzolkin* calendar combines 20 day names with the 13 day numbers to produce 260 unique days. Each day is numbered from 1 to 13 and then they begin again at 1. Each day has a name given in sequence from a list of 20 day names. They are used to establish the time of religious and ceremonial occasions and for divination. A different calendar was used when they wished to cover longer time period and to engrave calendar dates.

The Maya had yet another calendar system. This time-keeping method is called the Long Count which measures the days since a mythological starting point. This was needed so that their history would be recorded accurately. That early starting point corresponds to August 11, 3114 BCE. The Long Count could be extended to indicate any date far into the future or into the past. The cycles of the Long Count are independent of the solar year. Since the Maya calendar's starting date is August 13, 3114 BCE, all social events, wars and natural occurrences were dated based on the number of days that had passed counting from this ancient date. The information following the cosmic time period mostly include major historic events in the life of the ruler.

The Peten area, considered the center of Classic Maya civilization, noted the Long Count dates later; it first occured at Tikal on Stela 29 dated 292 CE. By the end of the Early Classic, most stelae inscriptions begin with Long Count period ending dates. The Long Count dates are usually inscribed with the Bak'tun first and the K'in position at the end, followed by the Calendar Round. Long Count dates are specific to the actual day.

The 13-month or *baktun* cycle of the Maya Long Count calendar counts 1,872,000 days or 5,125,366 tropical years. The Long Count was developed during the Late Preclassic and is found on many monuments of the Classic era. This count is one of the lengthiest cycles found in the Maya calendar system. This long cycle ended on the winter solstice, December 21, 2012. The date of creation for the Maya is equivalent to August 11, 3114 BCE in the Gregorian calendar.

Through these diverse and complex systems, the Maya kept track of time with precise accuracy. Time is the important marker of the Classic period determined by the span of time during which the lowland Maya were using the Long Count calendar on their monuments. They were using the Long Count consistently at the end of the third century CE. The Long Count was a complicated and clever calendric system that dealt with massive periods of time. From this time until the collapse of the Classic period, scholars have established a very closely dated archaeological sequence copied from inscriptions on stelae and other monuments and tombs. The Early and Late stages of the Classic are divided at CE 600 when upheavals occurred in the central areas; considerable cultural differences can be detected between the two halves.

At the Maya Collapse, one of the surprising development was the immediate stop of engraving Long Count dates on their monuments. The collapse took place close to the end of the Bak'tun 9; few Long Count monuments carried the date of the tenth Bak'tun. During the Postclassic period, the names of kings were rarely mentioned on monumental texts.

The question of the origin of the Maya's unique calendar concept has not been solved but possibilities have been offered. The calendar may have been suggested through the seasonal gatherings of mobile or nomadic groups, such as the Native American Pow Wow gatherings. This type of event took place in the lowlands of Mesoamerica during the dry season in February and March when farmers were freed from agricultural tasks. When major rituals or special events took place on certain dates of a solar calendar, those living in far off areas could use solar observations as a signal when to set off for the communal ceremonial centers without having means of communication.

Scholars debate the purpose behind the establishment of the unusual 260-day cycle, suggesting several possible motives. Many possible reasons have been suggested, including numerology, agricultural seasons, the human gestation period, and the time between solar zenith transits. Numerology would include ideas that add religious and cosmological meanings to particular numbers, which may apply to occurrences of special events in the social and natural world. The numbers 13 and 20, associated with human body parts, certain deities and cosmic levels were considered of utmost importance.

Experts have found growing evidence that the fixed 260 day from February to October represent the agricultural cycle, fixed by the trajectory of Venus, combined with observations of the Pleiades. They also consider the appearance and disappearance of the constellation of Orion. These celestial events were watched for more than a century before they were written into the Maya version of the almanac during the second half of the fifteenth century CE.

The careful observation of the heavens enabled the Maya to not only create their calendar system but to also utilize it to orient their main architecture. The gods had presented the calendar for their benefit, so that people could properly organize their society and create their successful agricultural programs. A structure that is oriented to sunset or sunrise was thought to be instilled with ritual symbolism with special calendric significance. The structure is therefore in harmony with the sacred dogmas of the calendric system. The essential civic and religious buildings created with calendric implications enhanced the ruler's prestige and authority; he would utilize these celestial connections to dramatize ritual performances. The community believed that their leader enjoyed close communications with the gods that inhabit the heavens and he worked to strengthen those divine illusions. Astronomer priests too used their calendric and astronomical expertise to increase their social and political positions. Accurate predictions of celestial phenomena granted them highly honored positions in their societies.

Maya seemed to focus on the Sun, Moon and Venus, with a handful of constellations and stars and the Pleiades star cluster making up their collective studies. The most important was the Sun, which was universally recognized as the giver of life. The Maya associated the god Ilhuicati Tonatiuh, a red eagle with a large and all-seeing eye with the Sun. Their observations were astonishing, given that they had to account for the tilt of the Earth's axis and the corresponding positions that the Sun would therefore be seen season to season. Study of their observations reveals a great accuracy in the calculated times for the Sun to rise and set, with records showing further calculations of 365 days to represent a solar year. A tropical year is actually 365.2422 days long an inaccuracy that would have grown more noticeable over time. The priest astronomers seemed aware of this, with evidence suggesting that they continually updated and amended subsequent records.

Maya Hieroglyphic Text

Another remarkable achievement of the Maya culture, which remained unchanged for close to 1000 years, was its intricate and detailed system of writing, particularly its use in mathematical numeration, astronomy and the calendar. The use of writing was considered one of the necessary constituents of civilization. Writing allowed the users to maintain permanent records and increased the transmission and amassing of knowledge. The Maya were able to record seasonal and astronomical information over long periods of time. These documents assisted in the development of accurate calendars and their advancement in understanding astronomical cycles. The Maya used a sophisticated mathematical notation that enabled them to write very large numbers in their carved inscriptions and in codices. They were among the first peoples, along with the ancient Babylonians, the Chinese and the Hindus of India, to develop the concept of zero and to use a symbol to express it (Fig. 3.3).

The Maya writing system was the most advanced in Mesoamerica; it consisted of pictorial symbols or glyphs that represent either entire words or syllables. Maya glyphs are the only Mesoamerican writing system that has been largely deciphered. Maya scribes recorded glyphic texts on buildings, monuments, stelae, altars, tombs and even on long stairways. The Lowland Maya created extreme versions of the standing stelae covering them with long texts describing the history, deeds, births, deaths and marriages of their kings. We will never know the great number codices written by scribes through the centuries but it must have reached many thousands.

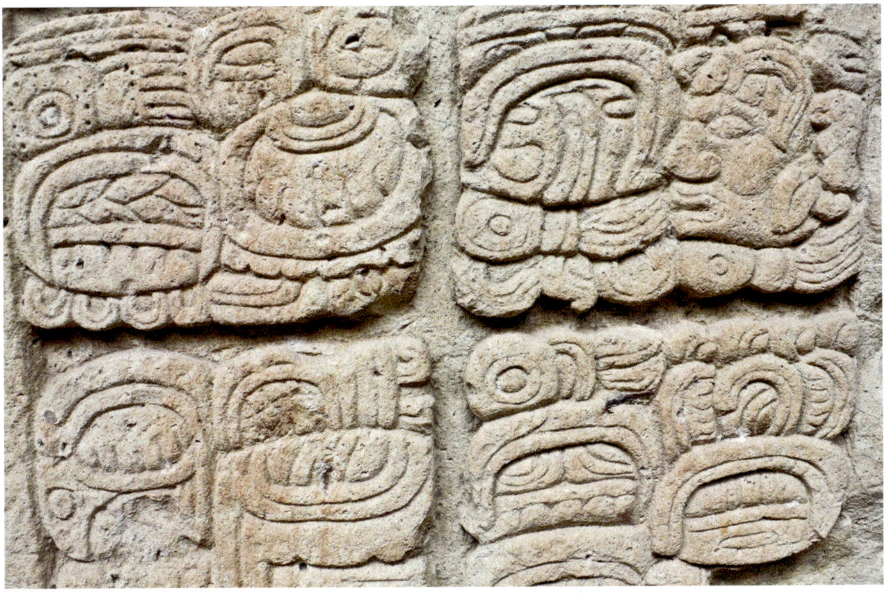

Fig. 3.3 Maya carved glyphic text from Copan (Credit Henryk Sadura)

Knowledge of Maya writing did not survive for long after the arrival of the Spanish. The glyphic script was considered a visible symbol of paganism. A Spanish priest wrote:

> We found a large number of books in these characters and, as they contained nothing in which there were not to be seen superstition and lies of the devil, we burned them all, which they regretted to an amazing degree, and which caused them much affliction.

The few codices that survive are devoted to astronomy, astrology and calendrics. Although not everyone could read, books were highly regarded. Scholars believe that literacy was not widespread among the general population. Scholars estimate that one in four of the Maya could read and probably fewer could write.

Maya writing was not static but continued to change and improve over time. A considerable portion of Classic Maya script was phonetic; that is, the written symbols stood for sounds or particles of speech that made up complete words. Maya writing had homonyms, synonyms and a grammar all its own. It differed from ours only in that it had over 1000 words rather than an alphabet to hold it all together. Epigrapher's recent advancements in their ability to translate Maya text has added new depth and understanding to their ancient writing. As an example a single word *kin*, signified time, day and the sun. Its meaning and basic glyphic form suggest that timekeeping was intimately connected with the practice of astronomy. The direction of the petals of the floral design that comprises the "day" glyph is thought to correspond to the standstill positions of the Sun at the horizon. A huge amount of information could be encoded into a single, small word.

Scribes wrote to recall important events of the past for present audiences and to record present day events for future generations. They tied the past, present and future to the social and timely events for which the texts were written. The Maya did not accept that terrible things could happen by accident. Every event was considered to be the fulfilment of patterns that could be read in the stars; they were activated by the gods. They thought that all things are closely connected and they have influence over each other, including people, gods, the natural environment and the heavens.

Mayan Codices

The Maya were the only culture of Mesoamerica that created a fully developed written language. Their hieroglyphic writing system was an advanced combination of pictographs representing objects and glyphs expressing more abstract ideas such as actions and syllabic sounds. It perhaps developed from the less advanced Olmec writing which only used literal images, pictographs; or it could have developed independently. Maya writing is thought to have begun around 300 BCE. Probably only the elite and scribes of the community could read, perhaps the nobility and priests as well. High status women were not excluded from learning to read and write. Some lower-status people may have been able to recognize the glyphs and interpret some common signs but were not able to read or write themselves. Writing was considered sacred and thus restricted and protected from the profane.

Although the Maya created innumerable screen-fold books filled with written text and colorful illustrations, few have survived the hot, humid tropical climate or Christian diligence. Advances in translating have reached the point where a large percentage of the surviving texts can now be understood. Scholars realize that the lost codices contained many more topics than are found in those few books that survive. We will never learn of the huge amount knowledge originally possessed by the highly-educated Maya. Most of the textual material that survives is not found written in books but carved on stones.

The textual information found on Maya monuments, although associated with astronomical time, is more concerned with dynastic history and political conquests. The codices, however, contain entirely different, more esoteric material and they are much more challenging to interpret. They were private and sacred documents holding complex, sophisticated information on the movements of the Sun, Moon, planets and stars, particularly Venus. The Maya codices were written with a precise accuracy which can be found in their architectural alignments as well.

The codices' writings and charts were private, for the priestly cult alone, not the general public, since they included divinatory methods, predictions and horoscopes. They contained information vital for understanding their astronomy, cosmology and spiritual beliefs. As written in an important quote, "it is an unsound approach to Maya astronomy to start from preconceived convictions about what the Maya could have known and what they could not possibly have known; one should, instead, draw conclusions only from the data as given in the inscriptions and codices".

The Maya astronomer-priests who wrote and illustrated their books maintained a precise calendar, predicted eclipses and aligned their pyramids and temples to incorporate this knowledge into their daily lives. Astronomer priests were guardians of the Maya calendar and of divine time. They watched and interpreted the heavenly movements keeping close track of exact times for festivals and ritual ceremonies. The priests recorded their beliefs and cosmology which was rich in symbolism; their religion was the central focus of their lives. Their basic astronomical knowledge was not reserved for the priesthood alone but was an integral part of the entire community's ritual and social life. In order to inform and communicate with the people, Maya priests traveled from town to town carrying codices, making predictions according to the positions of the stars, teaching myths and history and enacting religious rituals.

The Maya astronomer priests wielded enormous power in the communities and kingdoms since people believed that they could predict the future. By analyzing astronomical movements and past events on Earth, they assumed the same events would again take place in the future in conjunction with the heavens. Drawings in the codices depict astronomers using crossed sticks as a sighting tool for observations of the night sky. Even with little in the way of tools to assist them, their astronomical tables, eclipse charts and calendars were compiled with the greatest accuracy. They amassed an enormous amount of knowledge of the heavens.

One of the basic purposes for the Maya's diligence in collecting data and recording heavenly events was calendric. The Maya organized and conceived their world view by keeping an accurate calendar which extended into the past and into the future. Through the interlocking cycles of their complex calendrical system, the Maya astronomers could watch for planetary influences and predict events that might influence their politics, agriculture, or military affairs.

The Maya's astronomical knowledge was not reserved only for the use of priests and elites but was an integral part of the whole community's ritual and social life. In order to communicate with the people, Maya priests traveled from town to town carrying codices, making predictions according to positions of the stars, teaching myths and history and enacting sacred rituals. These sacred activities were synchronized with the motions of the heavens and as a result brought power, fertility and prosperity to their realms.

Unfortunately, there has been no direct evidence discovered as yet that provides information on the origin and early histories of the codices. Through careful analysis of codices, researchers have discovered hints of their history within the iconography, glyphic texts, stylistic manner and calendric material. Only four Mayan codices escaped the Spanish Conquest, the Dresden Codex, the Madrid Codex, the Paris Codex and Grolier Codex, now officially called the Codice Maya de Mexico. Each of the four will be discussed separately.

The Grolier Codex

When first discovered by Mayanists, the Grolier Codex was considered a fake, created by modern forgers. The damaged pages of the codex were said to be discovered in a cave in Mexico in the 1960s, found with a few other Pre-Columbian artifacts. It was shown to an archaeologist who displayed the codex at an exhibition in 1971 sponsored by the Grolier Club of New York, thus its name. The codex has been officially renamed the Codice Maya de Mexico and was moved to the National Museum of Anthropology in Mexico City in 2016. Even though once considered fake, numerous in-depth studies by Mexican specialists have demonstrated that the codex can be dated around 1230; most scholars now accept the Grolier as genuine. The archaeological studies and carbon 14 testing confirmed that it is actually the oldest book produced in the Americas. The codex was left in a dry cave as an offering.

The Grolier is incomplete, possibly a portion of a much larger text; only eleven damaged pages survive. It originally held twenty pages, the first eight and the last two have been lost. The leaves are typically 12.5 cm (4.9 in) wide. At the bottom of the pages 4 to 8, red framing lines that tell that the size was one much taller and the scribe left a space for text beneath the figures on each page. The codex would have extended for 250 cm (98 in), about the size of the Dresden Codex.

The Grolier contains an incomplete almanac that tracks the movement of Venus during the thirteenth century, along with an authentic and accurate Maya calendar. Like all Maya codices, the text and images were written on pounded fig tree bark sheets that the Maya call *huun*; their bark paper was first developed in the fifth century. This paper was even more durable and better for writing than papyrus. They covered the bark sheets with stucco and painted only on one side, even though both sides were prepared for text. Each recto page features the image of a standing god facing left, the left side of each leaf has a column of repeated day signs, thirteen all together. The images of the Grolier seem to be related to late paintings found at Chichen Itza; the human figural proportions are similar to Terminal Classic and Early Postclassic ceramics. This dating places the codex in the Early Postclassic period when both Tula and Chichen Itza were declining in power and Mesoamerica was collapsing.

The Dresden Codex

The Dresden Codex is considered the most important of the surviving books. It was rediscovered in the city of Dresden, Germany and entered the Royal Library in Dresden in 1739 after being purchased from a private collector in Vienna. The codex was carried to Spain by a missionary in 1618. The Dresden is now housed at the museum of the Saxon State Library. The Grolier Codex has been dated as the earliest Maya codex; the Dresden Codex appears to almost as old with a possible date of 1200–1250, the Postclassic period; other scholars prefer a late thirteenth or fourteenth century production date. Like the Grolier, the Dresden is also thought to have been created in Chichen Itza. Archaeoastronomer Anthony Aveni proposed that the Dresden Codex was most likely produced by astronomers at the Caracol building in Chichen Itza, often called the "observatory". The evidence is the specific orientation of the circular building and its windows and niches; they are all aligned to the important positions of Venus, especially its northern extreme. The observations notated in the Dresden may have been gathered by astronomers watching the heavens from the observation chamber of the Caracol's tower. The Dresden's astronomical tables hold Long Count dates that seem to be historically designated, dating from the eighth to the tenth centuries.

The Dresden Codex was also created from ficus tree bark and was covered with lime for a glossy finish. Its pages are 20 cm (8 in) high; the entire codex is 3.7 m (12 ft) long. Like other Maya codices, the codex is a screen fold, folded into 78 pages. The leaves have a decorative border of red paint but the pages have suffered deterioration due to their great age. The codex received some water damage during World War II. The book was composed by eight different scribes; each had their own individual writing style, glyph design and type of topics. The content of the Dresden is primarily related to astronomy, calendars, and noting the proper days for rituals, planting and prophecies, with references to sickness and medicine (Fig. 3.4).

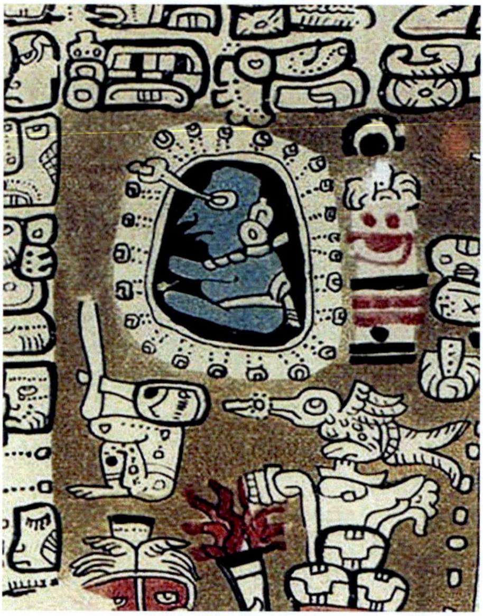

Fig. 3.4 Madrid Codex, Maya codex filled with astronomical charts (Credit S L Photography)

The Dresden begins with almanacs that measure time using only a 260-day calendar; next are almanacs of the celestial travels of Moon Woman and the planet Venus. Later the codex writes of the god Chaak. The Dresden provides background data for understanding the astronomical tables. The second portion of the codex is the Venus Table consisting of calendric sequences through 65 Venus Rounds. Each Round is divided and it notes the first morning visibility, last morning visibility, first evening visibility and last evening visibility. On the right side of five leaves of the Venus Table scribes related the prophetic meaning attributed to the first morning or heliacal visibilities. The Codex focused on the Venus almanac of five Venus cycles in eight solar years. During the Classic period, the Evening Star appearance of Venus was of most significance but the Dresden demonstrates that the Morning Star was thought to be of more importance during the Postclassic period. It also mentions an earlier text, dating to three or four hundred years earlier, that discusses local history and astronomical tables.

Six leaves of the Dresden are devoted to the 584-day cycles of Venus and the 583.9214-day synodic period of Venus. Tediously, they kept track of the amount of error that would have amassed and then developed a way for correcting their Table over very long periods of time. Amazingly, their corrections to the Venus Table were accurate to within 1 day in 6000 years. The Dresden Codex also contains a Venus-Mercury almanac. The synodic period of Venus (4×585) and Mercury (20×17) is very close to the 2340-day length of the Venus-Mercury almanac. One reason for this close interest in Venus positions was its correlation with the beginning and end of the rainy season at their latitude. The furthest northern extreme of Venus is reached in late April or early May and the southernmost extreme falls in late October or early November. The Dresden also charts the summer solstice and the *Haab' uayeb* ceremonies for the tenth century. A large number of gods and goddesses are discussed in the Dresden but the Moon Goddess is the most emphasized; she appears far more often than the other gods.

The Madrid Codex

The codex was first discovered in the 1860s in Spain; its two separate sections were found in two different areas. When the book was divided is unknown but the two parts were rejoined in 1888. Experts agree that it was probably sent to Spain from the colonies during the Colonial period, but how it got there is a mystery. The Madrid was one of the three codices dating to the Postclassic period (ca. 900–1521) and is now housed at the Museo de America in Madrid. Of the four surviving codices, the Madrid Codex is the longest, containing fifty-six leaves that are painted on both side, totaling 112 pages. The codex was created as a screen-fold book compiled from the bark of a fig tree. They coated the paper with a stucco wash and then painted over it with Mayan glyphs and drawn illustrations. The bark paper leaves measure about 23 by 12 cm (9 by 5 in).

Madrid Codex contains almanacs, divination material and astrological horoscopes; it includes images of religious rituals of human sacrifice and praying to the gods for rain. The book also depicts images of everyday life, such as hunting and pottery making. Some images portray a Maya god smoking a *sikar*, similar to a cigar. It also includes astronomical tables but they are less in number than found in the other surviving codices. The Madrid codex includes a number of astronomical tables and other celestial data in its almanac but has no Long Count dates. The astronomical information denotes locations of Mars, solar and lunar eclipses and solar positions at summer solstice and the vernal equinox. Five successive solar eclipses are recorded on pages 12b to 21b. The Madrid Codex tells important information about Maya gods and the rituals associated with the Maya New Year.

Some of its material may have been copied from older Maya books. Experts have determined that there may have been eight or nine scribes and they were possibly Maya priests because of its religious contents. They suspect that the codex was passed down from priest to priest to keep a continual record for their city. Although opinions vary, the Madrid was most likely created in the Yucatan Peninsula, determined from the language used. Scholars suggest its images were drawn in a manner similar to murals discovered at Chichen Itza and Mayapan. The dates determined for the codex range from 1250 to 1450. Evidence shows that the codex may have been acquired by a Spanish priest but it not known how it traveled to Spain or when.

The Paris Codex

Of the four surviving Maya codices, the Paris Codex is the most poorly preserved. It was discovered accidentally in a neglected corner of a Paris library in 1859 but is now held by the Bibliothèque Nationale de France, Paris. The codex consists of a bark paper strip measuring 140 × 23 cm (55 × 9 in) that is folded into eleven sheets that were painted on both sides, totaling twenty-two leaves. The edges of the pages are badly eroded, causing loss of the text except for the center area. The Paris Codex was probably written at a Maya city of the Yucatan, such as Tulum or Mayapan during the fifteenth century, perhaps around 1450, but earlier dates have been proposed. Mayapan is the favored origin point because its text is similar to inscribed stone monuments there. Evidence shows it could be a copy of a much earlier document from the Late Classic period.

The Paris codex includes a number of almanacs and astronomical tables similar to the Madrid Codex with a section that highlights *tun* and *k'atun* rituals. The celestial tables were created to mark solar and lunar eclipses, the rising and setting of Venus in the night sky and the placement of Mars in the zodiac. The codex also discusses Maya ceremonies, particular dates, historical information, and descriptions of the Maya gods and spirits. It shows patron gods and accompanying rituals for a cycle of thirteen *k'atuns*. The final two pages of the codex also tell of animals that represent astronomical signs along the ecliptic along with a scorpion suspended from a sky-band and a peccary. The Maya may have had a "zodiac", consisting of 13 houses; it may have been represented on pages 23 and 24 of the Paris Codex. If so, the first three signs or houses would have been Scorpion, Tortoise and Rattlesnake. These are the first three figures hanging in the middle section of page 24. The reverse side of the codex contains sections discussing a calendar cycle ruled by Chac, the god of rain. It also provides information on the prediction of rainfall and maize crop production. The codex tells of spiritual forces and the days of the *tzolkin* 260-day cycle.

The Maya created innumerable codices that do not survive; many burials contain traces of books that were left there as tomb goods. The books that the Maya produced contained much more than was found in those four surviving examples; even these four have many missing pages. Hints to the contents of lost leaves and lost codices can be found in texts and images that survive since they were painted or inscribed on more durable surfaces. For example, Sun-eyed Snake Jaguar of Palenque commissioned that information be carved on tablets on the periodicity of Jupiter, signifying that the planet was tracked by almanacs in books of that period.

References and Further Reading

Šprajc, et al. (2023). Origins of Mesoamerican astronomy and calendar: Evidence from the Olmec and Maya regions. *Science Advances, 9*, eabq7675.

Tedlock, E. (2010). *2000 years of Mayan literature*. University of California Press.

Wright, M. A. (2011). *A study of classic Maya Rulership*, Dissertation.

Part II
Astronomy and Architecture of the Ancient Maya

Chapter 4
Maya Astronomy and Architecture

Early Maya Cities and Architecture

Before the construction of temples and shrines, the Maya utilized places in nature, such as caves, mountains, volcanoes, cenotes and other special topographical features that were regarded as dwelling places of gods. Here they held ceremonies and recited prayers for the onset of necessary rains and the production of maize. In some areas present day Maya continue these ancient ritual activities at their traditional sacred locations, especially caves and mountain settings. Spanning several thousand years, the architecture of Maya kingdoms was tremendously impressive because of its monumental size and towering heights, its innovative designs and its exterior and interior sculptural artistic programs. Architecture of the ancient Maya cannot be studied independently; it must be looked at as a collective unit involving their astronomy, their religion, their symbolic artworks and their everyday life.

The accumulation of scholarly literature on Maya architecture and its incorporation of archaeoastronomy multiplies constantly. Recent research and a more precise deciphering of Mayan hieroglyphs has enabled scholars to gain a truer understanding of their architecture, their religious beliefs and their culture. The symbolic images and informative text carved on structures and monuments reveal accurate dates, purposes and the names of leaders who commissioned them. Their religious architecture placed an emphasis on extensive heights, adding prominence to the deities they revered. Their incredibly steep staircases climbed to the heavens, reaching closer to the celestial gods.

The Maya civilization was composed of multiple city-states many of which produced regional variations in the style of their architecture. When building their cities, they made close observations of the local landscape and integrated the contours and environmental features of the local area into their designs. In spite of local differences, the distinctive Maya characteristics are evident throughout the many scattered Maya regions. In some areas a hilly terrain would be incorporated to raise their

M. Dolan, *Decoding Astronomy in Maya Art and Architecture*,
Springer Praxis Books, https://doi.org/10.1007/978-3-031-61092-9_4

pyramids and temples to more impressive heights; decorative roof combs were added to reach even greater heights.

Much of the Maya area sits upon layers of limestone which was conveniently available to create their impressive structures. Unbelievably, their extraordinary and innovative architecture was constructed with only elementary tools, stones and mallets; they built without metal tools, the wheel or reliable pack animals. Once the large blocks of stone had been quarried, their only means of transport was over rollers pulled along the causeways. Some weighed hundreds of pounds, requiring an extensive labor force. Many had to be carried long distances and then raised atop the steep pyramids. Most stone buildings were built over strong foundations, using elevated platforms that varied in thickness from less than a meter to 45 m (147 ft), depending on the importance of the structure. The elevated base added prestige to their civic buildings and grand palaces. They brought even more attention to the most important buildings by placing them on raised platforms.

All Maya cities included open courtyards or plazas, terraced pyramids, temples, palaces, shrines, sweat baths and ball courts. Their cities grew by joining the great open plazas in the center with numerous raised platforms that surround it. The larger, more prosperous cities often erected an impressive royal acropolis that overlooked the main plaza with ball courts nearby. Through time and change of rulers, the existing structures were usually remodeled or over-built, growing more and more massive in size and height. The process of rebuilding over top of older ones is a common feature of Maya architecture.

At the beginning of large-scale construction, a fixed axis would be determined by Maya masons in alignment with the cardinal directions but they did not design the city with a formal overall plan. The large plazas in the city's center were then surrounded by the most important civic and religious buildings, providing space for the royal family, for elite nobles and for public rituals. Beyond the essential central structures were the palaces and houses of lesser nobles, smaller temples and individual shrines and altars. These less sacred and less used structures had privacy not obtainable at those more obvious buildings. Beyond this central area were the smaller, more modest wooden homes with thatched roofs occupied by the working people. The Maya's use of corbel vaulting, rather than the arch, limited the height of their buildings; as a result their interiors were restricted to only a few stories tall.

When constructing everything from basic houses to extravagant pyramids and temples, Maya buildings were designed with specific proportions according to the golden mean. They understood the golden ratio and applied it to their architectural designs. When planning their homes, the builders would start with a square and then use a cord or vine to measure from corner to corner used geometric proportions. By arcing the cord down, a new base length was established based on the previous square. The new rectangle's length would be the square root of two, a proportion found in nature. The Mayas' preference and close observance of specific proportions carried distinct religious significance; therefore their form of measure and proportion is called sacred geometry (Fig. 4.1).

The facades of their early pyramid/temples, public buildings and palaces were ornately embellished with impressive, stylized masks of the gods but the inner

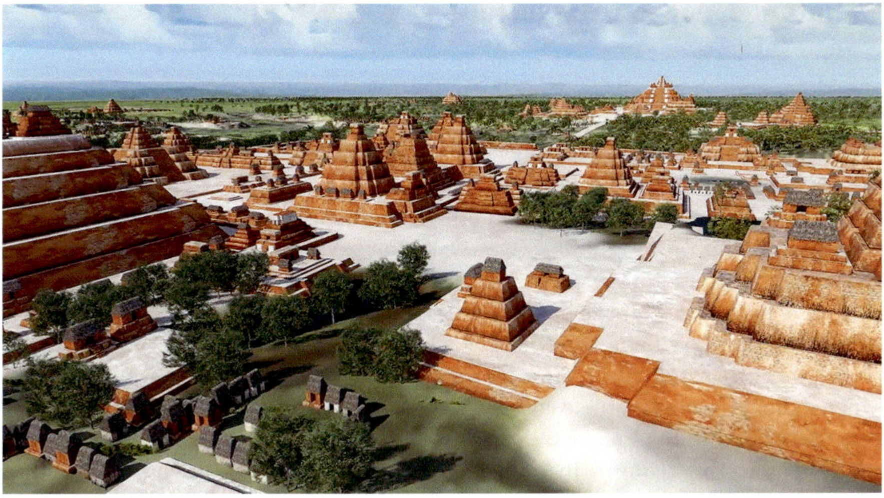

Fig. 4.1 Reproduction of early Maya city with pyramids, temples and raised platforms (Public domain)

rooms were often very simple, merely painted, sometime made cave-like. Builders enhanced the imposing power of their religious temples through elevating them atop huge pyramids and acropolises. Each was decorated with intricate sculpture; some contained small free-standing statues. The doors or gateways into the temples were referred to as mouths and from them would issue royal announcements and commands. Their doorways were often decorated in the form of an open mouth or maw, imitating a cave. Cities differentiated themselves architecturally and artistically by architectural features they emphasized and those they shunned.

As their cities expanded in distance from the central plaza, the Maya created raised, paved roadways to connect the large plazas with the more distant buildings. The raised road, called *sacbe* meaning "white road", connected different parts of a city and stretched to nearby, as well as far-flung, cities. Evidence survives to confirm that the roads were originally covered with limestone stucco or plaster which hid the stone and rubble fill of the *sacbeob* (plural).

Identifying the now hidden roads used by the Maya for communication and trade routes was almost impossible beneath the dense jungle canopy until the scientific techniques of radar, remote sensing and LiDAR have revealed a vast network of *sacbeo*b between the populated areas. In addition to the practical uses of trade and communication, the road systems of the Maya carried many layers of meanings; they were also symbolic connections of city centers, political markers and pilgrimage pathways. All *sacbeob* are thought to have ritual and religious connotations; some roads even trace mythological or celestial pathways. The explorer John Lloyd Stephens wrote in the early 1840s that some Yucatan Maya he witnessed still said a short prayer as they crossed a *sacbe*. Present day Maya call the Milky Way, *Sacbe,* indicating its importance.

Evolution of Maya Architecture

The Maya built their impressive architecture for more than 2000 years; during that time some structural programs changed but many, once formed, continued through the long era and throughout their wide spread territories. Their decorative programs and artistic styles showed more changes over time than the basic structures. Although each city-state was independent, the dominant structures carried small regional differences but almost all buildings were constructed with precise attention to their position and layout; the typical Maya features prevailed in each site.

By about 350 BCE the leaders of Maya cities began building elaborate architecture and pyramids that reached monumental heights. These ancient cities had already been constructing platforms packed with millions of cubic meters of fill which indicates a considerable control of economic resources and labor by the city's elite. Beginning in the Preclassic period, architecture was created with specific ritual designs having patterns that became universal in the Maya world. The settlements and buildings at that time were organized by certain alignments and also with strategic orientations and triangulations. Political and social powers were revealed in the architectural art that adorned all major structure facades. Altars, stelae, and panels, covered with stone and stucco engravings, were erected as historic and religious symbols. The portraits of rulers and textual recording of historic events became the main topics found on stone, ceramics and stucco.

These early settlements grew quickly with huge population increases which required the construction of dense residential urban areas. A duty of the inhabitants was to contribute in the constructions, except in the growing seasons. Temples, palaces, reservoirs, walls, agricultural terraces and causeways required millions of days of labor. Monumental art, especially portraits of the gods, decorated the facades of these larger and grander buildings and lined the staircases of pyramids, temples and civic buildings. By the end of the Middle Preclassic (ca. 600–400 BCE) both low and high relief artworks expressed authority and power through the representation of deities and supernatural forces that supported their royal politics and complex cosmology.

During the Classic period, the architectural structures and features that developed during the Preclassic were continued but they became larger, more elaborate and more highly decorated. Older buildings were over-built to renovate the structures and to mark certain time spans and to honor rulers. The city plans varied with the local terrain but the buildings were oriented along a north-south axis and positioned to align to celestial events or sight lines. There was no distinction between religious and civic buildings since their spirituality was a part of every Maya function.

Architecture of the Classic period experienced a "Golden Age" from CE 500 to 900, the result of a complex process of growth and change. The Classic temples found at Tikal, Copan, Uxmal and Palenque represent the zenith of Maya architecture and artworks. By commissioning ornate architecture, powerful kings promoted themselves and ensured their status and immortality. Their monumental pyramids, platforms, temples and sacrificial altars were used daily in their many religious rituals. The structural forms, the stylistic artworks, the glyphic texts provide the key to interpreting their religious beliefs and understanding their culture. The plan, location, and artistic embellishment of a building provide clues to how it may have functioned. The architectural formats and carefully constructed landscapes were an intimate part of the Maya everyday universe.

Maya Pyramid/Temples

Lofty pyramids were constructed in the blossoming Maya cities even before the Classic Period. El Mirador was already a major urban center in the fourth century BCE, the late Terminal Pre-Classic Period, but the majority of the majestic pyramids were raised during the Classic Period. The impressive city of Tikal constructed the largest number and the most impressive assemblage of pyramid that still peak through the forest canopy. Pyramids were identified by the Mayan word *witz* that also meant mountain; both were considered sacred. The mountains were revered throughout Mesoamerica; they were considered the home of deceased ancestors.

Most Maya pyramids include similar features, layered tiers declining in size as they rise, steep, narrow stairways and flat tops, upon which were placed small temples or shrines. Some believe the temples were architectural versions of the natural caves where offerings were placed during their ancient rituals. The flattened tops of the pyramid were used for important ceremonies but could hold only a limited number of participants; the stairways were used for religious processions. The steep incline of the stairs made the ascent a demanding task to reach the holy places above, especially when dressed in elaborate costumes (Fig. 4.2).

Although the upper platforms surrounding the temples were primarily used for religious rituals, they could serve as astronomical observatories as well. Their pyramids were often exactly aligned to the cardinal directions and functioned as a compass to follow the path of the Sun. The temple atop the pyramids was generally dedicated to a particular Maya god, often the Sun God, and was used to perform ceremonies that perhaps included blood sacrifices. Maya kings and priests would climb the narrow steps of the stairway, moving from the mundane Earth to the shimmering sky, conveying them closer to the heavens and the celestial gods. The earliest pyramids had deep tomb chambers used for royal and elite burials; that tradition lessened but never stopped through the centuries. Burials are rarely found in or beneath the temples.

Royal structures were often adorned with powerful symbols of rulership and illustrations of gods and sacred ancestors designed to validate their authority. These buildings were considered living entities; dedicated rituals would infuse them with a soul. Within the private chambers, leaders offered prayers and burned incense; they also performed solemn bloodletting rituals. The king's role and status were of supreme importance. The sacred glyphic texts, the iconography of the deities and

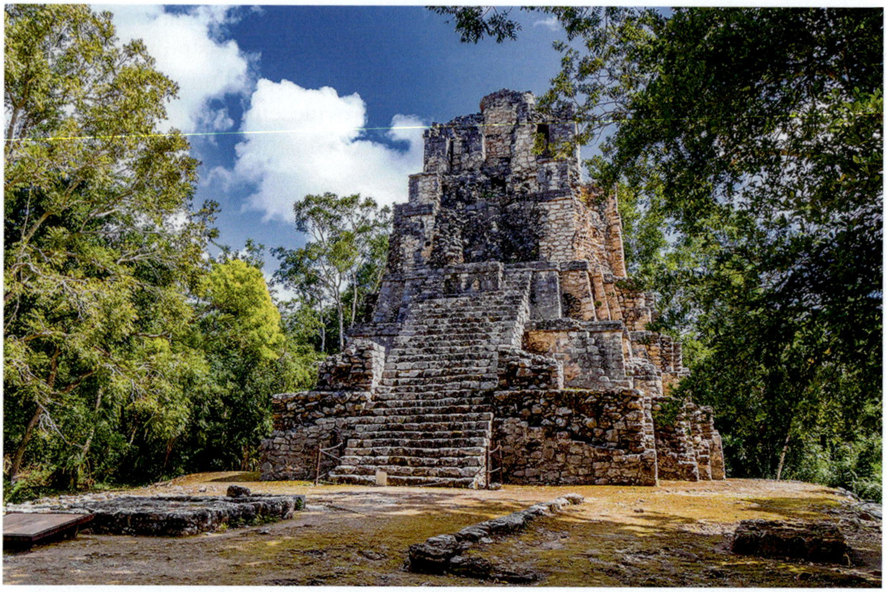

Fig. 4.2 Pyramid at Muyil Chunyaxche, Quintana Roo, site occupied from 350 BCE to CE 1200 (Credit Arkadiji Schell)

the consecrated rituals reinforced the different social ranks that existed between the king and the commoner (Fig. 4.3).

Nearly all Maya pyramid structures are decorated with a combination of revered mythological and spiritual tales and astronomical events that honored the gods. Elaborate roof combs decorated the tops of many Maya temples, making them even more impressive. The roof comb served no structural purpose; they were added merely as an ornate decorative device. The roof combs were comprised of two pierced framework walls that leaned on one another. Masons used a lattice of stonework to create the combs which proved to be strong and permanent. The roof comb was then covered with plaster and decorated with artistic depiction of gods or important rulers. At times they were painted with zoomorphic or anthropomorphic designs and rarely hieroglyphic texts were added. The beautiful pyramid-temples of Tikal display the best preserved roof combs.

Mayanists suggest that every 52 years, which relates to the Maya Long Count Calendar, many of their temples and pyramids would be remodeled or completely rebuilt. Some pyramids are stacked within each other like Russian dolls and one may encase up to seven inner buildings. This constant restoration system could also be linked with a new king or changing political conditions and served to revitalize and unify the people. Most remarkably, the North Acropolis at Tikal appears to be the result of 1500 years of revisions, rebuilding and improvements.

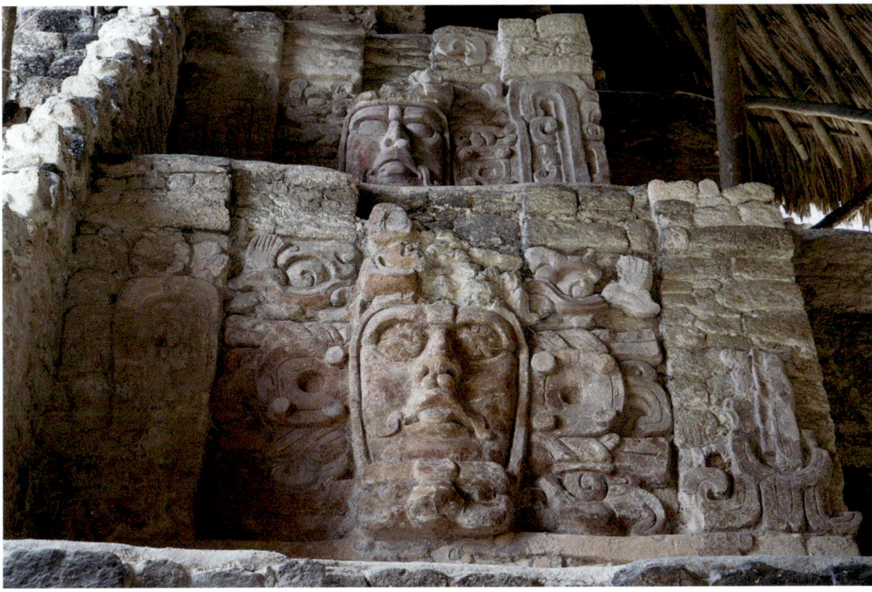

Fig. 4.3 Masks decorating pyramid at Kuhunlich, Quintana Roo (Credit Pe3K)

Maya Stelae

Around 400 BCE Maya sculptors began carving and erecting tall stone stelae (s. stela) to commemorate important calendrical or dynastic historical information. The stelae ranged from 4 to 5 m (13–16 ft) in height. Their production quickly spread throughout all areas of Meso-America. Every major city created a great number of monolithic stelae lining their buildings and decorating their plazas. The area with the largest number known was the great city of Calakmul, Mexico which has 166 stelae that survive, but their stone was of poor quality and is now badly deteriorated. Like most Maya structures, the megalithic stelae were quarried from limestone and were shaped and sculpted using only stone age tools: chisels, animal bones and wooden mallets. Most limestone found in Maya areas was easy to carve when first quarried and then became harder when exposed to the changing weather. The relief sculpture on the stelae covered all four sides and often featured life-size human figures, mainly kings, gods, and divine ancestors along with descriptive hieroglyphic text and accurate calendar dates. These monumental sculptures were vibrantly painted with the Maya-favored colors: mainly red, the color of the life force, with some yellow, black, and blue trim. Stelae are rarely found uncarved; probably, these monuments would have been colorfully painted originally, now washed clean (Fig. 4.4).

A stela at Quiriquá, Honduras stands over 24 ft tall, the tallest found so far; the base sunk into the ground adds 10 more ft. This colossal stela, weighing 59 tons,

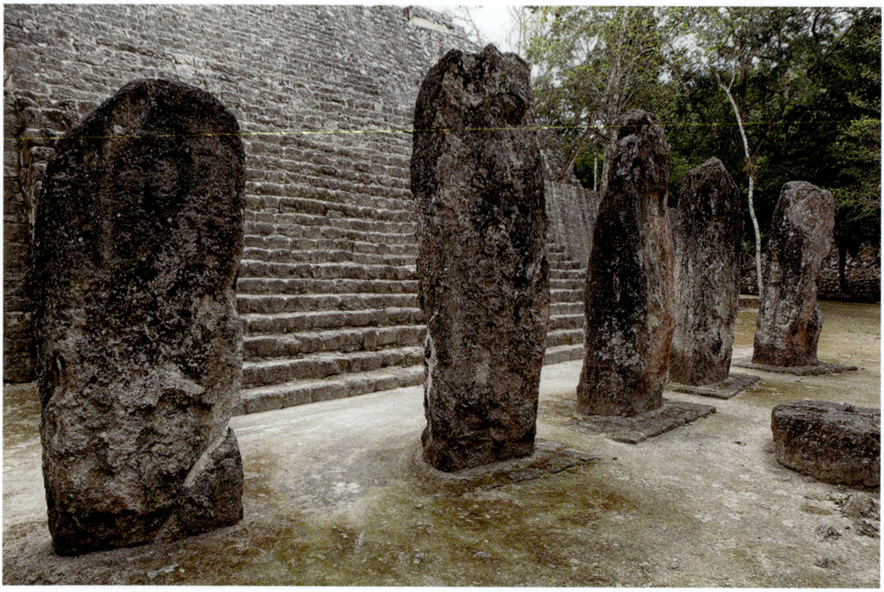

Fig. 4.4 Row of Stelae at city of Calakmul, Campeche (Credit Barna Tanko)

was dedicated on January 22, 771 CE to observe the end of the 16th K'atun, a period of 7200 days. These impressive stelae stood as effective symbols of royal power but were also erected to mark the end of their calendar cycles and record their understanding of time. Stelae most often honored and glorified the concept of divine kingship; the king would sometimes be shown standing next to his wife or his heir with his feet crushing the enemy. Stelae chronicled ritual moments in time as well as events in the heavens, such as important planetary conjunctions, and maintained them for eternity. These huge monuments also validated each ruler's reign and military triumphs and remembering his deeds. They also documented accession to power, lineage birth dates, special calendar dates, bloodletting ceremonies, dedication of buildings, marriage alliances. The stelae honored the heir to the throne and the capture of enemies and most importantly the deified ancestors.

The age of stelae carving and the supremacy of Maya kingship collapsed at the same time. The last known Maya stela was raised in 909. Presently, the many surviving stelae are vanishing due to looters; their great size and weight prevents them from being excavated intact. Instead looters try to remove individual sections which greatly damages the text and images; as a result the original context is lost. Portions of Maya stelae are preserved and studied in many museums of the world. The accurate drawings of Frederick Catherwood captured the original artwork on many stelae before they were eroded or looted. Although Mayanists have now deciphered the text and images on most of the known stelae, the question of exactly how the various stelae functioned in Maya city life is still being studied.

Maya Architecture and Archaeoastronomy

To accomplish these alignments, they looked to the stars, the home of the ancestral gods. The constancy and stability of the nightly lights traversing the heavens compared to the constantly changing conditions on Earth caused them to be viewed as living gods. By carefully observing their repetitive patterns, astronomers would come to know the behavior of the cosmic deities. This special knowledge encouraged sky watchers to seek omens to predict and order their daily lives. People needed assurance to determine the outcome of wars, the production of crops, and indications of the future. Reading the motions of the revolving sky acted as a text capable of revealing crucial information to those trained to ask the proper questions and interpret the answers offered.

Many buildings were aligned in accord with their astronomer/priest's interpretation of the orbits of the stars and the transits of the Sun, Moon and planets. Astronomical alignments of buildings played an essential role in Maya culture, defining and highlighting the key positions of the heavenly bodies. Observatories and temples were carefully positioned in order that the rising point of a celestial object, or some other celestial transit, could be precisely observed. An orientation calendar based on positions of sunrise on the equinoxes and solstices already existed by the Early Classic transition period (ca. 278 CE). These meticulously aligned structures with specific orientations were often placed in more private areas than around the central plaza.

The many years of archaeoastronomical studies in Mesoamerica have demonstrated that the non-uniform distribution of orientations in public and religious architecture is due to astronomical interests. No other purpose, such as climate, topography or defense, can account for the extensive and long-lasting alignment groups. The only plausible reason for the concentrations of alignments within particular azimuth ranges would be the rising and setting points of celestial bodies as reference points.

Archaeoastronomers have studied innumerable building alignments in the vast Maya area. Archaeologist Ivan Šprajc uses statistical analysis to discover the repeated patterns in the alignment of the principal architecture in Maya civic and ceremonial structures. The results of his huge, intense studies show the solar dates that were most important for Maya leaders, those who commissioned the architectural alignments. Studies found that most urban orientations appear in a zone 8° to 18° east of north with many focused to 14° east of north. Entire Mesoamerican cities were aligned to that narrow zone. The great city of Teotihuacan was designed on a grid system astronomically oriented 15.5° east of north. This variance from true north was not an error; the designers chose this position to align their cities and temples to unite them with the sky and the reliable motion of the celestial bodies. Other buildings' eastern orientation was directed at 25° south of east, which aligned them to sunrise on the winter solstice; those directed to 25° north of west are aligned to sunset on the summer solstice.

Maya monumental architecture became the setting for empowering leaders and impressing communities during rituals; connecting their huge monuments to the heavens was their ultimate achievement. Their ability in constructing and aligning their sacred architecture demonstrated knowledge of astronomical phenomena and accurate calendar-keeping through time. Buildings were even reoriented as the rising position of a designated star changed after a lengthy period. Grand public ceremonies took place to celebrate this intermingling of heavenly light with structures on Earth which helped to create a feeling of unity and a sense of awe. When these heavenly interactions produced light and shadow effects designed to occur on anticipated dates, they honored the ruling class and strengthened social accord. These special celestial alignment events reinforced the existing political order.

In addition to using their astronomical knowledge to orient architecture, the knowledge of astronomy enabled the planning and production of agriculture in agreement with the predictable weather cycles of the area. The dates recorded on buildings often denote agriculturally significant seasons; the intervals separating the solar dates are usually multiples of 13 and 20 days. These orientations enabled the use of observational calendars to arrange farming activities and the associated rituals and festivities.

Maya E-Group Complexes

An important innovation began during the Late Preclassic period, ca. 300–200 BCE, which was the beginning of triadic-style architecture, consisting of a large pyramidal platform that had three buildings on the top. These tripartite buildings had a dominant central structure with two smaller structures facing each other; this grouping became the most frequent architectural design during that early time. The development standardized architectural formats provided a solid, organized religious and political basis for a maintaining a sound and secure community. The Maya's uniform, consistent religious philosophy was an additional feature that stimulated and promoted their cultural sophistication.

This specific formation of buildings called E-Groups were constructed in a great number of Maya cities. Like pyramids, central plazas and ballcourts, E-Groups formed a core of Maya identity. The E-Group complexes were the first planned configuration of buildings in Maya settlements. Originally all E-groups seemed to be observatories for marking the motions of the Sun. It was thought that the astronomical alignments incorporated into all E-Group complexes were aligned to the Sun's solstices and equinoxes but this theory was soon proven wrong by archaeoastronomers. The latest astronomical studies have shown that the orientation of these widespread groupings are highly variable. Perhaps some E-Group alignments record significant dates important to agriculture as well.

The conventional E-Group has a symmetrical ground plan and includes four buildings; its central axis is directed roughly east-west. An E-Group format consists of four buildings and a raised platform that extends approximately north to south and is directed toward the eastern horizon and the rising Sun. A stepped pyramid is the main structure on the west side of a plaza facing the center of the extended platform that faces east. The three buildings are carefully positioned atop the long platform; the central building (E II) is much larger with a smaller replica (E I and E III) on each side. The central building is positioned directly opposite the main pyramid which has several levels and an eastern stairway leading to the summit where an astronomer-priest could sky watch from a prime observation deck. He could observe the Sun as it rose over the buildings; he could also watch the stars and planets as they transited the heavens throughout the night (Fig. 4.5).

The Sun's horizon extremes, north and south, and its equinox position can be observed from the western structure which is aligned to the three structures to the east. On the winter solstice if a building is properly aligned, the Sun rises over the southern-most corner of the southern structure (E-III) when viewed from the middle of the main structure (E-VII). At the equinoxes the Sun rises directly in the middle of the central building (E-II). Sunrise on the summer solstice occurs at the north corner of the northern building (E-I). The E-group complex seems to be an observatory for studying the Sun; the E-Group alignments were also thought to be involved in developing the Maya seasonal calendar necessary for agricultural success. The origin and purpose of this architectural grouping is still debated but in at least a few examples it appears that these building were erected for no other reason than for recording accurate astronomic data.

During the peak of Teotihuacan influence on Maya architecture, there was a hiatus of E-Group construction. But as Teotihuacan's influence faded, Maya cities began to look back to their own architectural history; they resurrected the creation

Fig. 4.5 E-Group Complex drawing (Credit Ivan Sprajc)

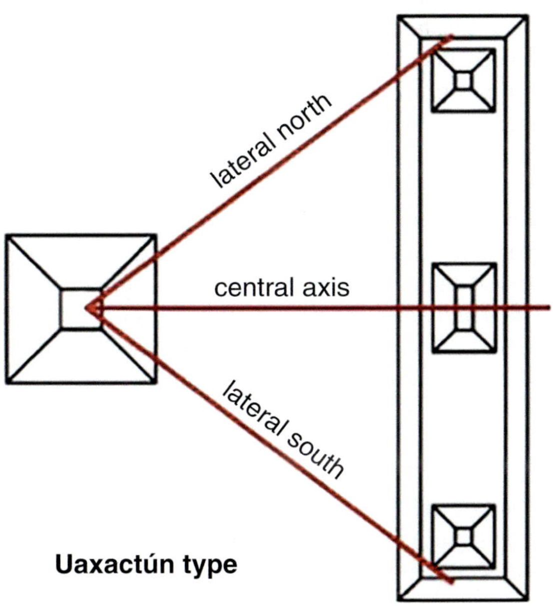

of pseudo-E-Groups. Their function changed from serving mainly astronomical roles to fulfilling principally social functions in the early Late Classic. The E-Group was part of the Maya identity; even small ceremonial centers erected E-Group complexes.

What started as an astronomically directed complex changed through time and evolved to accomplish other purposes. Like so many Maya ceremonial and civic structures, the E-Group architectural complex provided a variety of other ceremonial uses, along with the astronomical functions. The oldest known and earliest stage of E-Groups date to the Middle Preclassic period and were utilized throughout the Maya lowlands. This shows that they played a role in administrative and astronomical objectives in the earliest Maya ceremonial sites.

Most E-Group complexes were built in the Middle and Late Preclassic and Early Classic periods (1000 BCE–600 CE) and often had several construction phases. Later renovations and repairs often altered the original alignments of the complex shifting their accuracy over time. In 1924 the first example of a Group E-type complex to be recognized with solar orientations was discovered at the Maya city of Uaxactún, Guatemala. Dating to around 600 BCE the groupings astronomical alignments were measured and shown to be accurate to less than 2° north or south of due east. An even older documented E-Group complex, that was also astronomically oriented, was discovered at the site of Seibal (El Ceibal), dating to around 950 BCE. Dating to the Preclassic period, the E-Group complex at Tikal is also quite early, begun between 700 and 600 BCE. This complex was remodeled and enlarged between 500 and 400 BCE; then another renovation occurred sometime between 400 and 200 BCE. E-Group complexes are concentrated mostly in the central lowlands of the Yucatan Peninsula; they appeared in almost every city, even in the smaller settlements.

Over 160 archaeological sites with this same E-Group, four-building pattern, have been uncovered to date; only 20% have been excavated and reconstructed. Each E-Group complex seems to have been custom-made for the local topography adapted to the individual settings where they were built. Scholarly opinions vary; some think the E-Groups have a basic role in celestial observations; others consider them only symbolic or with only commemorative references to heavenly cycles with little to no real observational use. The astronomically significant directions, measured in the early E-Groups, assist experts in understanding one of the dominant aspects of ancient Maya architecture and urban planning.

The strictly astronomical theory of E-Groups has been challenged by archaeoastronomers since many of those studied have varied orientations and are not always aligned to the solstices and equinoxes. Experts suggest that E-Groups were also used for ritual and funerary functions; others propose that the artwork decorating some E-Groups relates to the mythological creation story, so that directed their usages. The groupings were reclassified as a “recurring architectural assemblage”.

When determining a building’s position, it is not difficult to find some astronomical alignment, since there are innumerable stars. But astronomers need to determine if that stellar position is merely by chance or if that particular orientation is significant. Was it repeated in other complexes? Scientific data supporting these varied theories is sometimes ambiguous or nonexistent.

In 2021 archaeologist Šprajc conducted a thorough analysis of seventy-one E-Groups in various Maya cities. After studying this large amount of alignment measurements and other evidence relating to E-Groups, he could find no consistent evidence of accurate astronomical alignments. The study indicated that E-Groups are found in practically all early cities in the central Yucatan Peninsula, but the celestial orientations embedded in the earliest E-Groups were later transferred to other types of buildings and architectural compounds. Some E-Groups were aligned with dates in March and in April which noted the auspicious time for performing ceremonies to ensure the start of the rainy season.

Šprajc argues with archaeoastronomers, asserting that the E-Groups were built mainly for ritual purposes but had no particular or prominent role in astronomical observations. Depending on the region and time period, there were large variations in the size and shape of the E-Group buildings. They had practical and symbolic importance but were not observational tools in the modern sense. He suggests that their significant role was to provide a stage for ritual ceremonies which imparted social and political implications to the E-Groups. Specialists now call the E-Groups compounds Commemorative Astronomical Complexes and "horizon calendar complexes". An accurate understanding of the complex's symbolic astronomical function is essential to gather knowledge of the fundamental principles of Maya architectural and city planning and their basic cosmological systems.

The E-Groups have an astronomical basis but with agricultural, ritual and ceremonial concerns. The observational orientations first embedded in E-Groups were later assigned to better suited structures. The primary function of E-Groups has now been shown as religious, residential or administrative; the rituals carried on were aimed at supernatural forces and were performed to secure divine assistance from nature and the heavenly gods. But their earlier astronomically and cosmologically significant directions are important in understanding basic aspects of early Maya architecture and urbanism.

The consistency of the Maya leaders through thousands of years of aligning the main observation building to the central axis indicates an astronomical basis to city planning. Astronomy was the original function but the distinctive E-Groupings adopted many more through its adaptable form, including ritual, residential, funerary or administrative. Many buildings in Maya cities had important astronomical functions but not only to determine solstices and equinoxes. Most civic and ceremonial buildings too were oriented to sunrises and sunsets on agriculturally important dates in their calendar. Maya rulers performed public rituals to mark celestial and agricultural cycles to reenact mythical creation events. The sheer quantity of E-Groups or pseudo-E-Groups implies that these architectural designs were embedded into the Maya conception of required elements in public architecture. The Maya's architectural orientations and knowledge of astronomy, along with their precise calendar, all worked together; on occasion the Maya astronomers directed these strengths to produce a hierophany.

Hierophanies in Maya Architecture

In Maya cosmology, hierophanies were manifestations of the sacred; a profound act, in communion with the divine forces that unite the natural with the divine, the sacred with the profane. The word hierophany is the combination of the Greek words for "sacred" and "to bring to light". Philosopher Mircea Eliade states that myths describe a "breakthrough of the sacred (or the supernatural) into the World – that is, hierophanies". A hierophany is a physical manifestation of the sacred; the sacred provides the world with value, focus and purpose. Those that believe this interaction of Earth and Heaven agree that all things need to imitate or adapt to the sacred models established by hierophanies in order to have true reality.

The Maya used architectural hierophanies to strengthen and emphasize solar events on days that were auspicious for political and religious purposes. The most important solar positions used to display light events were, of course, the solstices, equinoxes and zenith passages. During ritual ceremonies, these natural light displays, that occurred on architectural backgrounds, centered the viewer's attention, producing a perception of importance and intervention that dramatized an ordinary event to become an extraordinary event. Rays of sunlight striking specific locales, entrances or inner walls at certain times of the day, or when the Sun reaches key positions in its yearly cycle, greatly embellished the person honored or the ritual performed.

A hierophany alludes to a powerful meeting with spiritual forces; they may take place in public, in private, or both. This spiritual event can manifest an encounter with what is sacred within humans; it could also occur within a special stone, a tree or any object in nature. An entirely different experience takes place for those involved in the dramatic celestial phenomena. The object or person involved seems to be no longer a natural entity but becomes the holy representation entailing religious veneration. Creating heavenly hierophanies was an important ritual in most early cultures but the form they exhibit varies within the many different societies. They were evoked in behalf of royal, political, military and spiritual leaders. A great number are described in my book titled, *Decoding Astronomy in Art and Architecture.* Hierophanies were common within the ancient Maya communities, but they also occurred in most early cultures through astronomical correlations with the celestial deities, especially in Egypt, Greece and Rome. Egyptian pharaohs were thought to ascend to the heavens on a solar boat to experience their hierophany and become a star in the heavens.

As an example, the winter solstice sunset struck a special sculptural panel at Palenque to honor God L; an interbuilding alignment there also aligns with winter solstice sunset. These solar alignments at the winter solstice performed an essential role in emphasizing the turning point of the year as the Sun reversed its slow descent along the horizon and signaled the coming seasonal change. At Calakmul and at Palenque's Temple of the Sun, sunlight would strike a person standing at the temple doorway on the summer solstice sunset. Later the entrances of both temples were rebuilt to allow full sunlight to strike a person more completely and intensify the dramatic solar event (See Chap. 10).

These spiritual encounters with heavenly light have been known to transpire in relationships with not only material objects but in accord with certain actions, psychological functions, or even entertainment. As the carefully chosen object becomes spiritual, the object will be infused with a new meaning that is symbolic, referring to supernatural values and other realities. A hierophany removes the object that expresses the sacred from other things, from similar objects that have not been beatified; these remain profane.

For the Maya these astronomical epiphanies were associated with various facets of daily life: agricultural, calendrical, political, and religious. Through these celestial light interactions with their art and architecture, the king's social and political power was greatly enhanced. Controlled exhibits of directed sunlight would have been an effective technique for strengthening the notion that the ruler was the living personification of the Sun on Earth. Along with the light effects on special days of the year, participants might add music, chanting, speeches, songs, poetry or prayers to accompany the rituals and festivities when these sacred events were celebrated in public plazas. The event may gradually intensify until it reached its highest point, transitioning from secular to sacred in both space and time. Many will be described at their appropriate locations.

References and Further Reading

Almers, J. J., & Rice, P. M. (2006). *Astronomy, ritual and the interpretation of Maya "E-Group" architectural assemblages*.", Ancient Mesoamerica, 17. Cambridge University Press.

Aveni, A. F., & Hartung, H. (1989). Uaxactun, Guatemala, Group E and similar assemblages: An archaeoastronomical reconsideration. In A. F. Aveni (Ed.), *World Archaeoastronomy* (pp. 441–461). Cambridge University Press.

Awe, J., et al. (2017). Of apples and oranges: The case of E Groups and eastern triadic architectural assemblages in the Belize River Valley. In D. A. Freidel et al. (Eds.), *Maya E groups: Calendars, astronomy, and urbanism in the early lowlands*. University Press of Florida.

Dolan, M. (2021). *Decoding astronomy in art and architecture*. Springer Publishing.

Dowd, A. S. (2015). Maya architectural Hierophanies. In A. S. Dowd & S. Milbrath (Eds.), *Cosmology, calendars, and horizon-based astronomy in ancient Mesoamerica*. University Press of Colorado.

Freidel, D. A., et al. (Eds.). (2017). *Maya E-Groups calendars, astronomy, and urbanism in the early lowland*. University Press of Florida.

Šprajc, I. (2021). Astronomical aspects of Group E-type complexes and implications for understanding ancient Maya architecture and urban planning. *PLoS One, 16*(4), e0250785. https://doi.org/10.1371/journal.pone.0250785

Chapter 5
The Maya and Astronomy

Earliest Maya Astronomy

The ancient world of the earliest settlements can only be understood through their architecture and surviving artifacts since there are no texts or images from that early period. The earliest Maya structures could have been planned according to location, climate, defense or geomancy, but new research shows that the early architectural alignments reveal a nonrandom distribution which can only be explained by using the rising and setting points of heavenly bodies. The surviving structures from the early settlements are scarce but researchers are studying orientations to not only the Sun's yearly cycle but also the motions of Venus and the Moon. By joining the early calendar with astronomical observations, the Maya astronomers could predict the current date and keep an accurate calendar even in cloudy weather.

Large numbers of ceremonial centers found along the southern Gulf Coast that date from 1100 BCE to CE 250 have been analyzed for their astronomical orientations. These represent the earliest evidence of building construction in Maya territories. These studies demonstrate that the sophisticated astronomical knowledge that the Maya possessed during the Classic and Postclassic periods was already underway nearly a millennium before it could be seen in the written records. According to archaeologist Sprajc, the new lidar-based research covered a huge area consisting of 84,516 km^2 (32,631 mi^2) stretching from the core Olmec regions to the western Maya Lowlands. Within this wide area experts identified an unbelievable 33,935 architectural complexes and mound groups. The discoveries included 478 standardized complexes that dated to the Formative period. Numerous E-Group complexes were included among the architectural remains of the early cities.

Among those sites analyzed there was a large clustering of E-W orientations (89%) much, much greater than N-S azimuths. Since there was no evidence for orientations to bright stars, the buildings must have been aligned to the sun's movement along the horizon. (A random distribution would have 57% within that angle.)

M. Dolan, *Decoding Astronomy in Maya Art and Architecture*,
Springer Praxis Books, https://doi.org/10.1007/978-3-031-61092-9_5

A reason for this high proportion of narrow east-west orientations other than astronomy is scarcely credible. The most wide-reaching architectural group was aligned to sunrises on February 11 and October 29, separated by 260 days which are skewed south of east. This is a common orientation throughout Mesoamerica. This alignment appears most often in complexes that range between 1100 and 750 BCE. This study provides the earliest evidence of the 260-day calendar cycle.

In various parts of the ancient world, monumental structures were oriented to solar solstices and lunar extremes, even before agriculture was established. Once agriculture became important for survival, astronomical observations became more important and more complex. Maize became a staple in Mesoamerica between 2000 and 1000 BCE. The need for solstice directions continued through the Middle Formative period. Monument building increased significantly from 1100 to 900 BCE and extended over wide areas. At this same time Mesoamerican settlements were beginning to adopt a more sedentary lifestyle and a stronger commitment to maize agriculture. The architecture began to show a variety of astronomical orientations constructed in line with the 260-day calendar and its sections of 13 and 20 days.

The 260-day calendar of Mesoamerica is unique since no other world culture produced this kind of calendar. Experts have puzzled over the origins of the 260-day calendar for many years. These solar horizon calendars can only be made and maintained by making observations from a specific location. In order to record a long series of solar positions, a more settled lifestyle is required. These more permanent settled groups probably lived in larger centers, such as San Lorenzo. Understanding and keeping an accurate calendar would require specialists with advanced knowledge of astronomy and calendrical concepts.

Data from the recent study shows that this 260-day calendar was used from 1100 to 750 BCE and possibly even before. The Gulf Coast Olmec area and nearby regions were the primary starting points for the initial development of the Mesoamerican calendric system and astronomically aligned monumental architecture. This study provides important data on the relevance of archaeoastronomical principles that lie behind architectural design and urban planning; these discoveries have been mostly ignored by traditional archaeological literature.. The study demonstrated that orientations based on astronomical and calendrical bases were embedded in all civic and ceremonial buildings and they were duplicated by many nearby buildings, often dominating large portions of the city centers.

Astronomy and Archaeoastronomy of the Maya

When trying to determine the orientations of buildings in early Mesoamerican settlements, look to the sky for the clues. Studies by archaeoastronomers have shown that the more important civic and ceremonial buildings in Maya city-states were almost all aligned to sunrises or sunsets on significant dates, but the origin and range of their orientation practices are still unclear. Because the sky presented the basic references for alignments in space and time, the viewing of celestial patterns provided practical and useful knowledge. The obviously perfect and divine order observed in the heavens allowed various ideas explaining the role of celestial bodies in the heavens and their impact on earthly matters. The astronomically based orientations discovered in ancient architecture and urban designs were hugely influenced by both the celestial motions and their supposed influence on Earth.

The Maya integrated their astronomical knowledge into their unique world view. Their worldview was not compartmentalized into animate/inanimate or body/soul aspects as are western world views. The stars and planets were alive and motivated by the gods each for certain purposes. The Maya's complex and sophisticated worldview was not only holistic but also cyclical with the many aspects of their daily life integrated into a picture that was renewed by and in harmony with the cosmos. Isolating their art, architecture or astronomy from their religion, mythology, political and social structure is to consider only a small portion of their valuable legacy. Experts have carefully and thoroughly established that astronomy and celestial motions were an essential component in Mesoamerican societies. A series of temples on a raised platform was a sacred mountain range, while a forest of stelae represented trees. Numerous examples can be presented to demonstrate that astronomical and cosmological knowledge was incorporated into their rituals, myths, religious and political beliefs, in fact, their entire worldview.

The Maya created their unique foundation for understanding astronomy and the cyclical movements of the heavens. Celestial order, seemingly permanent and perfect, was considered superior to the earthly and human order. They attempted to reproduce the perfect, reliable order they found in the heavens in their daily lives. The Maya worked to establish stability in their community within the disorder and destruction of the constantly changing conditions they encountered on Earth. They conceived of a basis to formulate and understand the daily and yearly motions of the heavens. Their observations and analyses enabled them to create accurate charts of the heavenly cycles and then aligned their buildings, cities and agriculture to coordinate with those cycles. They aligned their sacred architecture to form meaningful designs that affirm and attune to these heavenly cycles to create a cosmic connection.

Archaeoastronomers recognized that solar and celestial observations were essential to the Maya's perception of the world. The aligning of their architecture in association with the cycles of the Sun, Moon, planets and stars are evidence of its importance. The designs of the Maya world were built and maintained not only through mapping but through myths, historic tales, stars and orientation of buildings. As in every area of Maya studies, huge advances have been achieved in archaeoastronomy in recent years, both in new discoveries and in rejection of formerly held theories. Scholars study not only architectural alignments but include codices, epigraphy, ethnology, iconography and calendrics to determine a deeper understanding of a building's purpose.

The Maya oriented their buildings to the heavens, channeling the celestial power of the divine gods into human built architecture. The points of sunrise and sunset on the horizon marked important moments of the year that became sacred. Their preferred alignments were the Sun's rising in the east and the Moon and Venus as Evening Star settings in the west. A great amount of data supports this conclusion.

The discoveries of archaeoastronomers provide an elemental basis to understand how our ancestors made observations, calculations and judgements that ultimately shaped the modern world of astronomy. Other than crossed sticks, the Maya had little in the way of observational equipment for astronomy; the majority of their work was carried out with the naked eye only. The astronomical orientations designed into the temples, pyramids and palaces were attempts to recreate and preserve the cosmic order in their environment. The responsibility for attuning the city, the buildings, the people and the activities with the cycles of the heavens above rested solely on the king or ruler who was considered a semi-god, in close contact with the gods.

Even without instruments, they excelled in their maintenance of accurate astronomical charts and their mathematical computations. The Maya conceived of the concept of zero and had a glyph for it by 400 CE; Europe did not understand the concept of zero until the thirteenth century, 800 years later when they adopted Arabic numerals. They applied their astronomical knowledge to the area of design and construction with numerous temples and buildings aligned to assist observers when watching for celestial positions. They used not only positions of structures but angled walls, doorways, windows, and small monuments to align with particular positions, mainly risings and settings of various celestial bodies. The Maya firmly believed that the Sun, Moon and planets and constellations had powerful influences on people and on nature. They understood that they needed to act in accord with the heavenly cycles. Two astronomical events that were particularly noticed and feared were eclipses and the first appearance of Venus as the Morning Star (Fig. 5.1).

The architectural orientations most often found by archaeoastronomer were to the solstice extremes of the Sun and to lunar standstills. The zenith orientations

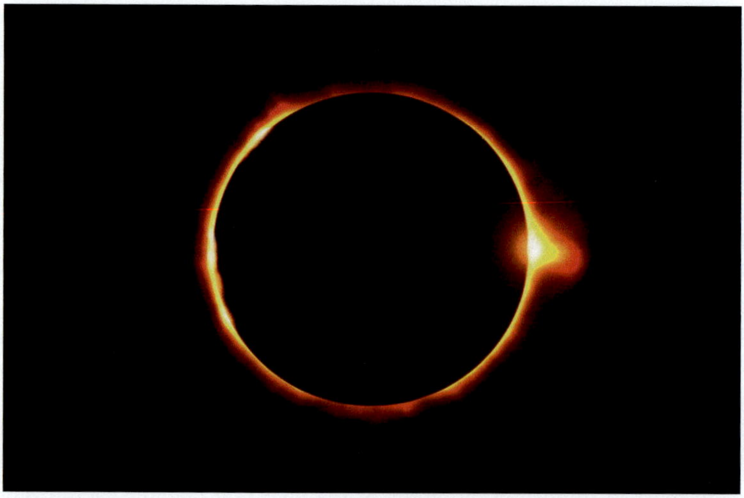

Fig. 5.1 Total Solar Eclipse, a time of fear and danger for the Maya, called "biting the Sun" (Credit Gabor Tinz)

were more often directed to the overhead position of the Sun rather than its rising or setting; these movements attracted less notice. Some evidence suggests that the first solar zenith passage each year could have been the most prominent day of the agricultural cycle, important because the zenith date nearly matches the start of the rainy season at their latitude. The less seen but more complex alignments were aimed at other celestial phenomena, such as heliacal rises of certain bright stars, or sunrises or sunsets at certain dates of the year that often had agricultural significance.

The buildings aligned to these astronomical events were not considered observatories because their main purposes were actually religious, residential or political, not skywatching. However, charting important celestial events could have been easily accomplished without building huge structures and orienting them precisely to the heavens. Accurately aligned sticks or raised stones could provide the same information. These orientations were combined with the Maya's beliefs in the structure and operation of the cosmos, and the functions of the city-state or community. They organized the universe compounded with their own myths, cosmologies and world views. They designed their myths to incorporate and retain important astronomical information .

There appears to be no direct connection to the type of building, temple or pyramid and its orientation. The sunsets at the equinox are noted more frequently than equinox sunrises. The orientations to the Sun found in religious and civic buildings are usually separated by intervals that are multiples of 13 and 20 days. This pattern has been found in most Mesoamerica cities. The 13-day and 20-day time periods were basic units of the Maya calendar system; these orientations were easy to manage in keeping their calendar. These dates were important for timing agricultural tasks and related ceremonies. These observation were important for the 260-day calendrical count where the cycles of 13 and 20 days were interconnected.

Overall, the Maya related the east with the Sun and the dry season and the west with the Moon, Venus as Evening Star, water, maize and the rainy season. This is appropriate since the Sun rises in the east every day and the Sun's light is more powerful in the dry season. Archaeologist Šprajc states the interest in aligning to the equinoxes was slight. Instead he believes that the Maya were more interested in the "quarter days" that occur 2 days before and after the spring and fall equinoxes. The quarter days divide each of the two halves of the year into two almost equal parts of about 91 days each. These dates mark the mid-points in time between the solstices which is closer to the true intent of the Maya astronomers.

The astronomically aligned building had a practical function of pointing out the moments in the seasonal year that were most important. They reinforced the Maya concept of uniting time and space. The orientations to Venus and lunar extremes were not useful for timekeeping but instead showed the symbolic significance of these celestial bodies that were essential to their religion and worldview. The building alignment did not stand alone as a concept but intertwined with the city design, with other aligned buildings and with the surrounding landscape.

Because the Maya lived in independent city-states with no overall political affiliations between them, each city-state acted in its own best interest. There are no architectural characteristics that apply to every settlement; many variations are found in the numerous Maya cities; each area has its own particular manifestation and preferences dictated by the ruler. No rules of alignments or spatial relations can apply to all areas and every time period. Each city-state had its own relationship with the heavens and followed their individual preferences in alignments, but every city had numerous alignments with the heavens. The leaders utilized these alignments to create special light effects by capturing solar and lunar rays that would illuminate their artistic carvings and their buildings. The timed appearance of the light beams could be used to spot light the king on specific occasions to enhance his power and position, creating a cosmic sound and light show. The special light effects emphasized the blessing and approval from celestial gods bestowed upon the divine ruler.

Both ancient and modern Maya cities were generally designed as microcosms of the four-quartered world with symbolic directions, a replica corresponding to the macrocosm, a miniature of their concept of the universe. The four corners and bearers of the sky were positioned at the four solsticial points on the eastern and western horizons. The movements of the heavens were not considered abstract mathematical concepts but meant to relay messages. The Maya believed that stars and planets were living beings; powerful forces inspired by the gods with particular purposes. The king held the highest position and asserted himself as the principal authority that controlled the cosmic order. Displays of power and control performed before the public were the methods used for social cohesion. A close relationship between Maya kings, chief deities and favored celestial bodies can be readily seen in iconography, historical sources, glyphic inscriptions. It can also be seen in the décor of noble residences and in civic structures. These universal features are brought together through the astronomically aligned architecture; together they exemplify their cosmological beliefs.

Motions of the Sun

The Sun was by far the most important celestial body for the Maya; it was the basic unit of the Maya day which was marked by the Sun rising from the Underworld at dawn and crossing the sky to be swallowed once again into the Underworld. All blessings came from the Sun. Most are aware that throughout the year, the Sun's rising point shifts along the horizon between two extremes, the northern solstice (June 20) and southern solstice (December 20) where it stalls a few days, but it passes through the equinoxes or half-way points (March 20 and September 23) rather quickly. Because Earth's orbit is elliptical, the days between the two solstices are not exactly the same.

The Sun appears to travel its path against the background of specific groups of stars, the twelve constellations of the zodiac. The orbits of the Sun, Moon, Earth and planets travel along this plane called the ecliptic; it spans about 8° of arc. Our Sun progresses through the starry background about 1° each day; but the Sun and constellations together move in a westward direction that opposes its daily motion. In reality, the center of mass of Earth revolves on an ellipse, with the Sun at one focus as it rotates on its axis. The Earth's axis is not perpendicular to the ecliptic but inclined by 23° 30′.

When the Sun, Moon and Earth revolve in their orbits within the ecliptic plane, they cross paths; at times the three line up exactly and a solar or lunar eclipse occurs. Another important element in astronomical measure is the celestial equator, an imaginary celestial sphere that projects the Earth's equator on the sky. Because of Earth's axial tilt, the celestial equator too is inclined by about 23.44° with respect to the plane of Earth's orbit. Throughout the year, the constellations appear to shift in position, relative to the Sun. Those constellation stars that appear just above the eastern horizon as the evening sky darkens rise a few minutes earlier each day; as a result, new constellations will begin to appear. The first moments of appearance and disappearance of a star or planet is called its heliacal rising or setting. Through history these sightings remained one of the most utilized and reliable timekeeping methods.

The zenith point of the Sun, when it is directly overhead and casts no shadow, was not noticed in western time keeping because of its restricted appearance. It only occurs in the latitudes between the Tropic of Cancer and the Tropic of Capricorn. In contrast the zenith point of the Sun was of utmost importance to the Mesoamerican people. Archaeological and ethnographic discoveries show that the first solar zenith passage of each year was thought to be the most important day of the agricultural year. Noted because the zenith marks the beginning of the rainy season at many latitudes of the Maya lands. At some Mesoamerican settlements small shafts were situated within specially placed buildings to allow the zenith Sun to send a ray down the tube to officially mark the correct day of zenith passage. During the twentieth century important rituals associated with the solar zenith, still performed by Maya priests were recorded by ethnographers. Their studies showed that the observance of the zenith was even now important and honored. Archaeologists often examine ethnographic records when available to find which features of the sky were important to the living Maya and then looked for archaeological correlates.

In Maya beliefs, the Sun was the most powerful natural force; it was given form as the cross-eyed Sun of the day and as the Jaguar God of the Underworld; the Sun protected them and guided them. The Maya ordered their lives in accord with the movements of the heavens. The motions of the heavenly bodies were thought to be reenactments of cosmic mythical events. They changed natural features and built architectural programs to designate significant points on the horizon, particularly that of the Sun. The glittering celestial bodies were perfect and eternal; through time they became personalized and deified, thought of as alive. The stars and planets inspired extraordinary artworks, as well as poetry, song and myth.

Solar ideology is not a separate practice, event or image; it is a correlated set of beliefs, ritual events and symbolism centered on the daily appearance and yearly transit of the Sun. Its motions were a validation of order and security in Maya daily lives. The ancient Maya kings looked to the Sun to reaffirm their authority and their right to rule. The Sun god was often depicted as a king sitting tall on a throne or as a ruler carrying the two-headed ceremonial bar. The king usually took the name of the Sun as part of their royal name, as a symbol of divine authority. The ruler wore iconographic symbols that connected him with the Sun and with other gods. When the king wore the attire of a god, he became one with that god. The Maya king could be integrated with the Sun god. Ancestral Maya kings were shown joined with the Sun God; they were sometimes shown while vertically descending from the solar zenith. The solar aspect of a king appears to suggest apotheosis and life after death; it could also be considered a pathway to the Underworld.

The Motions of the Moon

The Moon was of utmost importance to Maya astronomer priests; they viewed, analyzed and predicted the Moon's movements through its monthly cycles with amazing accuracy. Watching the lunar phases and the first appearance of the crescent Moon were of utmost importance in the daily ritual lives of the Maya. The initial crescent of the Moon first appears in the western sky soon after sunset. The Moon's conjunctions with planets and constellations were also carefully observed. The Moon's phases marked the times for planting and harvest and they announced the rainy and dry seasons. The Maya understood and kept accurate records of the Moon's synodical cycles, periods of visibility and invisibility, and the eclipse cycles. They also recorded the extreme points on the northern and southern horizons reached by the Moon's rising and setting.

The movements of the Moon display greater variability and are clearly more complicated than those of the Sun and it travels much more swiftly. The Moon revolves around Earth, tidally bound as Earth revolves around the Sun, always showing the same face to those on Earth. The Moon completes its orbit of Earth in relation to the fixed stars in a sidereal month or about 27.3 days. The cycle of the Moon's phases from one new Moon to the next comprises its synodic month which transpires every 29.5 days. A further complication is the Moon's axial tilt; this measures only 1.5422°, much less than the Earth's 23.44° slant.

The Moon's size appears to be almost the same as that of the Sun since the projected diameters of both are 30° as viewed from Earth, but the Sun is actually about 400 times the lunar diameter and distance. This expedient similarity allows the Moon to block out the brilliance of the Sun during eclipses, permitting important visual perceptions for astronomers. Some information on the motions of the Sun can be gathered during a partial eclipse, but a total eclipse allows the disc of the Sun to be entirely blocked, cutting off its direct light. At those few minutes of totality astronomers have the opportunity to observe and photograph the Sun's corona which is normally obscured by its intense brilliance. Only during an eclipse can astronomers view and photograph the Sun's corona, the solar atmosphere, and the brightest stars near the Sun that are ordinarily invisible.

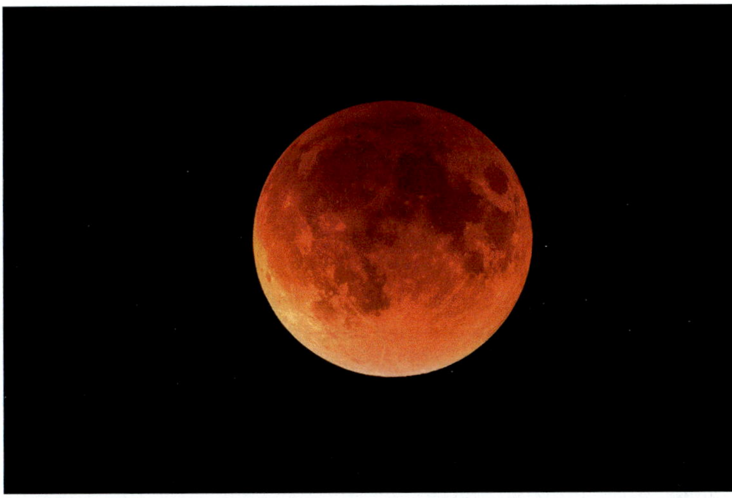

Fig. 5.2 Total Lunar Eclipse, a time of sacrifice and prayer (Credit Chris Harwood)

There are two points where the orbit of the Moon intersects the ecliptic, called the lunar nodes. The ascending (or north) node is where the Moon moves into the northern ecliptic hemisphere. The descending (or south) node is the point that the Moon enters the southern ecliptic hemisphere. Solar eclipses at the ascending node occur in March, solar eclipses at the descending node in September. Therefore, lunar eclipses at descending node occur in March, lunar eclipses at ascending node in September. Many megalithic monuments of England and Ireland appear to designate those particular times (Fig. 5.2).

The Moon deviates from the center line of the ecliptic plane by 5° with a maximum deviation on either side but does slowly return. The Moon travels in a zone about twenty moon-breadths wide, half the year north of the ecliptic and half south of it. Tied to the Sun, the Moon travels to its northernmost and southernmost limits along the horizons along with the Sun. This varying range between the extreme points of its declination, called its standstills or lunistices, changes over the course of about two weeks or one-half a sidereal month, 13.66 days. Declination is a celestial coordinate measured as the angle from the celestial equator, analogous to latitude. The declination of the celestial equator is 0°. One major or one minor lunar standstill occurs every 18.6 years due to the precessional cycle of the lunar nodes at that rate. Examples have been found that show ancient cultures marked alignments to the moonrise or moonset on days of lunar standstills These complex motions of the Moon's cycles make gathering accurate data and demonstrating lunar alignments much more difficult for archaeoastronomers than for verifying solar alignments.

Since the Moon appears in a new place of the sky every night and at different times, Maya astronomers realized that they could use the Moon as well as the Sun to measure time. Their lengthy records of the Moon's positions over long periods of time is called the "Lunar Series". Their lunar charts enabled Maya priests to predict eclipses and they can still be read in the Dresden Codex. The phases of the Moon continued to be an essential factor in other areas of daily life, especially scheduling the yearly agricultural activities along with the associated ceremonies.

Inscribed on a monument in Palenque, scholars discovered the textual evidence that the Maya knew that there were 405 moons in 11,960 days; the text also gave an average lunation period of 29.53086 days. Their sophisticated knowledge and degree of accuracy is quite remarkable. The actual lunation measured by current scientific methods is 29.53059 days. Some of their measurements differ from systems of today. Currently, astronomers consider the new Moon to occur at the conjunction of the Sun and Moon when they have the same ecliptic longitude. But the Maya considered the start of the lunar cycle as the first day that the very thin crescent waxing Moon could be seen, which is actually about two days after the actual astronomical new Moon. Because the full Moon always rises around sunset and sets at sunrise, the extreme positions of the Sun and Moon are viewed on diametrically opposite sides of the horizon. Due to celestial mechanics, the full Moon brightens the night for the longest period exactly in the part of the year with the shortest days and is hidden providing no light on the longest day of the year.

Lunar Architectural Alignments

Research of archaeoastronomers in the Maya Lowlands has discovered architectural alignments that match the major and minor extremes of the Moon, also called lunar standstill positions, on the horizon. A large group of alignments related to the major lunar extremes were revealed in recent quantitative analyses. Most of the structural alignments were discovered at cities along the northeast coast of the Yucatan Peninsula which verifies that they were oriented intentionally, not randomly distributed. This data is supported by the fact that an important lunar cult to the Moon Goddess Ixchel was located in that area during the Postclassic period (See Chap. 6).

Many studies have confirmed that throughout Mesoamerica the alignments in civic and ceremonial structures display a distinct pattern and group around certain azimuthal values. This wide-spread appearance of particular orientations shows that astronomical bodies on the horizon were used as reference points. As we have seen, most structures were oriented to sunrises and sunsets on specific days; many structures were aligned to Venus as well, but most other celestial bodies were considered less important and were less studied. Current research has revealed a significant number of buildings that show connections with positions of the Moon along the horizon. Every month the Moon can be observed as its rising and setting point moves along the horizon between its most northerly and southerly points. The Sun's transit covers almost the same span of the horizon in a year, solstice to solstice, but the lunar extremes are further north and south than the Sun. This signifies whether the building is oriented to the Moon or the Son.

The patroness of the Moon Goddess, whether Ixchel, Goddess I or Goddess O, is still debated, but the widespread popularity of the lunar cult cannot be argued. The Moon cult devotions drew massive pilgrimages to her cult centers on the Isles of Mujeres and Cozumel during the Late Postclassic period. The coastal setting surrounded by water would emphasize dramatically the rising and setting of the Moon and perhaps arouse specific emotions and attract wider attention to the Moon. The natural watery environment would provide the perfect location for rituals, ceremonies, prayers and processions to the various temples and shrines honoring the Moon and the goddess.

The major and minor lunar standstills reoccur at 18.6-year intervals but at these times the Moon is not always in the same phase, new to full. The full Moon extremes always occur near the solstices; this is close to where the Sun also reaches the extreme points of its transit along the horizon. Due to celestial mechanics, the Sun rises and sets at its farthest southern point when the full Moon reaches its northern extreme which always happens around the December solstice. The full Moon rises and sets at its farthest southern point when the Sun reaches its extreme northern rising and setting points around the June solstice.

In the Maya Lowlands analyses of structures with lunar orientations discovered that their alignments functioned toward the west horizon and the setting of the Moon, not the rising of the Moon. The solstice orientations, that were measured, linked with alignments to major lunar extremes tend to show a preference to the Sun's December solstice sunrises, rather than with June solstice sunsets. This indicates that they functioned mainly to the east. Orientations to the minor lunar standstills are much less common than major standstills, but they have been discovered. The most apparent location of lunar minor standstills alignments would be suspected to appear on the northeast coast of the Yucatan which is exactly where they have been found. Not surprisingly, the Late Classic Maya city of Tulum on the Yucatan northeast coast has two structures aligned to the Moon's minor standstill.

The sophisticated astronomical knowledge held by the Maya astronomer priests included the complex motions of the Sun, Moon, planets and stars. They also understood the periodic oscillations of the Moon's extreme rising and setting positions. This advanced knowledge was soon lost after the Spanish Conquest when the Maya were subjected to the severe Christian indoctrination. In spite of the attempts to remove all beliefs of their "pagan" past, many of the ancient beliefs and associations with the heavenly motions have persisted, although diminished and altered, among the present-day communities. The motions and phases of the Moon still remain useful considerations in organizing the agricultural calendar.

In addition to the spiritual concepts behind orientations to lunar standstills, recent studies have provided evidence that a correlation exists between tides, rainfall patterns and temperatures and the lunar nodal cycle of 18.6 years. There exists no direct evidence that the Maya were aware of these weather-related connections to lunar cycles, but they associated water and fertility to the Moon Goddess. The possible weather correlations could be another incentive for closely tracking the lunar cycles.

The alignments of a substantial number of civic and ceremonial buildings in different sections of Mesoamerica relate to major lunar extremes on the horizon. Any other interpretation of their placements than astronomical associations with the Moon could not be justified because there is no other celestial object that fits these orientations. Another assurance of lunar correspondence is the fact that they are mainly located in the area where the cult of the Moon goddess was most popular and celebrated. The positions of the lunar standstills always occur near the solstices; therefore other factors are considered, such as textual evidence or associated concerns, to assure the alignment was to the Moon not the Sun. There are numerous alignments directed toward major lunar standstills and many toward minor lunar standstills, but the minor standstills are less certain since the reference point may have actually been the Sun.

The Motions of the Planet Venus

When considering all of the heavenly bodies that pass overhead in the night skies, the planet Venus certainly received the most interest, was the most highly observed, and most studied by Maya astronomers. After the Moon, Venus can often be viewed as one of the brightest objects in the nighttime sky. It was compelling to the Maya not only due to its radiance, but also due to its unusual rhythmic cycles that fascinated them. The special ability of Venus to switch from a morning to evening star led the Maya to both honor it and fear it. The rays of Venus were considered particularly dangerous at its heliacal rising, when it reappears as the morning star after its 8 day disappearance in front of the Sun.

The most essential planet in the artworks, mythology, and calendar of the Maya is undoubtedly Venus. The Venus glyph appears consistently in their artistic rendering and its representational icons were carved into their walls, monuments and stelae. Maya sometimes illustrate Venus as a god hurling spears to the Earth. Different attributes of Venus embody seasonal and timely concerns. Like the Moon, Venus has numerous personalities and identities. Venus was not depicted and understood in only one aspect; it actually had many meanings and pictorial forms. Its aspects changed with the seasons of the years and in its significant position in its cyclical orbit. Some traits of Venus appear to be connected to the rainy season and agricultural fertility; others reflect the dry season and warfare.

Maya astronomers kept records of Venus' entire planetary cycle which averages 584 days; five Venus cycles are almost exactly equivalent to eight Earth years. Each position of Venus, once noted, will reoccur at about the same calendar date 8 years later. Their daily life activities were often influenced by this celestial convention. The first and last appearances of Venus in its heavenly transit were quite important to the Maya. Venus seldom moves out of the plane of the ecliptic; always rises and sets near the rising and setting of the Sun. For about 9 months of a solar year, Venus is positioned in the morning sky above the spot where the Sun will rise. As the days pass, the planet slowly moves away from the Sun until it reaches its maximum declination of 46° from the Sun. Then it reverses and gradually moves back toward the Sun. At that position Venus becomes too near the Sun and its bright light is overwhelmed by solar rays, except for a few fleeting moments in the morning sky. When Venus passed behind the Sun at superior conjunction and between the Earth and the Sun at inferior conjunction, it is invisible. When Venus and the Sun are blended in images, it suggests that they are in conjunction. The brilliant star disappears as the Evening Star and then magically reappears as the Morning Star approximately 8 days later. It then repeats the same performance as an evening star. The cycle of Venus lasts for 583.92 days but it can vary up to 6 days.

Venus' northern and southern extremes are the same as the Sun's but over longer periods. The period of ascent when a Morning Star is very rapid, about 80 days and the Evening Star also makes a speedy descent of about 80 days, when it becomes lower in the sky. The period of most brilliance is during its ascent as the Morning Star. Each aspect of Venus and its celestial cycles are illustrated by specific imagery in Maya art. Scholars have suggested that the Maya astronomers observed Venus' maximum elongations, that is when Venus is furthest from the Sun. Since Venus extremes do not occur annually and do not always occur on exactly the same dates of the year, they cannot be especially useful for time keeping. The retrograde motion of Venus was important but only during the Classic period.

The Dresden Codex is filled with close observations of Venus, noting its heliacal rising, its first appearance in the pre-dawn sky, which marked a significant time period. The Venus tables in the Codex are arranged in rows, each column registers five periods of 584 days. Their Venus charts were extremely accurate, off only 1 hour in 400 years. The Classic Maya astronomers appeared to focus more intently on Venus as the Evening Star since 70% of dates of star glyphs studied refer to that aspect. The Dresden Codex tells that the Morning Star instead received greater attention during the Postclassic period. Drawings in the Dresden illustrate signs of Venus as a god striking his victims with arrows, which symbolize the first rays of light from the planet. Speared victims and associated ritual ceremonies are also shown. This star chart devoted to documenting the Venus cycle indicates its prime importance for Maya astronomers. The greatest emphasis was focused on the Venus Almanac of five Venus cycles in eight solar years .

The Maya associated Venus with the major god Quetzalcoatl, the Feathered Serpent, called *Xux Ek*, meaning Great Star. The Venus cult of the Feathered Serpent god was closely connected to the city of Tula; it also had great influence in other cities of the Maya world. The Tlaloc cult too is connected with Venus. Most building orientations associated with Venus correspond to the major northerly extremes of the evening star. The particular attention paid to these northern extremes is noted because they announce the start of the rainy season.. The Maya considered Venus responsible for the climate change. The predominance of this orientation undoubtedly reflects the overwhelming significance of this agricultural season. The Venus glyph appears consistently in engravings in codices and on stone carvings.

Maya astronomers kept accurate calendars not only for time keeping but also for recurring rituals and feasts. Their star charts were necessary for many purposes: astronomical, astrological, numerological and agricultural. The number eight is associated with the Maya maize god which is the number of days Venus spends in the Underworld before its rebirth as the morning star. The 8 years required for the Venus cycle to harmonize with the agricultural year, once again recalls the maize god's patronage of the number 8. These connections indicate deep relationships between Venus and the rains and between Venus and maize. Therefore, the Venus-maize-rain complex does reflect a cosmological reality. All of these Venus cycles may be evident in Maya dates and images.

Other Planets and Stars

The Maya believed the universe was designed according to a sacred pattern that could be understood through the cycles of the stars in connection with time, through its units of hours and days. Priests and astronomers kept amazingly accurate track of time using the motions of the Sun, Moon, stars and planets as their measure. The Maya were well aware of the five planets and watched them closely. Jupiter and Saturn are two of the brightest celestial bodies. When the Earth passes these two superior planets, they appear to stop moving in the usual direction of their transit and back up for a period of time. They seem to stop for a short time; then they resume their usual path along the ecliptic. This action is called apparent retrograde motion.

The Maya named those star groups that are the most easily seen in the night sky or that mark changing seasons in some way. They recognized certain star patterns as connected to patterns in their lives. They named the planets and attached omens that could be applied to the fate of the city-state and to their daily lives. The celestial events occurring overhead were thought to be mirrored by events on the terrestrial plane. As above, so below. The omens or messages from the heavens were viewed as commands to motivate human actions. The heavenly signs were the basis of their complex astrology. Understanding of the constellations recognized by the Maya is still unclear.

Pleiades consists of a small but bright cluster of stars in the constellation Taurus, called the Seven Sisters in western astronomy. The asterism appears to have only six very bright stars, but actually contains more than a thousand stars. The group glows with the blue light of hot, young stars and are quite distinctive and easy to spot in the star-filled sky. They were watched for the start of the farming season and for sea navigation from the earliest times, even noted by Hesiod and Homer. The rising of the Pleiades during April signaled the most significant day in the Maya year, the time for planting maize. The Maya called the Pleiades *Tianquiztle*, which means marketplace (Fig. 5.3).

The Milky Way is an enormous collection of stars that glows in the night sky; the Maya called it a white roadway or celestial river but it is actually a spinning disc of stars, dust and gas. The Milky Way Galaxy is composed of a central core of young, newly formed stars orbiting its central nucleus with four main spiral arm bursting with more stars. The hazy road is the edge-on view of the galaxy; it spans the night sky from one side to the other. The dark areas within the milky band are interstellar dust clouds that hide the stars around them. The Maya pictured the Milky Way as an

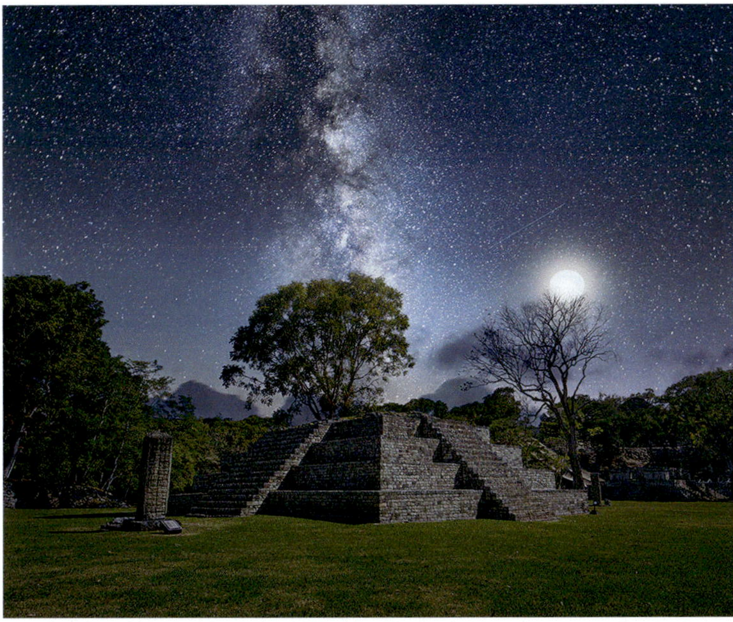

Fig. 5.3 Brilliant stars of the Milky Way and Full Moon over radial pyramid at Copan (Credit Soft Light)

enormous two-headed snake with one head representing life and the other death. The great sky serpent was thought to be an embodiment of the supreme Maya god Itzamna. From one of the serpent's mouths came the rising Sun in the east, the bringer of life. From the second mouth in the west came the consuming of the Sun and death.

Comets were named the "tail of the star" and were considered evil omens causing disease; also called "windy tail" and "fire-ill-omened star". They were frequently represented by a stellar image on a blue background emitting streams of smoke and often are considered an omen that a foretold that a ruler or noble in Maya society will die. Curiously that same interpretation appears in medieval manuscripts. Eclipses too have a variety of descriptions; the iconography and names for various stages and types of eclipses and the Eclipse Goddess are overwhelming. Comets were also objects of foreboding, called 'a star that smokes'. Over time, these links between celestial and terrestrial events observed since the earliest Maya settlements became deeply implanted into the Maya belief system and mythological stories.

References and Further Reading

Aveni, A. F. (Ed.). (1992). *The sky in Mayan literature*. Oxford University Press.

Aveni, A. F. (2003). Archaeoastronomy in the ancient Americas. *Journal of Archaeological Research, 11*(2), 149.

Garcia, A. M. (2014). *Astronomía, geometría y arquitectura en Chichén Itzá*. Academia.com, ISSN: 0719-4579.

Milbrath, S. (2017). The role of solar observations in developing the Preclassic Maya calendar. *American Antiquity, 28*(1), 88–104.

Sprajc, I. (2009). Astronomical and cosmological aspects of Maya architecture and urbanism. In J. A. Belmonte et al. (Eds.), *Cosmology across cultures ASP conference series* (Vol. 409).

Part III
Maya Gods and Mythology

Chapter 6
Deities of the Maya

Introduction

The Maya considered all Mesoamerican lands and natural features as sacred; the landscape, mountains, water and caves were alive with supernatural beings. The sky too was filled spiritual beings; the Sun, Moon and planets were believed to be living gods closely involved in the physical world. There was no distinction between the natural and the supernatural realms, they were thought of as a unified whole. But some locations were more potent with spiritual energy than others. The presence of especially sacred areas and pilgrimage centers shows that there were similar belief systems among various settlements throughout the Maya world; the energy of the sacred places attracted visitors with like beliefs. Some sites whether manmade or natural settings are so exceptional that priests and pilgrims would travel far distances to perform rituals and prayers there, especially the Sacred Cenote at Chichen Itza and the Temple of the Moon Goddess at Cozumel.

The Maya considered the world to be imbued with dangerous sacred powers that demanded ritual ceremonies which often included bloodletting. Offering sacrificial blood to the gods was a religious duty, required to maintain the cycles of life and of the heavens. The Maya passed on through the generations not only their society's history but their religious practices, sacred stories of the gods and goddesses and their essential creation myths. Understanding the vast variety and attributes of the Maya gods is complex and not fully agreed upon. Experts disagree and offer differing interpretations of texts and images. Maya gods cannot be defined in one particular category since on occasion many can merge with other deities.

The Maya gods were not abstract notions but beings that could be contacted in the form of masks or effigies. They could also be reached in dreams, in visions or imagined in smoke rising from sacred offerings. The gods were not like humans; they were superior beings capable of influencing the natural forces and performing a role in human destiny, for either good or for bad. Maya gods were sometimes

M. Dolan, *Decoding Astronomy in Maya Art and Architecture*,
Springer Praxis Books, https://doi.org/10.1007/978-3-031-61092-9_6

depicted in a fully human form when they could not be discerned from humans. The Maya rulers performed as intermediaries between humans and the deities; they were believed to have the power to control the supernatural order. Rulers claimed divine status through using the names of the gods as their personal names. Some royal portraits show leaders merged with gods. Maya artworks at times depict leaders attired as gods or perhaps they were considered gods themselves. The Maya trusted that their sacrificial offerings would impel the gods to carry out their requests.

The Maya considered the Earth as a flat, four-cornered surface and the sky above was held by four sky bearers or four world trees. In the center stood a blue-green ceiba tree with roots descending into the Underworld. The tree's trunk rose from the surface of the Earth or from a cave and its branches reached up toward the sky. The Earth's surface was similar to the back of a crocodile or turtle floating in a sea of water lilies; the balancing feature in the celestial realm was a two-headed serpent. Time was not thought of as linear but as cycles that brought continual creations and destructions. The Maya thought that the world was ordered by time which was measured by movements of the Sun, Moon, planets and stars. They amassed accurate records of celestial cycles and developed various calendars. Before the present world existed, several other worlds had preceded it. The current Maya world cycle began on August 13, 3114 and they thought it would complete its Long Count cycle on December 12, 2012. The cycle ended and the world moved on, although some doubted that it would.

Astronomical Gods of the Maya

The chief gods of the Maya pantheon were astronomical gods. The star gods were beseeched in Maya art and architecture for well over a thousand years and they were thought greatly involved with human affairs. The movements of the heavens were closely watched for they carried messages from the deities. Many events of Maya were planned to coincide with preferred celestial moments. In fact, a war might be delayed until the gods were in the proper position or a ruler might ascend to the throne of a Maya city only when a particular planet was visible in the night sky or in the desired position (See Chap. 10).

The Sun, Moon and Venus were the most well-known astronomical gods in ancient Mesoamerica. The Milky Way too was considered a god, while the constellations were visualized as celestial animals or special locations. There were many more heavenly bodies invoked in Maya worship but they are not as commonly mentioned. The Sun and Venus were thought of as male gods; the Moon carried both male and female attributes. At times the same deity could embody both male and female aspects.

The Maya honored a rich display of astronomical deities called "lights in the night". The number of Maya gods is quite large but not accurately known; the identities of some deities changed through time and in different areas. Later manuscripts speak of 166 deities but others tell of closer to thirty. Rather than a single, supreme being, the Maya concept of god was not one individual force but could be four or five different individuals. Their gods may hold multiple aspects of humans as well as animals. A god may appear as carrying not only male and female traits, both young and old, or kind or wicked. The gods were often visualized in temples, on monuments and on pottery.

The ancient Maya pantheon included an abundant and complex array of deities. In Maya mythology Itzamná was the ruler of heaven, day and night who was thought to live in the heavens. With his wife, the older version of Goddess Ixchel (Goddess O), they were supremely powerful, the parents of all other gods. Itzamná was a fantastic mythological being that embodied the opposites of our world (earth-sky, life-death, male-female, light-dark). The Mayan word Itzamná means caiman, lizard or large fish. Typical of Maya gods, he was known by many names; scholars refer to him as God D. He was a creator god thought to have created humans, along with the art of writing, the sacred calendar and was the patron of medicine. He was associated with the Sun God and God K. He also has four aspects assigned to world colors and the directional points of the compass and their assigned colors: east-red, north-white, west-black and south-yellow.

Itzamná was linked with the sacred World Tree which the Maya believed joined the sky, the Earth and the Underworld together. Illustrations of Itzamná usually depict him as an old man sitting on a throne with other deities around him. He may be seen as a wise priest with a hooked nose and large square eyes, wearing a tall headdress with a beaded mirror. He is also shown as a two-headed underwater serpent or with a combination of human and caiman features, called the Cosmic Monster. The two heads of Itzamná mark the two crossroads that are the intersection points of the Milky Way and the ecliptic. Maya chronicles imply that Itzamná has a quadripartite body with four colors that relate to the four celestial roads. The roads create a cosmic diagram showing the intersecting paths of the ecliptic and the Milky Way.

In Illustrations, it is thought that his white body indicates the rainy-season side of the Milky Way; while his black body evokes the underworld side of the Milky Way connected with Xibalba. God D, God K and the Sun God all became aspects of Itzamná when they are located in the Milky Way. Itzamná also manifests as the Cosmic Monster and holds celestial gods in his mouth; he carries the Sun, Moon and planets as they traverse the Milky Way. Various features of Itzamná show an association with the Sun, the Moon and the planets, probably symbolizing the time these celestial bodies were seen in the Milky Way. Important star groups are also related to Itzamná when they are where the ecliptic crosses the Milky Way.

Solar Deities

The Sun God is referred to as K'inich Ahau and is identified with jaguars, decapitation, fire, and rulership. The Sun was of utmost importance and K'inich Ahua was one of the most powerful in the Maya pantheon, considered an aspect of Itzamná.

The Sun God would shine in the sky all day and then transform himself into a jaguar at night to pass through the Underworld, Xibalba. He was also known as God G in Postclassic codices. The Sun and Venus were recognized as male gods; the Moon possessed both male and female traits.

The representation of the solar deity K'inich Ahau is depicted as youthful in Classic Maya art; but in the Postclassic codices, he is pictured as an aged, bearded male with a sunken jaw having only a few teeth. The beard was meant to resemble solar rays. The older version of the Sun is associated with the Lord of the Night. The Sun has many different animal avatars; a few are the red macaw, the hummingbird and the puma.

The Sun's path through the sky is illustrated as sky ropes or twisted cords. The Sun's motion represents the four corners of the universe, meaning the four extreme points of the Sun's passage at the solstices rising and setting. Thus he was also associated with the number four. The most important solar symbol is the glyph *k'in* meaning "Sun" or "day" and is found on his head or body when personified as human. His name was "Lord Sun" during the Classic and Postclassic periods and he was connected with the north. Ancestors are sometimes honored by showing them apotheosized in the form of the Sun. The Sun can also be pictured as cross-eyed, a sign of beauty to the Maya. When illustrated full figure, he usually wears male clothing. At times he shows spots on his cheeks which could be death spots that link him with the Underworld, but they are more often identified as the spots of a jaguar, indicating his nocturnal associations. The nightly disappearance of the Sun was considered its death and then its resurrection followed each morning (Fig. 6.1).

Lunar Deities

The Moon, almost as important as the Sun, carried great significance for the Maya. The Moon was represented by a female deity who had a powerful influence on terrestrial events. The Moon carries multiple aspects, including a male characteristic that are evoked at times, which includes the warrior and the priest, or male ballplayer, usually when the Moon is full. The lunar deity changes gender with the death and decapitation of the Moon Goddess; this transformation of the Moon's gender takes place during the conjunction of the Moon with the Sun. The Moon was thought to grow old as it progressed from full to new. Maya mythology associated the Moon with a maiden, an old woman and a rabbit. The Maya dynasties at times claimed to be descended from the Moon, as well as from the Sun and planets.

The primary Maya Moon goddess was Ixchel, the powerful woman who struggled with the Sun and forced him to descent into the Underworld each night. Even though formidable, she was the patroness of women, pregnancy and fertility. Another aspect of the moon goddess was known as Goddess I, who was discussed in some of the Mayan codices. Goddess I appeared as young and beautiful; she was perhaps Ixchel in her youth or in a different apparition; she was associated with the waxing Moon. Goddess I in a death attribute represents the eclipsed full Moon. The

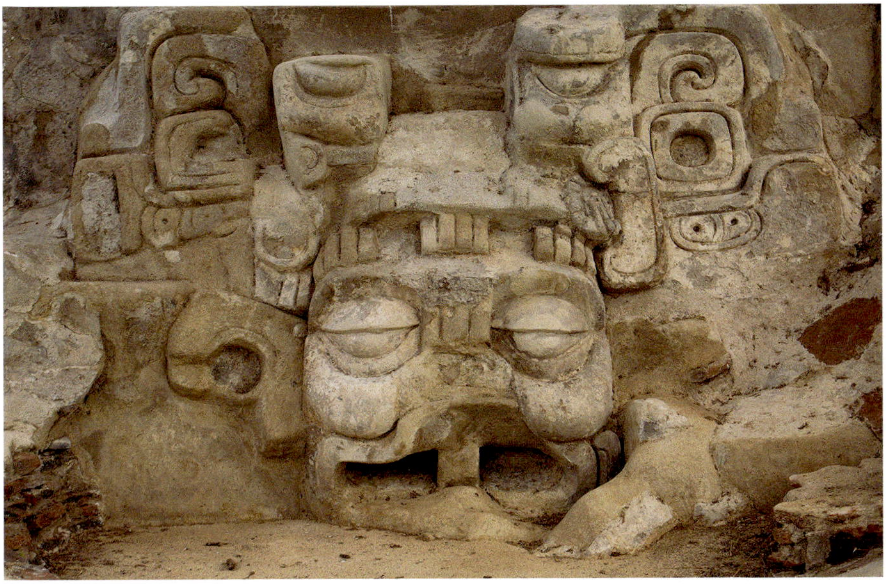

Fig. 6.1 Sun God Kinish Ahau, Temple VI at Comalcalco, Tabasco (Credit Matt Gush)

waning Moon was considered to be an older female goddess who ruled over childbirth and weaving. Goddess O represents aspects of the Moon Goddess as the aged Grandmother.

Most scholars agree that Ixchel was the Moon Goddess who was associated with pregnancy, medicine, divination and weaving and that she was the same deity as Goddess I. But others disagree and claim Goddess I is a deity with similar attributes but she was not connected with the Moon Goddess. In her exhaustive study of astronomical deities, Susan Milbrath claims that there is little evidence to connect Goddess I to the Moon. She believes that Ixchel is instead associated with Goddess O who is also a lunar deity. She believes Goddess I and Goddess O are related not to the Moon itself but to different phases or aspects of the Moon.

Ixchel's nature is shown in a dance held in her honor on the sixth day of Zip with a dance that translates to the "dance of the moon". In the Dresden Codex the Moon Goddess is shown with numerous names and different imagery; she was pictured in love scenes or carrying burdens or making offerings that show maize symbols and fish. Ixchel is also illustrated as the archetypal mother with small deities in her arms or strapped to her back. Ixchel was highly honored in particularly dedicated temples and shrines found in various times and areas of the Maya World.

Scholars account for the large number of roles attributed to the Moon because of the many characteristics that it takes on through its quick transition from new to full. The Moon is constantly changing its position in the sky and then as the crescent Moon decreases in size; it eventually disappears for 3 days when it becomes conjunct with the Sun. The Maya understood the Moon's cycle as the process of aging, with the new Moon considered as a newborn. In some imagery the Moon is thought to change its gender during the cyclic transformation from new to full. Both the male and female aspects of the Moon have important connections to the attributes assigned to the Moon.

In Maya artworks, female depictions of the Moon are dominant but male components appear in Maya images as well. She can be pictured with a child strapped to her back or carrying small deities in her arms that symbolize her divine children. The lunar crescent is often shown on the ball player's belt, indicating its male aspect. Some accounts show the Moon as bisexual showing both male and female features. The Moon Goddess has two aspects, that of the Young Moon Goddess and the Old Moon Goddess. A waxing Moon had the traits of a beautiful woman, while a waning Moon was considered to be an older female who ruled over childbirth. Lunar goddesses can also be linked with snakes, even the Plumed Serpent and sometimes wears a serpent headdress. During the late Classic period she is associated with women, marriage, sensual love and with weaving.

The Moon is also connected with the maize deity and maize agriculture and producing rainfall during the rainy season. The Maya also saw a link between the Moon and Earth; they are joined in images representing lunar conjunction. Since the Moon appears to vanish into the Earth for several days each month, the Earth and the new Moon are supposedly united. The fertility of the Earth needed a bisexual concept; therefore the Moon within the Earth was also bisexual.

Venus Deities

Venus was certainly the most important of all the planets, appearing quite often in Maya art, cosmology and codices. Similar to the Moon, Venus can be seen with many personifications. The planet can be connected with the rainy season and agriculture, or with warfare and the dry season. In the Yucatan Venus often references the number eight that recalls the 8 days of inferior conjunction or the 8-year cycle in the Venus almanac. The Maya had a significant interest in the first and last visibility of Venus as it passed through its cyclic pattern. The wind aspect of Venus, Ehecatl-Quetzalcoatl is connected with round temples. The multiple names of Venus was due to its numerous aspects. When inside their homes, the Maya would block their chimneys on certain days, so that the evil light from Venus could not enter (Fig. 6.2).

Venus is commonly associated with feathered serpent images popular throughout Mesoamerica. The Maya believe Venus has a dual nature as well, with differing roles according to its time of visibility. The Morning Star was written about more often than the Evening Star and was often identified with the Sun since their paths seem to be close. When Venus is pictured as the feathered serpent, the tail appears to be the Pleiades, which is called the "rattlesnake's rattle" in the Yucatan today. Venus' anthropomorphic counterpart is known as Quetzalcoatl in Central Mexico and Kukulkan in the Maya areas. The Maya god Quetzalcoatl or Kukulkan is

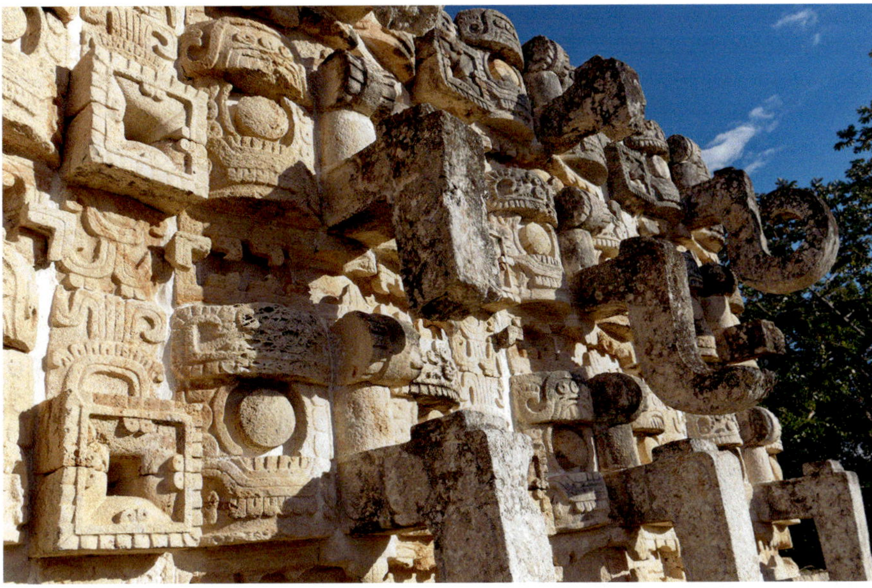

Fig. 6.2 Rain God Chaak Masks, Kabah, Yucatan Peninsula (Credit Belikova Oksana)

visualized as the Plumed or Feathered Serpent; his connections and connotations are quite complicated. There are two characters involved with Quetzalcoatl, the divine aspect and the human aspect. His godlike qualities include great creative power, bringer of rain, a supernatural being with whom powerful rulers and priests could merge. During the Postclassic the Feathered Serpent became the source of royal power. The historic Quetzalcoatl was thought to be a highly honored and successful Maya ruler, leading the Toltecs, but some experts question whether he was real or legendary (See Chap. 11).

There has been a massive amount of data written on the importance of Venus but very little information on the other planets, individual stars or constellations. Considering the enormous interest in Venus in artworks and in building orientations, one would expect that the other planets would be of interest as well, featured in Maya myths and folk astronomy. This does not mean that the Maya were uninterested in the other planets, just that the knowledge has not been handed down or has not yet been discovered.

Other Astronomical Deities

Information on the godlike aspects of the other planets is sparce. After Venus, Jupiter and Saturn are two of the brightest objects in the heavens. Only a few names have been discovered for Jupiter; God K (K'awil) has been identified as Jupiter by

Maya scholar Susan Milbrath. She has also tied together the synodic cycles of both Jupiter and Saturn with the katun cycles of the Long Count. Her work demonstrated a connection between God K imagery and dates matching with Jupiter's stationary points when in retrograde. She states that K'awil is actually the god of the retrograde cycles of Jupiter and Saturn. Names for Saturn do not appear and information on the red planet, Mars, is scarce. Mars was recognized as a planet and astronomers charted its celestial cycles and its retrograde period of around 78 days, but its appearance in Maya iconography is still cloudy. There is a glyph called the Mars Beast that appears on some occasions. Mars is suspected to be associated with weather changes.

Few articles have been published on the presence of Mercury in Maya artworks, perhaps because it is not easy to recognize Mercury, both in the heavens and on artworks. Some works thought to be Venus personification could actually refer to Mercury instead. Mercury has a particular name and identifying glyph when it reaches its inferior conjunction, but a totally different name at superior conjunction. One name is used for its time as a morning star but has a different name when it becomes an evening star, and yet another at its heliacal rising and setting. At certain times in its speedy cyclical orbit around the Sun, Mercury can be called 9 Wind, 1 Reed, 4 Wind, 8 Twisted and 9 Rain.

Mercury seems to be related to owl imagery. The Maya would have known about Mercury since its cycle is similar to that of Venus; but may not have observed Mercury regularly since its nighty appearance is always quite short. It is difficult to observe precisely. Mercury was thought of as the messenger of Xibalba, the Underworld. Since Mercury seems to be invisible when in conjunction, the role of messenger to the Underworld seems appropriate. God N is possibly connected with Mercury. Future research may produce more valuable information on the identities of the other planets.

The Milky Way was also considered a god and the constellations were envisioned as celestial animals or locations. It is actually the edge-on disc of our own galaxy; it is a 10° wide band of light that crosses the sky. Its most significant feature is the great dust cloud that creates a dark rift in the southern and western sections. No almanac of the Milky Way has yet been discovered. As expected the Milky Way has many different associations; it was visualized as a river, a place of death, a causeway or a road of ice, which was quite common throughout the ancient world. It was also depicted in Classic Maya art as a celestial serpent. The serpent was imagined with its mouth to the south and its tail to the north. The Milky Way has also been illustrated as an aged goddess dressed in white with a starry skirt or as a male god with his face painted black and surrounded by stars. The Maya visualized the Milky Way as a path or axis intersecting the ecliptic.

The Milky Way as a "Cosmic Monster" was pictured with a sky band body carrying symbols of the Sun, Moon and Venus. The Milky Way and its dense belt of stars were carefully watched to predict seasonal changes; the Moon's position crossing the Milky Way was also noted. It has been suggested that the Maya associated the Milky Way with the dead. The Milky Way could symbolize the road of the dead with its innumerable stars representing the souls of the dead. The rotation of the Milky Way was plainly meaningful in Maya iconography.

Stars and Star Groups

Strong evidence exists that Maya astronomers organized certain star groups into patterns as constellations. These constellations include a turtle, which represents stars in Orion, a rattlesnake constellation that includes the Pleiades and a fish-snake consisting of stars in Sagittarius. Surprisingly, the constellation presently known as Scorpio, the scorpion, was also recognized by the Maya as a scorpion constellation, joining the same stars,. The brightest star in Scorpius is Antares; it was called the "heart of the scorpion". A variety of different astronomical associations was applied to this constellation in ethnographic literature. A Maya zodiac of thirteen constellations, with five animal constellations on a sky band, depicts star groups at the intersection of the Milky Way and the ecliptic.

Like farmers in all parts of the world for thousands of years, Maya farmers observed the Pleiades for timing their agricultural tasks. The position of the Pleiades would predict the moment of the Sun's passage over the zenith point around the proper planting time. In mid-May the Pleiades rose at dawn, just before planting and at the start of the rainy season. The evening setting of the Pleiades would mark the planting season for maize in the higher altitudes; while the conjunction of the Pleiades with the Sun in May would announce the time for planting in the lower altitudes. The conjunction of the Sun and the Pleiades matches the zenith passage of the Sun and the onset of the rains.

The Big Dipper (Ursa Major) was related to the number seven since it has seven bright stars; the Maya linked the same seven stars in their unnamed constellation. Some areas warn that counting the stars will cause the death of one's spouse. Neither the Little Dipper (Ursa Minor) nor the North Star (Polaris), the first star in the handle of the Little Dipper were frequently recorded by the Maya astronomers. Constellations positioned at meeting points of the Milky Way and the ecliptic appear to be particularly important in Maya cosmology.

The three bright stars in Orion are associated with a peccary (a medium-sized pig); three stars of smaller magnitude are the peccary's piglets, positioned close to their mother. Orion is also linked with the number three because the three bright stars found in Orion's Belt are easy to find, even when the full Moon is near. Some call Orion's Belt the "three kings". The Maya joined Alnitak in Orion's Belt with other stars of Orion, Rigel and Saiph, to create a celestial triangle called the "three hearth stones" which will be discussed in chapter seven. Few Maya sources provide information on other astronomical phenomena, such as comets, meteors and supernovas. Future studies may produce more information on the astronomical deities of the ancient Maya.

Other Deities of the Maya

One of the most honored deities of the Maya pantheon and the most depicted is Chaak, the god of rain and lightning. Chaak can also be identified with war and human sacrifice. With its multiple manifestations, Chaak is one of the god that is most difficult to interpret. The rain deities have been among the most lasting and commonly highlighted components of Maya religious activities. They are still prevalent in modern-day Maya villages. For 25 centuries the rain deities have been recipients of prayers and offerings from believers to assure successful agricultural production (Fig. 6.3).

The Maize God appeared widely, as the Maya held maize to be sacred. The Maize God was personified often in Maya artworks and artifacts, with an idealized appearance, ritualized gestures and a pose of contemplation. The Maize god was of great importance and often depicted as statues, on temple walls and on ceramics. The god is pictured as young and handsome with attributes of maize plants. Often his image seems to dance and sway in a gentle breeze, similar to the corn plants. Identified as youthful and kind, the image of the Maize God was highly honored by royalty, who often dressed in his iconic attire. Like the maize plant, the Maize God

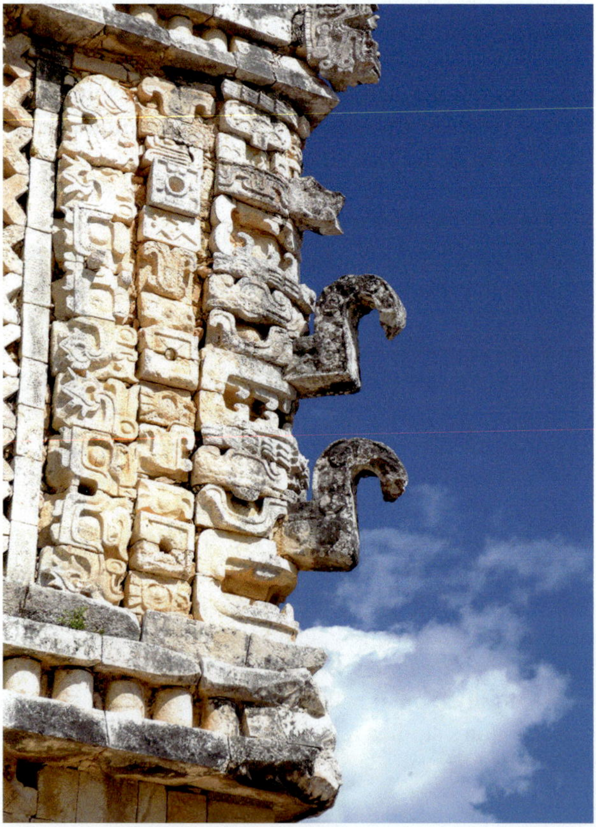

Fig. 6.3 Rain God Chaak Masks, Uxmal, Yucatan Peninsula (Credit Nahdjo)

moves through the life cycle; during one season he is a handsome young man, alive and active; but in the next he becomes decapitated with his head on a plate as an offering. But the young Maize God emerges again to guarantee successful corn production.

The Jaguar God of the Underworld embodies the night Sun, apparent when the Sun sets in the west and travels through the Underworld to rise again in the morning. The Lord of the Underworld is connected with the jaguar and the owl, both are linked with caves and the night. He is also related to death and destruction as well as abundant riches. The Lord of the Underworld was sometimes pictured as an aged god with a screech owl and wearing a brimmed hat. He has usually has a black body and jaguar characteristics; he is sometimes shown smoking a cigar. The Wind God represents the breath spirit and is the god of music; his number is three. The glyph for the wind is the *ik'* sign. The Water Lily Serpent signifies terrestrial waters, including rivers, lakes, cenotes and the sea. This god's number is thirteen and has close association with Chaak. During the Classic period rulers often impersonated the Water Lily Serpent.

God K or K'awill is the god of royal dynasties and elite lineages; he was considered very important to the Maya. He is also the god of celestial lightning. God K is often depicted in the manner of a Mankin Scepter, an axe held by the rulers. Another female called the Old Goddess is related with opposite forces, creation and destruction. She is also connected with divination, healing, childbirth and weaving. She was often illustrated with a red body and a serpent headdress. She was fiercely depicted with claws and fangs and wears a skirt with crossed bones and death. At times Goddess O was connected with actions related to weaving. Because weaving was by tradition a dry season activity for the Maya; that imagery of Goddess O is thought to represent a dry season aspect of the Moon. The numerous gods of the Maya each have numerous depictions and numerous roles to play.

References and Further Reading

Ishihara, R. (2009). *Deities of the ancient Maya*. Dumbarton Oaks Research Library.

Milbrath, S. (1995). *Gender and roles of Lunar Deities in Postclassic Central Mexico and their correlations with the Maya Area*. https://www.academia.edu/2762949/

Milbrath, S. (1999). *Star Gods of the Maya: Astronomy in art, folklore, and calendars*. University of Texas Press.

Milbrath, S. (2014). Deities in ancient Mesoamerica. In C. L. N. Ruggles (Ed.), *Handbook of Archaeoastronomy and Ethnoastronomy*. # Springer Science+Business Media New York. https://doi.org/10.1007/978-1-4614-6141-8_55

Milbrath, S., & Dowd, A. S. (2015). An interdisciplinary approach to cosmology, calendars, and horizon based astronomy. In A. S. Dowd & S. Milbrath (Eds.), *Cosmology, calendars, and horizon-based astronomy in ancient Mesoamerica*. University Press of Colorado.

Milbrcath, S. (1999). *Star Gods of the May: Astronomy in art, folklore, and calendars*. University of Texas Press.

Miller, M., & Taube, K. A. (1993). *The gods and symbols of ancient Mexico and the Maya*. Thames and Hudson.

Chapter 7
Maya Mythology, Cosmology and Art

Introduction

Mesoamerican cosmology is based on the principles of time and space; their beliefs are exemplified on inscribed monuments, with astronomically aligned architecture and in illustrated codices. Favored themes of the Maya were the myths of sacrifice, cyclical regeneration, the symbol of maize, the cave of creation, divine cosmic order and ruler accession. These mythic themes were all associated with astronomy in different ways and embodied over and over in the art and architecture of the Maya.

The ancient Maya world was conceived of as having three layers: the starry expanse of the celestial dome, the stoney Middle-world of Earth and the dark waters of the Underworld below. Originally there were only two regions, the heavens and the Underworld but the creator gods decided to raise the sky and form the earthly region in between, establishing the area for souls to experience the Middle-world of Earth. After death souls would descend along the roots of the World Tree to the Maya Underworld, called Xibalba, the place of fear, decay and disease. In this place the soul would be challenged and undergo tests before they could arrive at a place of eternal rest in the afterlife.

This conception of the world can be dated back to the Preclassic period. These three regions were not separate layers but interrelated and alive with spiritual powers that must be appeased and honored. The area of human existence was a sacred place and the four cardinal directions presented a basic grid for the surface of the Maya world. The main axis for the Middle-world was the daily path of the Sun as it traveled from east to west. Each of the four directions had a particular tree, bird, color and special gods linked with each region and sacred rituals linked with each god. The directional colors were affirmed to show that east is red, west is black, north is white and south is yellow.

M. Dolan, *Decoding Astronomy in Maya Art and Architecture*,
Springer Praxis Books, https://doi.org/10.1007/978-3-031-61092-9_7

The World Tree

Their concept of the world also had a center and passing through the center was the world axis that they conceived of as a tree that "raised up the sky". The World Tree symbolized the essential axis that existed through all three levels; it too was associated with a particular color which was green. The great tree's trunk was positioned in the Middle-world; its roots reached into the waters of the Underworld and its branches stretched up to the upper levels of the celestial realm.

The World Tree embodies the four cardinal direction, a symbolic axis mundi; its image with both directional and central meaning is pictured over and over in Maya art and in mythology. Its most well-known appearance can be found on the sarcophagus lid of the tomb of Palenque's revered King Pakal. The axis mundi was a place where one could access manifestations of the divine or spiritual power in the form of a hierophany, where or when communication with the divine was made possible. The central World Tree was thought to be a ceiba tree; it also represented the starry band of the Milky Way. The World Tree was thought to center the cosmos both vertically and horizontally. Manifesting a living axis mundi, a Maya ruler became a living expression of sacred space, wherever he stood became a holy place. The king's control of this cosmic world axis would be a powerful communication between the ruler and those he ruled.

The Vision Serpent

The serpent was a revered symbol in Maya mythology, both the Vision Serpent and the Feathered Serpent. The fact that snakes can shed their skin makes them a symbol of rebirth and renewal. Serpents can also represent the means by which the heavenly bodies cross the heavens. The Feathered Serpent, a snake with feathered neckpiece, was considered a primary religious icon, one of the most honored gods (Fig. 7.1).

The Vision Serpent was one of the highly honored mythological deities. The serpent was usually bearded with a rounded snout. The mystical serpent was sometimes drawn with two heads or with a god or a holy ancestor emerging from its jaws. According to experts the Vision Serpent corresponds with the constellation of Scorpio. During bloodletting ceremonies, the person involved usually experienced hallucinatory visions during which they could interact with the gods or the honored ancestors. The Vision Serpent was also a feature of other religious practices and its frightening form appeared on ceramics and as sculpted reliefs. Early Maya concepts connected the Vision Serpent with the World Tree which created the central axis that enabled the king to communicate with the spiritual world and use the sacred action to enhance his power and successful reign.

Myths have as their objective the explanation of the universal and humanity's material and spiritual reassurance. Myths are realities in the inherited cultural consciousness of a community or a people. They narrate events that occurred at the

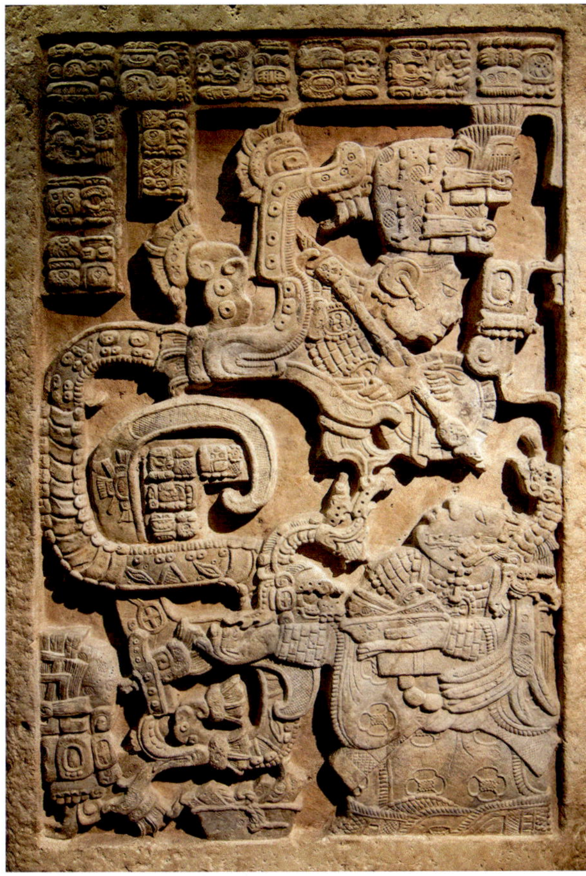

Fig. 7.1 Lady K'abal Xook conjures Vision Serpent during bloodletting; she looks up at a god emerging from its mouth. The serpent corresponds to the stars of Scorpio, fron Yaxchilan, Chiapas, mid-eighth century (Credit Lucas Uher)

beginning of time and they tell how reality came to exist through the acts of supernatural beings. Myths explain the act of creation as a sacred process of transformation, as to how the world and humans were brought into being. All aspects of Maya life include astronomy, especially their mythology, cosmic stories and creation myths. Some scholars even suggest that the artworks at Palenque, Copan and Tikal and other major cities actually symbolize maps of the sky on the dates associated with both the date of Maya Creation and with important events in each city's history.

Maya mythology, like all cultural mythology, attempts to show the origins and the relationship between humans and the environment, including the heavenly bodies, especially the Sun, Moon and Venus. Many myths have been reassembled by scholars from analyzing the surviving artworks and their symbolic iconography that illustrate the essential features of the ancient, symbolic tales. The myths often relate epic stories that personify the powerful forces of nature with the deities, supernatural entities and heroic figures that interact with them. One of the most important Maya myths that survives intact from the sixteenth century is the Popol Vuh.

The Popol Vuh Creation Myth

The Popol Vuh, the Sacred Book of the Quiche Maya People, is the ancient sacred book of myths, history, traditions and the Maya story of how the world was created and the history of the Quiche Maya, a later group of Maya who lived in modern day Guatemala. Experts estimate that the Popol Vuh was first fully composed and written between 1554–1558 but the book incorporates thousands of years of oral history passed down by the Maya people. The sacred book stands as the foundational narrative of the Quiche people. The Popol Vuh was originally written as a single, long poem; it combines folklore, mythology and historical accounts.

When initially written, the manuscript had no divisions but was later divided into four books when translated into European languages. The Popol Vuh was probably derived from ancient writings along with long-recited oral traditions. After the Spanish Conquest the unknown author thought he must write down the myths and history of his people before they were totally lost. The Quiche author implies that he had to keep his writings hidden from Christians who would have destroyed them, as they did to so many Maya codices.

The mythical books of the Popol Vuh say that the world begins in darkness with only the principal creator deities; it then describes the birth of the Sun and the Moon. It relates the story of the creation of humans and animals. The attempt of the gods to create humans was a long process of trial and error; they tried using wet clay first and then wood, but nothing was suitable until they tried maize. Then finally, using maize they were able to form the perfect humans who were able to plant crops and praise the gods.

The Popol Vuh includes the stories of many different characters and that of the young Hero Twins, named Hunahpu and Xbalanque. The Twins were magically conceived after their father's death. The Hero Twins succeeded in undertaking difficult conquests and they conquer several different beings, sending them to the Underworld. They visit Xibalba to avenge the deaths of their father and uncle. There they face many trials during their numerous adventures in the Underworld where they manage to trick the Lords of Xibalba over and over. In the Underworld some deities plan the defeat of the Hero Twins through playing ball, their sacred sport, but the twins ultimately win by defeating the mighty Lords of Xibalba.

The books describe the exploits of the brave and daring twins in the Underworld and their triumph over all obstacles, including death. The Hero Twins eventually destroy the evil forces and then ascend the World Tree, finally reaching the sky where they are transformed into the Sun and the Moon. Recent studies argue that when Hunahpu and Xbalanque emerged from the interior of the Earth, they already are the Sun and Moon. They rose on the day of the zenith transit. Their rising now brings forth the time of planting and germination; this is an event that coordinates well with the very first zenith passage.

The Popol Vuh is replete with stories and incidents that refer to the celestial bodies and their cycles. The reference to four roads in the Popol Vuh relates to a cosmic diagram describing the intersecting paths of the ecliptic and the Milky Way. Depictions of individual scenes derived from the Popol Vuh appear on temple walls, stelae and various Maya structures. The artworks offer evidence that scenes from the Popol Vuh were pictured on Maya walls; some images were discovered even in the Preclassic period which hints to the ancient age of the myth.

Creation Myth of the Three Hearthstones

It is well known that the Maya were excellent astronomers who identified constellations and formulated a zodiac similar to the western one; their Scorpio is the same as ours. The Maya formed many constellations from small star groups that we would call asterisms. For example, the three stars of Orion, the stellar hunter's belt, are now called Alnitak, Alnilam and Mintaka, most individual stars were named by Islamic astronomers. These three stars are only a small part of our constellation of Orion, but for the Maya these three stars formed the constellation they called the turtle. The three Hearthstones of the heavens are three different stars of Orion. They form a triangular asterism; these three stars are Alnitak, Saiph and Rigel. The three stones are part of Maya constellation that sits just below Orion's belt. This triangle of stars symbolizes the three stones that form the triangular hearth found in Maya homes from the Preclassic period to the present day (Fig. 7.2).

One of the fundamental cosmic myths of the Maya is the story of the Three Hearthstones. For the Maya still today, the three hearthstones symbolize a place of origin, foundation, center, beginning and the source of all that follows. The three hearthstones are always placed to form a triangle. A modern Maya tradition when building a home includes this ancient ritual known as "planting the stones". The

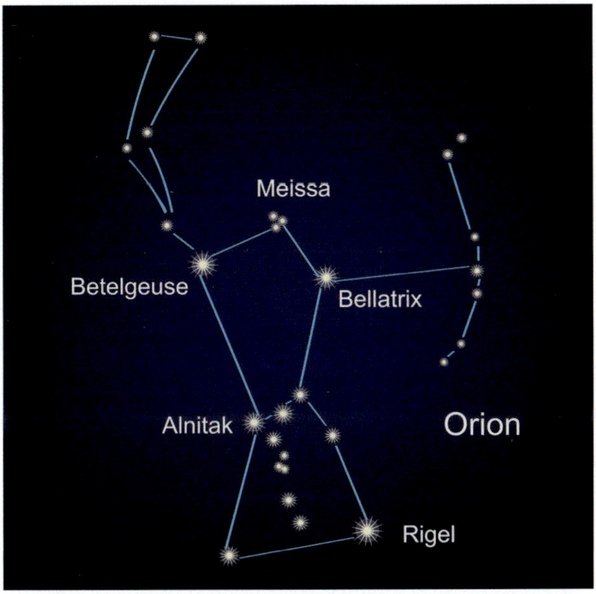

Fig. 7.2 Constellation of Orion which contains the Three Hearth Stones of Maya myth (Credit Polii)

Maya tradition requires building a hearth at the heart of their home by setting three large stones. The planting of three stones as described in the myth is symbolic of building a new home for humanity.

The Maya story starts by explaining how the first stone was set by Jaguar Paddler and his brother, Stingray Spine Paddler. In myths the brothers were responsible for paddling the canoe that transports the dead across the rivers of the Underworld. They planted a stone called Jaguar Throne in the House of Five Sky. The second stone was called the Snake Throne that was planted in the Earth Place. This stone had been planted by an unknown god called the Black House Great. The third stone known as the Water Throne is thought to be set in the Lying Down Sky or the Edge of the Sky. This stone was planted by Itzamna, the important deity of the Maya who was linked with learning and books. He was said to tie the stones at the edge of the sky, which became the new place of the stones. The location of this event refers to the condition of the pre-creation universe as black since the sky had not yet been lifted from the Primordial Sea.

According to scholar Linda Schele, the three physical hearthstones enclose the cooking fire and define the center of the Maya home, while the three stone thrones of Creation created the center of the cosmos and allowed the sky to be lifted from the Primordial Sea. In addition, a hazy nebula (called Orion Nebula M42) appears in the center of this triangular asterism which the Maya identify as the smoke from the central fire. Another asterism that we call Orion's Sword is a collection of nebulas and open clusters of stars; the brightest star in this group is called Na'ir al Saif, "the bright star of the sword" in Arabic. Perhaps the Maya considered this hazy group as puffs of smoke that rise from a heavenly hearth fire. The Maya believed that the beginning of the current age occurred when the Milky Way, Orion's belt and the Three Hearthstones were in a certain celestial position. The Madrid Codex presents an illustration of a turtle that is carrying the three hearthstone stars in a triangular form, the hearthstones symbolize the center of the Maya universe. This location is very close to the position of our solar system in the Orion Arm of the Milky Way.

The Classic period myth of creation was more than a collection of unusual tales provided by the Maya religion. The myths were based upon the bodies and motions of the heavens that timed their daily lives, including public ceremonies, pageants, feasting and political life. Maya artworks often represent images that they gathered from the sky. Public ceremonies and royal rituals were carefully timed in coordination with heavenly events. Leaders and nobles wore symbolic clothing and accessories based on the myths and deities of the sky. Their interpretation of the heavens provided a structure of meaning that all Maya people could understand and follow; the myths of the heavens kept them in accord with the forces of nature.

Art of the Maya

Reaching back to our earliest history, humans have had an innate desire to beautify their surroundings, their possessions and even themselves. Neanderthals were known to fashion jewelry and create sections of the amazing painted caves that date back to 35,000 BCE. The oldest ceramics uncovered in the ancient China, Middle East and Egypt were already decorated with geometric patterns to differentiate them and to enhance their appearance. Preserved bodies from ancient eras display tattoos, pierced ears and jewelry created with beads and shells. The earliest settled humans decorated their dwelling places, such as Catal Huyuk, with murals, paintings, sculpted images and bull heads. The early Maya elite too enjoyed beautiful possessions and lived well; they embellished their homes, temples, palaces and courtyards with sculptures, altars and paintings. They adorned themselves as well with elaborate decorative costumes, headdresses and elegant jewelry; the elite were buried with painted vessels and great quantities of luxury goods. The artworks of the ancient Maya were designed not only to enrich their environment but to augment their religious beliefs and their spiritual lives.

Decorative and symbolic art flourished during the 600 years of the Classic period; Maya artists produced a multitude of outstanding and imaginative creations. They integrated their astronomical knowledge and belief systems into their unique worldview. The Maya did not compartmentalize their outlooks, separating animate/inanimate or body/soul aspects, but united all aspects of daily life into a complete model that was in harmony with the cosmos. To attempt to isolate the art and architecture from the culture's religion, mythology, political and social structure would be to view only a segmented portion of their legacy.

Large numbers of artworks survive, such as sculptures, figurines and murals of their royalty, deities, myths, and military successes. The most important subject of monumental art was of course the city's king, whose title was *k'uhul ajaw* (holy lord). Maya artisans recorded the city's history through sculpted images of their rulers, molded figures and narrative scenes that decorated their buildings both inside and out. The numerous blocks of glyphic inscriptions added essential details by specifying accurate dates and historic details for later generations. Only a small group of educated Maya were thought capable of reading the complex glyphs but everyone, young and old alike, could easily recognize the stories and myths pictured in the art displayed on buildings and monuments. Extraordinary events that occurred in the heavens were documented in stone, including eclipses, conjunctions and comets along with their explanations. Descriptions of their essential ceremonies often combined their sacred myths and their artworks with reality, as if both were actually real.

Even though Mesoamerican artisans possessed no metal tools and worked with simple stone implements and abrasives, they produced expert engravings. Various materials, including wood, bone, shells, limestone, jade and obsidian were shaped into sophisticated designs. Although almost no wood survives, historians believe that many wooden artworks would have been carved; but, if not destroyed as idols by the early Spanish, the humid, rainy climate would have eroded them. Architectural facades were almost always decorated with stucco relief carvings and usually finished with a thin plaster covering and then brightly painted. In the Late Preclassic giant stucco masks, often bearing Sun symbols, were used to adorn temple and pyramid facades; masks sometimes lined the pyramid's stairway as it climbed to the temple at the top. This practice of applying appropriate masks to honor a deity or a favored ruler continued into the Classic period.

Even at the earliest Maya settlements, symbolic art appeared in wall murals and in carved sculptures. Excavations of richly painted murals at San Bartolo, northern Guatemala, dating to the Late-Preclassic, between 300 and 200 BCE, show that the Maya had a long tradition of splendid mural painting. Tomb walls too were often colorfully painted. The best preserved painted murals were discovered at Bonampak in the state of Chiapas, near the Guatemala border and date to the Late Classic era, c. 580 to 800. The brightly painted murals celebrate royal and military events (Fig. 7.3).

Fortunately the sculpted artworks have survived in larger numbers and in better condition than the murals and paintings. They are rarely preserved in the warm, humid climate of Mesoamerica but those examples that do survive offer essential information on the area's cultural history. Most major artworks focus on the royal dynasties and nobles clad in their finest ceremonial clothing, commemorating important occasions. In complex works of art the king may be shown alone or with others, but the size, posture and placement of the ruler always indicates that he alone is most important, the central subject of any scene.

Maya architectural facades were built up from stucco and then molded into the finished form; when creating human forms, they were shaped first and then their costumes were added later. The final finished patterns were then colorfully painted. Color performed an essential part during sacred Maya rituals and public occasions; each color contained its own symbolism. Rulers, priests and nobles were adorned with bright costumes and brilliant plumage of exotic birds. Even sacrificial victims wore particular clothing and at times were painted turquoise. The use of boldly colored paint on objects and buildings enriched the messages of the divine, dynastic or military that were directed to the inhabitants.

In key Maya cities, extraordinary stone stelae were created by the thousands. Tall standing stelae decorated the central plazas and lined the exterior of important buildings, pyramids and temples. Portraits of kings and deities were intricately carved from great slabs of limestone, which were completed with complex patterns

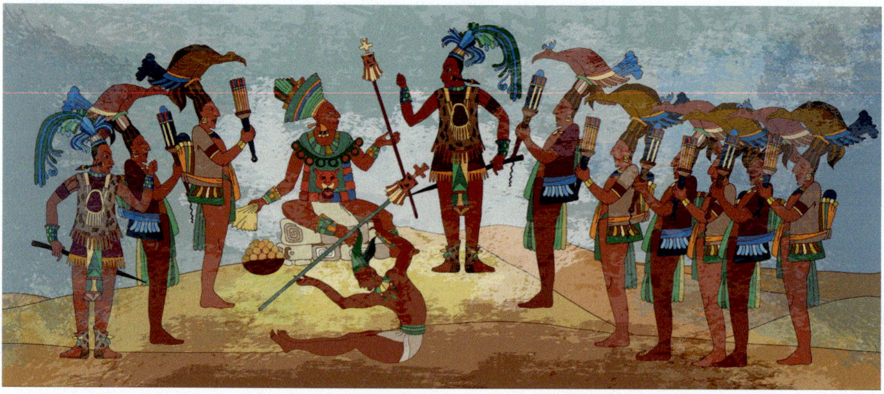

Fig. 7.3 Maya Nobles wall mural at Bonampak, reproduction (Credit matryoshka)

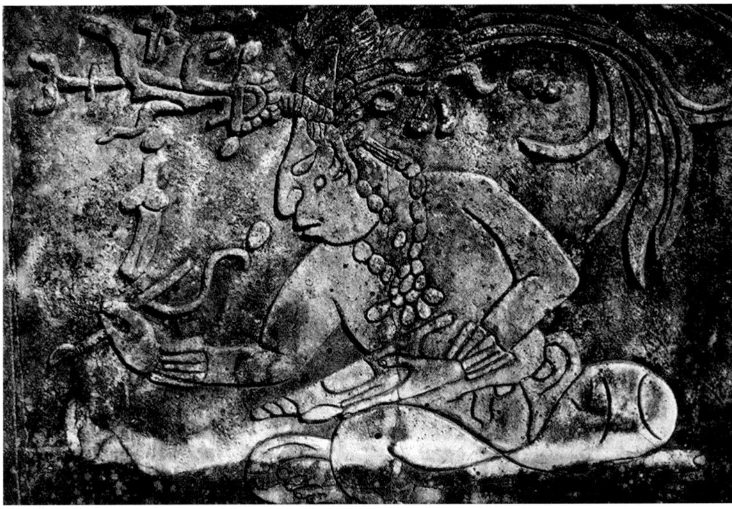

Fig. 7.4 Bas relief carving of Maya artist at work painting on a wall (Credit Lebedev)

and engraved with rows of glyphs explaining the dates and events being honored. Stelae narrate the ritual events of Maya elite; they promote the institution of divine kingship. The inscriptions can also include events of supernatural and ancestral significance, tributes to gods and kings; sometimes they boasted of great military achievements. Dedication of stelae often relate to ritual ceremonies of accession and ceremonial bloodletting sacrifices. The stelae that still stand today bear only a hint of their original color but they were always brilliantly painted with symbolic colors to further emphasize the impact made by the huge megaliths (Fig. 7.4).

The most commonly surviving artworks are smaller, less formal paintings found on ceramic vessels, cups, bowls and figurines. These everyday items are not worthy of the king but are more likely to reveal information on priests, scribes, craftsmen, musicians and the average Maya population. Maya screen-fold codices made of fig-bark included numerous small scale paintings as well but very few codices have survived. The artifacts that have survived in the best condition are artworks found buried in tombs and those hidden in sacred caves.

Symbolism and Personification in Maya Art

The identification of solar, lunar and planetary personifications that are carved on stelae or buildings or drawn in codices can be bewildering in their multiplicity and complexity. By comparison, in European art the symbols and personifications of the Sun, Moon, planets and constellations appear very uniform, standardized and easily recognized. The iconography established during the Hellenistic era was very consistent and carried on through the Renaissance. The powerful Sun was pictured as

Helios or Apollo, a handsome young man with rays of light surrounding his head, usually directing a quadriga pulled by four horses. The Moon was always portrayed as young female, identified by a crescent Moon symbol or headdress. Mercury was represented by a young man with winged feet or sandals who carried a caduceus entwined with snakes. The goddess of beauty and love, Venus, was always an exquisite young woman in various stages of undress. In medieval manuscripts Venus was called Lucifer (bringer of light) when appearing as Morning Star and Vesperus when appearing as an Evening Star. The planet Mercury was the messenger with winged sandals and Mars was always a military figure, clutching the accoutrements of war. Mighty Jupiter was depicted as the majestic ruler of the sky with long curling hair and beard; Saturn was easily identified as an older man with white hair and a long beard, personifying time as an aged and weary figure.

In contrast to the simplicity of the Hellenistic system of astronomical iconography, the art of the Maya pantheon is much more detailed and multifaceted. The celestial pantheon does not have a single or standardized format for any of the celestial bodies; every planet has many names and personifications. The sophisticated Maya system retains and conveys much more specific astronomical information in their personifications. The planet's specific name can tell its position in the sky or the phase of the Moon. They have at least two and sometimes numerous representative figures portraying the same god, planet or constellation; many are complete opposites in their appearance. For example, the Sun is usually represented by a four-petaled glyph referred to as the *Kin* sign, but the Sun is also associated with the jaguar.

Many of the heavenly deities appearing in artworks during the Maya Classic period are presented as bisexual; they were illustrated in codices, painted on colorful vessels, and carved on stelae and temple walls displaying both male and female attributes. Some celestial gods appear as totally male or female, but others seem to be ambiguous, with a dual nature. An image of a god may look to be clearly male even though having female attributes or a female may exhibit male traits.

In Maya symbology lunar characteristics are quite complex as well, with many faces and many personalities. Some characteristics appear to be connected with gender roles. The identification of Venus is complicated as well since its symbols are quite plentiful. Venus is referred to as Lord of the Place of the House of Dawn or the Morning Star when rising before the Sun but has different name at inferior conjunction, Five Flower; another at superior conjunction, Nine Dog or Nine Snake; at heliacal rising, *haab-cuc*; at the autumn equinox it is called One Flower. Venus was also called *noh ek* (Great Star), *chac ek* (red star), *sastal ek* (bright star) and *xux ek* (wasp star). The Venus of Mesoamerica was not the beautiful woman, the essence of love and beauty, but a symbol of danger, death and destruction in a male guise. The rhythmic cycles of Venus were carefully recorded by the Maya, for each significant position in its eight-year cycle predicted times of danger, drought or conflicts. On inauspicious days, certain rituals had to be performed to counteract Venus' negative influences (Fig. 7.5).

Maya sculpture and paintings contain abundant examples of symbolism and religious iconography and images alluding to their rich supernatural world. Omnipresent is the human form, representing humans and gods in astonishing lifelike activities.

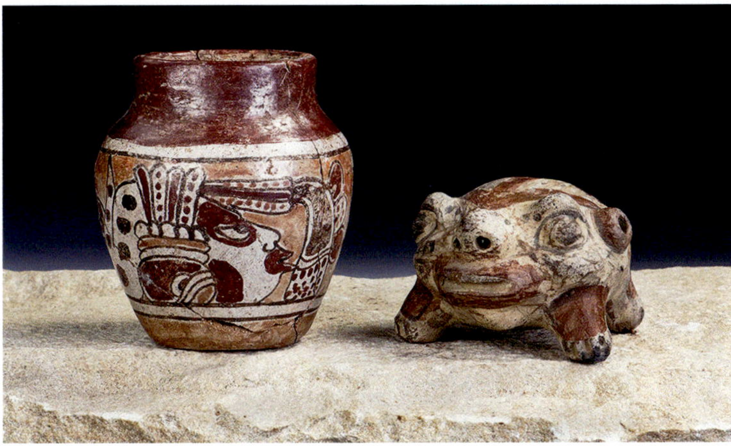

Fig. 7.5 Maya ceramic vase and turtle rattle, CE 600–1000 (Credit W Scott McGil)

The rulers or kings of a city were often depicted in the costume of a god, so it is difficult to understand the true meaning of the symbolic figure without an inscribed text. Living creatures were an important aspect and carried many layers of symbolism. Most common were jaguars, snakes, eagles, and crocodiles which all had religious and cosmological connotations. Another influential symbol appearing in Maya art was the serpent since they could easily transverse between the Earth and the Underworld, connecting the two spheres. Because of their ability to shed and renew their skin, snakes represented the cycles of creation and destruction. The deity Kukulkan was usually depicted as a serpent with a feathered neck; the added feathers enabled him to reach all three realms, the Earth, the Underworld and Otherworld or heavens. Eagles too were divine in Maya beliefs since they could soar to the heavens; they were associated with the Sun God, Kinich Ahau (Fig. 7.6).

Another prominent animal symbol in Maya art is the jaguar which was an image of strength and bravery; some gods and goddesses are shown with jaguar features, called zoomorphs. Kings and priests could shape-shift and transform into zoomorphic shapes that were half-human and half-jaguar. By adopting jaguar aspects, kings could then absorb the jaguar's abilities of stealth and agility. Sculpted images of zoomorphs with religious implications were jaguars, snakes and crocodiles; the turtle represented a constellation and relates to the Maya Creation Myth and acted as an important supernatural symbol in art as well.

Works of Art in Jade and Turquoise

In the world of the ancient Maya, jade was more valued than gold, not only for its rich, pure color and beauty but also for its treasured symbolism, and its religious and spiritual importance. Jade was one of the hardest and most durable stones found in

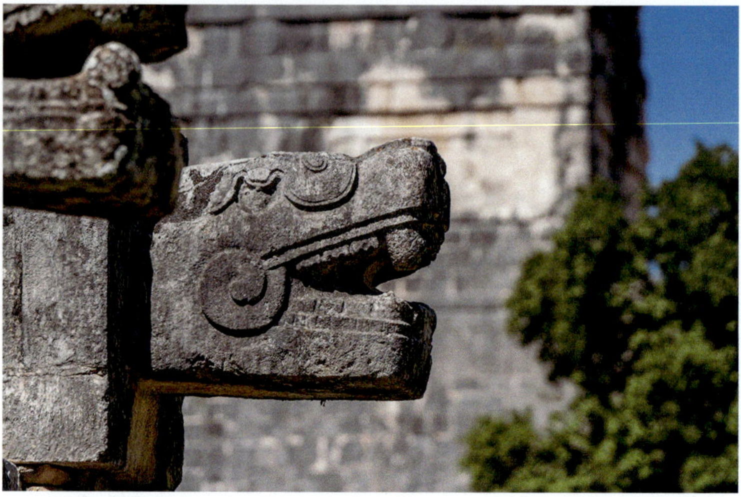

Fig. 7.6 Serpent Head of Kukulkan, symbol of Venus on the Venus Platform or Temple of Venus (Cedit Nik Waller Productions)

Mesoamerica. The rare jade stone was accessible for artisans' creative works throughout the Classic period since it was mined in Guatemala, one of the few sources for the mineral in the West. The typical color of jade or jadeite ranges from pale apple-green to deep emerald-green but is also found in white and white with green spots and a few other rare shades. Masks, jewelry, figurines and artifacts of jade, no matter their size, were considered luxury items and confined to royalty and elite members of the Maya communities. Pieces of jade were sculpted into depictions of deities, rulers, animals, plants and abstract forms with masterful artisanship.

The Classic Maya attributed various meanings to precious jade stones, including rulership and life-giving maize, since the color of jade was similar to the leaves of their maize plants. Jade was also associated with water and considered the embodiment of wind and the breath spirit. Jade was regarded as essential in funeral rituals and during rites seeking communication with the gods or the ancestors. Earspools of jade carved as flowers were believed to be supernatural sources or pathways for the breath spirit, often shown as a bead or a serpent emerging from a jade flare. Some Maya nobles practiced dental modification and inserted the valuable jade into their teeth.

In Maya artworks, jade is closely linked with depictions of Classic Maya rulers and a king is rarely fully dressed without a full ensemble of decorative jade jewels. When depicting captive enemy elites, part of their disgrace was being stripped of their treasured jade adornments. Jade is identified with antiquity because their ancestors passed down their precious heirlooms through their dynasties; they used jade to contact their deified ancestors (Fig. 7.7).

When placed in their tombs, Classic Maya rulers were usually adorned with many precious objects made of jade; valuable beads of jade were often placed

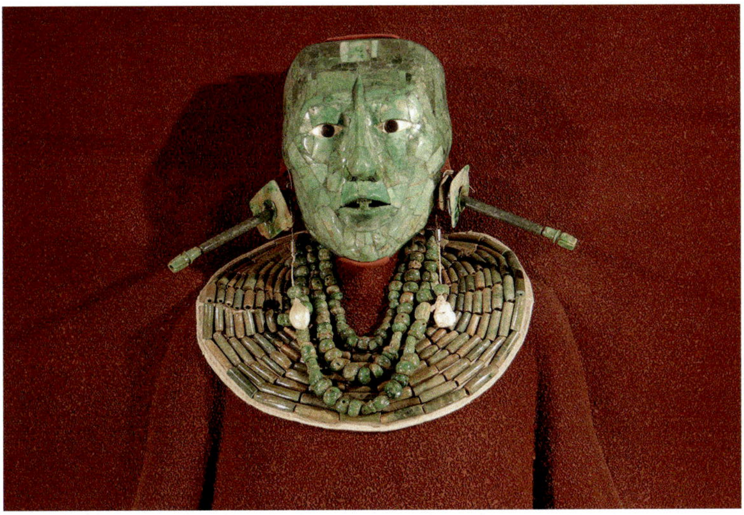

Fig. 7.7 Jade death mask of King Pakal with jade jewelry, seventh century (Credit S L Photography)

within the mouth of the honored dead. The most famous king honored with a vast array of jade treasures was the Late Classic ruler of Palenque, K'inich Janaab Pakal. The lid of his great sarcophagus pictures Pakal as the Maize God at the base of the World Tree. His funerary mask is a stunning mosaic of jade and he was laid in his coffin with jade stones oriented to the Earth's four quarters and one at the world center. Four large jade beads were laid at his hands and feet as well, enthroning him as the world axis, the personification of Maya cosmology (See Chap. 10).

For the Classic Maya jade earflares symbolized rain-making wind, as well as the breath spirit; their hollow, flared form resembled flowers and cave-like entrances. Elite persons adorned with jade articles were perhaps thought of as able to connect with spiritual beings. Pairs of massive earspools placed at the end of ceremonial bars were utilized when conjuring gods and ancestors. The durable nature of jade allowed these cherished objects to be used and reused for centuries as heirlooms, relics and reminders of previous events. Jade stones were considered precious living material that linked the living Maya to their revered gods and ancestors.

Another valued gemstone, turquoise, was only mined in southwest United States and northern Mexico; it arrived late in Mesoamerica trade centers. The gemstone was greatly admired by the Maya. Turquoise jewelry was first worn by Toltec rulers and then soon after appeared at Chichen Itza as well. For the Terminal Classic Maya, turquoise carried very different symbolism than jade which was associated with maize and water. Turquoise instead was considered a 'sky stone' and they related it to the Sun, fire and meteors. Although green and blue are quite different colors, the Maya used the same word for green and blue. Like turquoise, gold too had a late arrival in Maya cities; it was not until the tenth century CE when metallurgy arrived in Mesoamerica introduced from South America. The Maya learned the skill of metallurgy but limited their metal production to small objects in gold, silver and copper, mainly found as offerings in the sacred cenotes of Chichen Itza.

Art, Symbolism and Royal Power

The Maya used art, space, and architecture not only as reflections of their power but to maintain that supreme power. The great honors shown to the king demonstrate the hierarchical differences between the ruler and other nobles, and the even greater differences between the ruler and commoners. The information and artworks carved on stelae and temples includes truths of their cosmologies and the hierarchies of power. Plazas in Maya cities were used in many ways, for public rituals, for processions, for markets and other daily activities; images of the ruler in public plazas were displayed to inform and influence the largest and widest audience.

The Maya produced both public and private art; the public art was meant for all inhabitants and visitors while the private art inside temples and palaces was meant only for the few nobles or priests allowed to enter the sacred, ceremonial and residential structures. The audience for the information inscribed on stelae in public plazas is quite different from those few people who have access to the information on lintels or panels inside a sacred temple. Ballplayers were also honored with their images, wearing all the regalia necessary for playing, carved on walls and on plaques (Fig. 7.8).

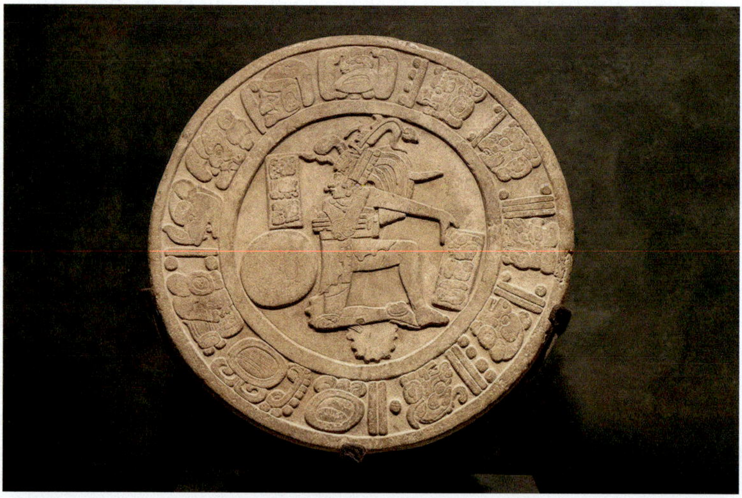

Fig. 7.8 Maya Ballplayer, carved disc with player hitting the giant ball with his hip. Early Classic c. CE 591, Chiapas, Mexico (Public domain)

Ruler portraits were dominant at the ground level of monuments, while mythological images were dominant in the more private upper levels. Public rituals took place in the central plazas, but the commoners in the audience just witness the start of private royal rituals. While those of higher status are allowed to ascend the steps of the pyramid or building. The stairways would have been decorated with dynastic histories that assisted one to look back through the array of ancestral rulers, reaching the dynastic past and into the supernatural realm as well. Once there they could see the symbolic art on the temple first hand and enter the ceremonial space above. During that process they would enter a space that was dominated by cosmic imagery as they symbolically transitioned into the supernatural world. The advantaged participants left the terrestrial world of the plaza filled with images of the ruler to enter the supernatural world filled with cosmological images. Some buildings had top tiers or areas that acted as stages on which the less exclusive rituals were acted out for the audience in the plaza below.

The interior spaces of Maya architecture were not large, especially sacred temples, but the special place provided spiritual context for religious ceremonies. The inner rooms in the upper levels of the buildings were the settings for private rituals where the king interacted with deities and ancestors, assuring his power. The symbolic sculpted images emphasized the ruler's position with the supernatural. Relatively few elites were allowed to enter the ceremonial sites and only when they were participating in the limited rituals. The specially directed symbolic images of deities and myth conveyed specific messages to the privileged audience. The king was placed within a cosmic setting where his power was enhanced by his intimacy with the supernatural. To assure his position above the powerful elite, the king was recognized through his actions, costume and the surrounding carved and painted monuments.

The ruler was shown communing with the supernatural and performing acts that all others could not. Through their participation in the ceremonial rituals, the nobles of the city-state acknowledged and accepted the ruler's superior position and approve his ultimate power. The ceremony of bloodletting and the ritual objects used were shown in private settings; they were never displayed to the public. For the Maya time is associated with cosmic order and the ruler communicated his power over society through his understanding of their history, customs and of the cycles of time. Their royals' precise knowledge of time was acquired through their long, careful study of heavenly phenomena.

References and Further Reading

Burdick, C. E. (2010). *Text and image in classic Maya sculpture: A.D.600–900*. PhD dissertation, Academia.edu.

Grofe, M. J. (2014). *The Copan Baseline: K'atun 9.11.0.0.0 and the Three Hearthstones in Orion*. Academia.edu.

Hopkins, N. *Not just any Turtle: The Maya constellation of Orion's Belt*. Academia.edu.

Milbrath, S. (1995). Gender and Roles of Lunar Deities in Postclassic Central Mexico and Their Correlations in the Maya Area, Academia, (62) Gender and Roles of Lunar Deities in Postclassic Central Mexico and their Correlations with the Maya area, Academia.edu.

Milbrath, S. (1999). *Star Gods of the Maya: Astronomy in art, folklore, and calendars*. University of Texas Press.

Miller, M. E., & O'Neil, M. E. (2014). *Maya art and architecture*. Thames & Hudson.

Sanchez, J. L. (2005). Ancient Maya Royal strategies: Creating power and identity through art. *Ancient Mesoamerica, 16*(02), 261–275.

Taube, K. A. (2005). The symbolism of Jade in classic Maya religion. *Ancient Mesoamerica, 16*(02), 23–50.

Taube, K. A. (2012). *The Symbolism of Turquoise in ancient Mesoamerica*. https://www.academia.edu/4069017

Part IV
Maya Preclassic and Classic Cities

Chapter 8
Early Cities in Mesoamerica

Monte Alban, Oaxaca

One of the earliest groups of buildings constructed with astronomical alignments was found at the enormous site of Monte Alban in the southern Mexican state of Oaxaca. The ancient city sits atop an artificially leveled low mountain range overlooking three important valleys in the central area of Oaxaca and dates to around 600–500 BCE. The site was first settled by the Zapotec people who created an important economic and political center that thrived for over 1000 years. By about 100 BCE the city had become the capital of the area and interacted with other Mesoamerican city-states, especially the great city-state of Teotihuacan. Teotihuacan was also astronomically aligned from its original layout, including its street grid and pyramids. By about 300 BCE Monte Alban held a population of about 5200; by about 200 CE the population had grown to 17,200, one of the largest cities in Mesoamerica at that time. Most of the monumental structures of Monte Alban date from about 300 to 700 CE when it reigned as the principle ceremonial center of the Zapotec civilization (Fig. 8.1).

Currently, the hilltop ruins comprise the second largest ceremonial center in Mesoamerica, only the immense city of Teotihuacan is greater in size. Monte Alban was not only a successful fortress city; the land in the valley below was a rich agricultural region needed to support the ever-growing population. The selection of a steeply walled plateau, 400 m above the valley floor, to build their city is an enigma that had puzzled scholars. Through their research, archaeologists have discovered a variety of different motives for selecting the inhospitable location; it was not chosen for convenience. Although the site was difficult to access, it provided unobstructed views in all directions and was easily defended. Surprisingly, the location has no natural sources for water, nor for quarrying strong stone for construction projects. It required immense human effort to haul sufficient water, food and building blocks up

M. Dolan, *Decoding Astronomy in Maya Art and Architecture*,
Springer Praxis Books, https://doi.org/10.1007/978-3-031-61092-9_8

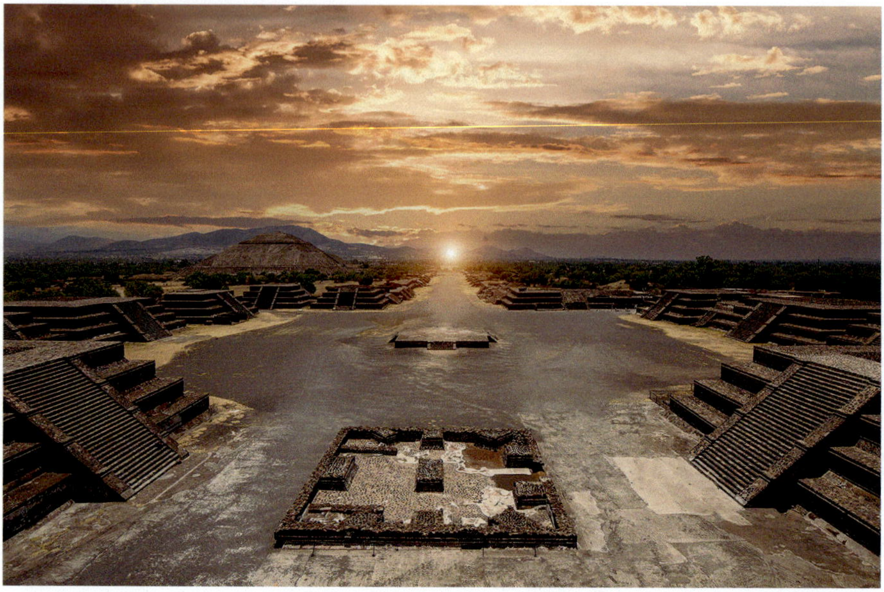

Fig. 8.1 City of Teotihuacan, Street of the Dead solar alignment (Credit eskystudio)

the mountainous slope necessary to operate a major city. From its strategic elevated position, Monte Alban became a fortress city able to control other settlement groups in the area.

Architecture and Astronomy at Monte Alban

Scientific excavations and reconstructions of Monte Alban's ceremonial center began in 1931. The monumental center of Monte Alban is the main plaza that measures about 300 m (984 ft) by 150 m (492 ft) and was large enough to hold the entire population of the city for sacred rituals or civic ceremonies. This great plaza construction project was an enormous feat, requiring the leveling of huge rock outcrops and filling deep cavities. The city core consists of huge plazas, numerous pyramids, a ballcourt, underground passageways, and over 170 tombs. For the taller and larger structures, the builders used thick walls and sloping sides as protection against earthquakes. The most essential buildings and the elite residences are located around the main plaza or in close vicinity. These buildings were originally covered with white plaster or stucco.

Even though many elite palaces have been found, archaeologists have found no evidence that the general population lived at or near the city center. The elite homes were built with stone and consisted of three or four rooms, around an inner patio and a sub-patio tomb that could be reached by a stairway. The tombs of the noble families were walled with stone and often decorated with painted murals; these burials contained some of the most elaborate offerings found in Mesoamerica. In contrast, the poorer population of the Zapotecs resided on artificial terraces created along the slopes of the mountain below the central plaza. The lower level residences were made of adobe brick, not stone, and held only one or two rooms with a central, partially enclosed patio. Another ambitious architectural project was the city's defensive walls that measured 9 m (29.5 ft) tall and 20 m (65.6 ft) long; they were built surrounding the settlement, demonstrating the power of the elite class (Fig. 8.2).

The plazas and buildings of Monte Alban were not designed in a symmetrical manner; the distances between the structures vary considerably from building to building. The center of Monte Alban consists of a series of buildings set atop artificial terraces and facing almost squarely onto its central plaza with clusters of raised architecture along the ridgeline and surrounding areas. Monumental mounds were found within the city also and seem to be evenly spaced throughout. A row of mounds lay north to south in the center of the plaza. The orientations of the architecture changed as the city expanded. On the western side of the plaza, early structures were rotated south of east; later buildings were aligned closer to the cardinal directions. The majority of the city's temples faced east or west, aligning with the Sun's path. These temples were constructed with a two-room floor plan; a shared porch was built at the front that was joined to a private sanctuary at the back.

The selection of Monte Alban's inconvenient geographic positioning was perhaps due to its observational abilities. Archaeoastronomical studies at the site have revealed many important celestial connections. Of all the remarkable architecture at Monte Alban, Building J, dated to 250 BCE, stands out as extraordinary; it was built

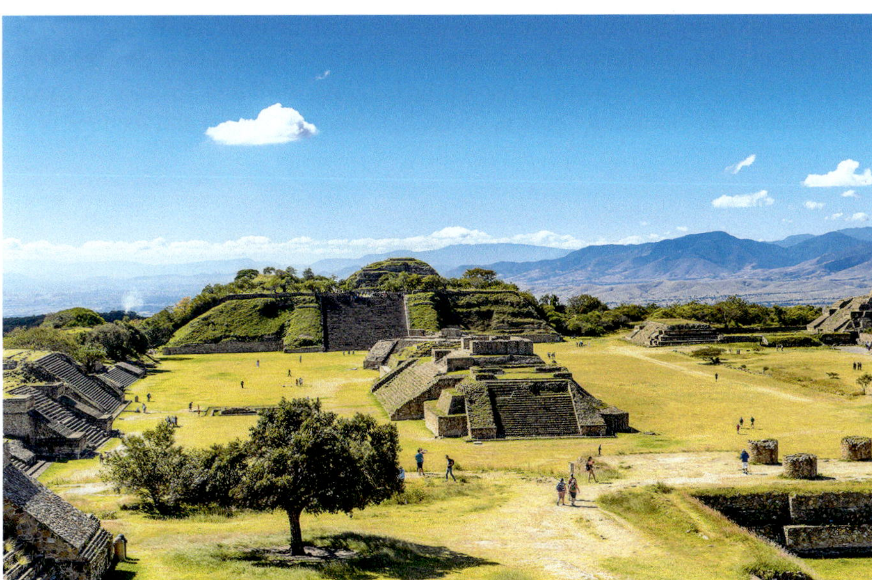

Fig. 8.2 Monte Alban central plaza area ruins (Credit Anton Ivanov)

at a completely different orientation than the surrounding buildings. Although one of the early monumental buildings, its odd shape and juxtaposition to the other plaza architecture must have been designed and oriented for some particular purpose. The building is aligned 47 ½° east of north and its peculiar pentagonal form points like an arrow directly toward the southwest. The unusual shape of the structure plus the odd tunnels found inside implied it must have an astronomical function. Archaeoastronomers conclude that the perpendicular to its doorway is deliberately designed to align with the sixth brightest star Capella, in the constellation of Auriga, at its heliacal rising about May 2. That same date corresponds with the transit of the Sun at its zenith (Fig. 8.3).

There is also a precise alignment of the doorway Building P with the stairway of Building J. Three-quarters of the way up the stairs of Building P there is a small square hole in the middle of the staircase. This thin shaft leads to a chamber 6 ft below the stairs. The shaft, called a zenith tube, is aimed directly at the zenith point overhead. At noon on May 6, weather permitting, a shaft of sunlight would flash directly down the zenith tube; a small ledge 2½ ft above the floor of the chamber would receive the burst of light, dramatically and accurately announcing the time for festivities celebrated at the beginning of the growing season.

Structure P with its exact alignment to the zenith Sun could also act as a calendrical devise. By counting the days that had elapsed between zenith passages of the Sun as it headed north, and then returned, the astronomer-priest of Monte Alban could calculate the length of the year quite precisely. Apparently this cosmic event was so important to the Zapotecans that they designed Building J in such an unusual manner and skewed its orientation to direct attention to the zenith tube in Building P. What seems to be asymmetric to us was to the astronomers of Monte Alban a symbol of harmony between their place on earth and the heavens. In addition, low relief carvings on Building J, showing a crossed stick sighting device, which strongly suggests an astronomical use of the building. Building J is also aligned to the position of the Southern Cross in 25 BCE. Other odd angles and features of the

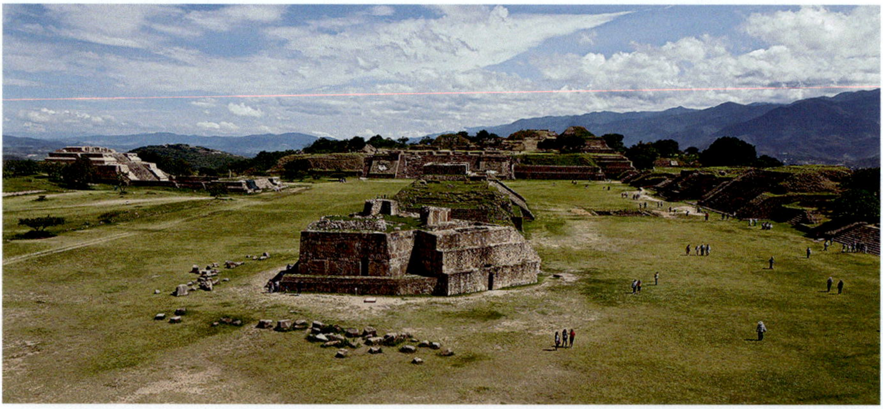

Fig. 8.3 Monte Alban Building J, astronomically aligned (Credit J Vadim Petrakov)

peculiar observatory building may be decoded in the future to supply additional astronomical information. Monte Alban displays the large amount of knowledge regarding celestial motions that Mesoamericans already possessed in the Formative period.

Monte Alban was not only known for its early astronomical alignments; the city was one of the earliest in Mesoamerica to have written texts. Evidence of early writing has been discovered at Monte Alban, but the Zapotecs are not considered the inventors of the first Mesoamerican writing system or the first recorders of time in Mesoamerica. Undeciphered hieroglyphs found at Monte Alban are among the earliest writing systems in all of Mesoamerica. Extensive texts in a glyphic script have been found carved on a large number of stone monuments around the central plaza. Some glyphic signs are recognized as calendric but their script has not been deciphered as yet. The foundations of calendar and numerical notations were widely used in various areas of Mesoamerica before the middle of the first millennium BCE. The Zapotec text that survives is only found on stone, but early scribes may have also used more perishable writing surfaces, such as wood, cloth, bark paper or deerskin. The Zapotec script was no longer used by the late Classic period; it was replaced by another form of script that developed into the later Mixtec and Aztec writing system.

Monte Alban was an important ceremonial center during the Late Formative and lasted through the Classic period; but the city was impeded by its inconvenient geography. The mountain city of the Zapotecs lessened as it was surpassed in size, population and wealth by the enormous site of Teotihuacan. By the end of the Late Classic, 600–750 CE, the city's influence inside and outside the Oaxaca Valley worsened. Soon Monte Alban was totally abandoned. Through the later centuries the site was reoccupied at various times and surviving structures were reused into the Colonial period. The once powerful Monte Alban was replaced by dozens of competing smaller cities, which continued until the Spanish conquest.

The Olmec Civilization

The very beginnings of the Maya civilization are difficult to establish with certainty. Historians claim that the earliest populations first gathered into settlements in the Maya lands sometime between 7000 BCE and 2000 BCE. The Olmec are thought to be the first inhabitants of Mesoamerica to give up the nomadic life style and establish permanent residential clusters. Around at least 4000 BCE, perhaps even earlier, settlers began the cultivation of corn or maize, the elemental crop of the Maya, which enabled the inhabitants to adequately feed themselves and for communities to grow.

Archaeological studies have shown that the Olmec culture established its roots in the early coastal village traditions located in the swampy lowlands of the Gulf of Mexico, similar to that of the Pacific shores. The inhabitants had a broad array of resources: fishing, hunting, gathering, and farming. The word "Olmec" means "rubber people", which refers to their practice of extracting latex from the rubber trees abundant in the area. The name was given to them much later; the name they called themselves is unknown. Often called the "Mother culture" of Mesoamerica, the Olmecs developed the foundation and cultural pattern that were continued by the Maya civilization.

The Olmecs were the earliest stratified society in Mesoamerica; they date back to 1500 BCE but reached their height between the thirteenth and fifth centuries BCE. The Olmecs were flourishing by the Formative period and by 1200 BCE social stratification was becoming more complex. Settlements in the tropical Gulf lowlands, such as San Lorenzo, La Venta, Tres Zapotes and Laguna de los Cerros in southern Vera Cruz had formed into principal cities. The Olmec city of San Lorenzo was a low, hilltop ceremonial center and showed remains of its earliest occupation, that was before 1200 BCE, such as pottery with incised designs. Recent discovery of remains, ritually deposited at a shrine there, have pushed the site's dating back further to at least 1600–1500 BCE.

The richly fertile environment promoted an increasing population which soon inspired the rise of an elite class. The elite class was responsible for the production of the symbolic and splendid artworks and monuments that identify the Olmec culture. The Olmec constructed large earthwork platforms, pyramids, and plazas which were drained by extensive stone-line conduit systems. San Lorenzo became a major center that was densely populated; the leaders began constructing even more elaborate projects, such as huge temple complexes that must have required hundreds of workers. San Lorenzo was becoming deserted by about 900 BCE around the same time that La Venta rose to its heights.

After the decline of San Lorenzo, La Venta grew to be the dominant Olmec center, from about 900 BCE to about 400 BCE, when the city was abandoned. La Venta developed into a temple town containing monumental earthen platforms and sculptured monuments. The town's hilltop was leveled and shaped and then became crowded with numerous monuments, building platforms, stelae, including the famous colossal stone heads. The Great Pyramid of La Venta was the largest structure in Mesoamerica at that time. Even today it reaches 34 m (112 ft) after 2500 years of erosion. Artifacts from La Venta include large mosaic pavements, votive offering of polished jade celts, figurines, pottery, monumental heads, and 1000 tons of smooth serpentine blocks. So far, seventeen colossal basalt heads have been uncovered in Olmec areas with a height up to 11 ft; the largest weigh up to 50 tons. Their production, subject and purpose is still debated by scholars (Fig. 8.4).

The most manifest remains of the Olmec culture are found in its creative sculpture. They are particularly admired for their artwork, carved altars, statues, and etched celts. The Olmecs were first identified by their distinctive art styles; their small-scale sculpture and large monuments were quite different than other Mesoamerican artifacts that were appearing in the modern art markets. The appearance of exotic materials used to create their luxury items indicate that they had access to an extensive trading network for valuable resources, such as jade, obsidian, basalt and magnetite. Some materials were transported from mines more than 200 miles away.

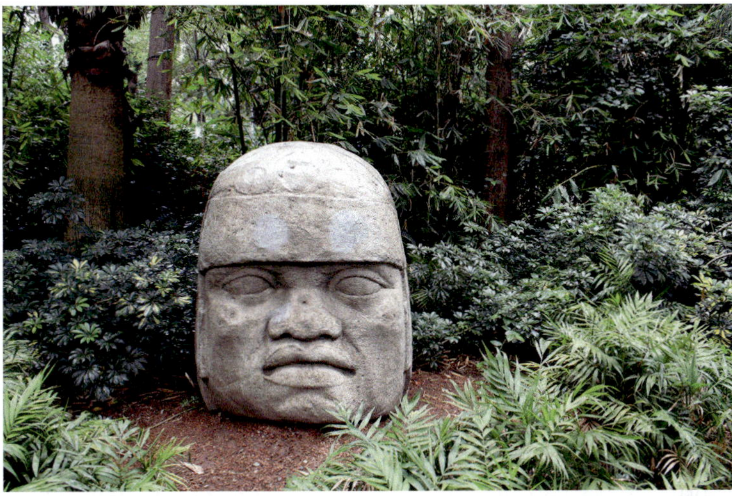

Fig. 8.4 Olmec Colossal Head of early leader, c. 8 tons, ca. 900 BCE (Credit Zbig)

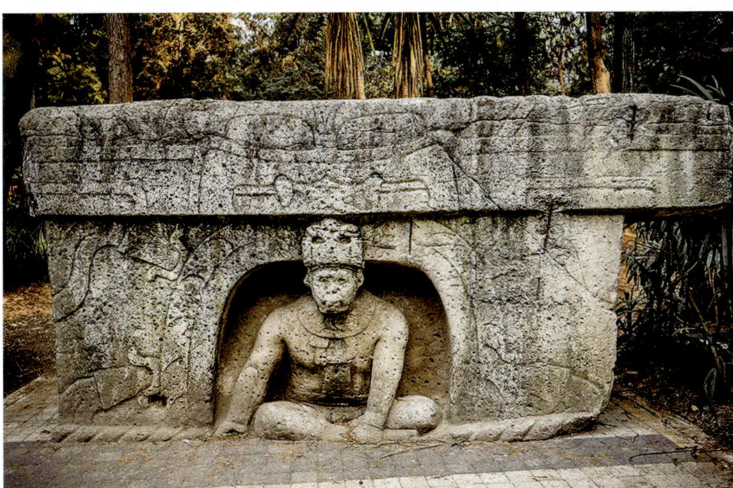

Fig. 8.5 Olmec sculpture, seated royal figure from La Venta (Credit Lev Levin)

Like the Maya, the Olmec sculpture consists of three major thematic groupings: mythic images, effigies of supernatural beings and human figures. The distinctive traits of Olmec art were half human and half animal, such as a human body whose head had feline traits. They are known for the jaguar-like masks, figures with a grooved or notched forehead, prominent teeth, and small nose. These distinctive traits defined Olmec sculpture in carved panels, and in both small-scale and monumental sculpture. The unique characteristics and artistic techniques that the Olmec refined remain unsurpassed in Mesoamerican art (Fig. 8.5).

Scholars have long suspected that Olmecs had a 260-day calendar dating hundreds of years earlier than the Maya; they are almost certain that the Epi-Olmec used the agricultural calendar. Olmec centers, such as San Lorenzo, may have been prime locations for the development of solar calendars. The early ritual calendars were connected to the Mesoamerican numerology of 20 and 13; they used the Sun, Moon and stars for timekeeping which was reflected in the alignments of some early architectural complexes. Initially their calendar was a tool designed to mark critical dates for farming. The earliest confirmed confirmation of the 260-day and the 365-day calendar was actually found in the Zapotec capital of Monte Alban in Oaxaca. At that site, Stele 12 has a carved date confirmed as 594 BCE. Through the Preclassic period there were about 60 different calendars invented in Mesoamerican societies and some communities throughout the area still use versions of the ancient calendars.

In addition to all of these Olmec practices inherited by the Maya, experts are investigating whether the Olmec gods were the same gods worshipped by the Maya. Contacts and interactions with the neighboring Olmec civilization is thought to have encouraged the Maya into initially constructing monumental architecture, which soon expanded into religious ritual complexes to appease the gods. The Maya inherited many of their practices from the neighboring Olmec. The Olmecs were the first culture to practice ritual bloodletting; they were also first to play the Mesoamerican ballgame. The Olmec had the base-20 numerical system, an accurate calendar, use of the concept of zero, and hieroglyphic text. The numerical system consisting of dots and bars probably dates back to the Olmecs as well. It was formerly believed that the Olmec did not create a writing system but new discoveries in Mesoamerica include decorated stone tablets dating to the Olmec period. The tablets were engraved with inscriptions that are thought to be a writing system used only by the Olmec but the scholarly opinions still vary.

The oldest artifacts pertaining to the calendar in Mesoamerica date to the end of the Olmec period, the Epi-Olmec period. The calendar-oriented artifacts were found outside of the Maya lands which indicates that they were created by the Olmec civilization. No positive evidence has been uncovered as yet, but archaeoastronomers believe:

> the 260-day cycle materialized in Mesoamerica's earliest known monumental complexes [and was used] for scheduling seasonal, subsistence-related ceremonies. We cannot be certain exactly when and where it was invented.

These numerous advanced practices developed by the Olmec continued in the advanced Maya societies. We now know that the Maya more closely resembled the earlier Olmec civilization than scholars previously thought.

Unfortunately, written records describing the history, practices and beliefs of the Olmecs do not exist. Many of their architectural sites have not yet been explored. Their known history has been pieced together through the remains of their buildings, their ceramics and their small and monumental basalt sculptures. Their many accomplishments, their intelligence, and their aesthetic refinement has shown their high social level. They had a way of life, thought, and aesthetic expression more advanced than other early agrarian cultures.

Collapse of the Olmec

Historians are puzzled by the eventual extinction of the Olmec civilization. Between 400 and 350 BCE, the population in the eastern half of the Olmec territory diminished dramatically but the site remained sparsely occupied until the nineteenth century. The cause of their decline is thought to be the result of dangerous environmental changes that affected farming conditions. Another problem was that nearby rivers changed their courses. A third theory for population decline was attributed to volcanic eruptions that could have covered the lands with ash and forced the Olmec to move their locations to fresh territories. While the Olmec civilization was in decline, the epi-Olmec culture arose in the remains of the city of Tres Zapotes.

The Epi-Olmec culture is considered a critical link between the Olmec and the Maya. The Epi-Olmec was limited to the central region of Veracruz and lasted from about 300 BCE to CE 250. The Epi-Olmec was not an entirely new, unique culture but a gradual transformation of the Olmec traditions and political systems. Even though Epi-Olmec did not reach the achievements of the Olmec, they did gain sophistication in their writing system and calendrics. They never attained the size and importance of the grand Olmec cities that thrived before them.

The great advancements achieved by the Olmec in architecture, artworks, societal structures and cosmology were never lost but were assumed by the Maya who shaped them into their own particular forms. The astronomical knowledge acquired by the Olmec through the astronomer-priests that watched the night skies was surely passed down to Maya skywatchers as well. The Olmec and the Zapotecs were predecessors and contributors to the culture of the ancient Maya but they were not the only two; there were multiple cradles of civilization throughout the vast area of Mesoamerica.

Maya City of Nakbe

Located in the Mirador Basin in the Peten, Guatemala, Nakbe is one of the earliest and oldest ceremonial centers of the Maya civilization yet discovered. The Mirador Basin is an area of Maya sites of unusual dimensions and architectural complexity, many of its archaeological sites date to the Preclassic period. The largest buildings discovered in the Basin have been dated to the Late Preclassic (350 BCE–CE 150). Even at the early time, the Maya were creating architecture in the "triadic architectural form" which is a huge, raised platform with three buildings atop it.

The Maya city of Nakbe dates to the Early Formative period with habitation beginning around 1400 BCE, only El Mirador, about 13 km north, is larger for that time. Nakbe was discovered later than many Maya cities; it was first noticed by aerial photography in 1930 but was not thoroughly excavated until the 1980s. When the ancient site was unearthed, it revealed surprising information that revised the existing theories on the development of the Maya civilization. The primary archaeologist at the site today is Richard D. Hansen, who has excavated numerous Maya cities of the Mirador Basin and has made amazing discoveries at Nakbe.

The archaeological ruins of Nakbe were deeply buried in the thick tropical rainforest of northern Guatemala. The very origins of settlement in the Maya lowlands began here when the Maya carved a major city from the jungle and first began to construct monumental architecture. The primary period of occupation and construction can be dated to the Middle Preclassic, based on radiocarbon dates and ceramic styles. Nakbe's civic and social core was a rather small area divided into an eastern and a western group; the area containing the larger architectural structures is not expansive. During the first half of the Middle Preclassic (1000–600 BCE), the architecture of Nakbe was small with low walls, similar to other early sites. Within a few centuries a complex civic, political and religious ceremonial center emerged. A major transformation took place during the latter half of the Middle Preclassic (600–400 BCE) when a massive base platform and a group of terraces were constructed more than 13 m (42 ft) tall. This was a time when experts believed only simple villages consisting of elementary housing existed. Four of Nakbe's tallest buildings were created ca. 400 to 200 BCE. Studies of Nakbe's architecture demonstrated that the early Maya had organizational, artistic and technical capabilities centuries before the Classic period.

A huge pyramid called Structure 1 was unearthed in the 1990s, the most impressive building at Nakbe. Structure 1 was constructed within the Western Group of major architecture. Excavations conducted on Structure 1 have provided valuable new information related to the nature and form of the architectural evolution of a construction which ultimately became the dominant building at the site. The Structure 1 was built atop a Middle Preclassic raised platform. The grand pyramid was topped with three roofed structures created in the triadic architectural form; all three buildings of the triadic group were built at the same time. The tall pyramid reaches 48 m (157 ft) over the forest floor. The building was adorned with two giant stucco masks illustrating favored Maya deities; they had been intentionally buried by later versions of the masks and with panels that were added during the last construction phase in the early Late Preclassic period.

During the early Late Preclassic architectural artworks appeared and began to decorate more of the architecture of Nakbe in the form of deity masks. They were discovered on each side of the central staircase of various buildings, but most survive in poor condition. Excavations in the main plaza of Nakbe's eastern architectural group revealed a surprising stela, called Stela 1. The stela had been smashed into forty-five small fragments in antiquity but archaeologists were able to reconstruct the original monument. When the pieces were reassembled, the experts realized that the carvings on the stela presented a most important find. The image pictured on the stela showed two males who wear a very early style of Maya clothing and stand face to face. The iconography of the two figures indicates that they were the key characters from the Popol Vuh. They appeared as the supernatural twins Hunahpu and Xbalanque with Xbalanque pointing to his father's head, acknowledging his connection to a royal lineage. There was no indication that the stela was engraved with relevant text but its images of the creation twins at such an early date was most important and quite unexpected.

One of the essential assets of Nakbe was its extensive limestone quarries, the crucial necessity for the construction of monumental architecture. Buildings at Nakbe were studied to learn the techniques and tools available to the ancient necessary to cut and transport the innumerable stone blocks needed to build their enormous pyramids, palaces and temples. The tools available for the Maya when quarrying were minimal, such as bifacial stone picks and axes, hammer-stones, and flake cores that were chipped from chert which probably had some type of handle attached. Cutting the blocks from the matrix was tedious and back-breaking work, after which the huge stones had to be moved to the building site. The architecture uncovered at Nakbe prove that the Maya builders and architects were quite competent even with only minimal equipment.

Discoveries at Nakbe have revolutionized the former concepts of the early development of Maya civilization but their settlement was rapidly eclipsed by the rise of El Mirador. Nakbe declined in population and then collapsed during the Terminal Formative period (CE 100–200) about the same time as El Mirador. The city was deserted until Late Classic period when a new group of Maya people arrived and reoccupied the city for a time. The newcomers left examples of fine Classic ceramics, but they built no new architecture or monuments.

Mirador Basin Architecture

Important studies are now underway in northern Guatemala headed by archaeologist Richard Hansen, the current director of the Mirador Basin Project. The discoveries made by him and his team have led them to believe that more than forty-five newly mapped sites uncovered in the Mirador Basin formed the earliest well-defined political state in Mesoamerica. Many of these archaeological sites are the earliest and largest in the Maya world. Fourteen have been studied and about thirty more still await excavation surveys.

As a graduate student, Hansen made an amazing discovery that changed the prevailing understanding of Maya history. He discovered potsherds lying scattered just as the early inhabitants left them; he knew that the type of potsherds he found dated to at least 200 BCE. "I realized at that moment the whole evolutionary model for the economic, cultural and social history of the Maya was wrong," he said. "What you had here was the first state-level society in the Western Hemisphere, a thousand years before anyone suspected."

Before his discoveries, scholars believed that almost all of the great monuments, pyramids, painted murals and stone carvings of the Maya dated to the Classic period, not before CE 250. They thought that Maya civilization reached its peak during the Classic period and that any earlier sites would naturally be smaller, more primitive and less impressive than the amazing Classic cities, like Tikal, Copan and

Palenque. They were astonished to learn that their long-established understanding of the development of Maya civilization was totally wrong.

Hansen has excavated, mapped and explored a total of fifty-one Preclassic cities in the huge Mirador Basin with different dates in the Preclassic. Archaeological evidence proved that this area developed the first state-level society in the Western Hemisphere, now dated a thousand years before archaeologists expected to find this sophisticated level of organization. Scholars were surprised when realizing that Preclassic Maya settlements advanced from simple clans and chiefdoms to complex societies with class hierarchies and an organized ideology very quickly. The early Maya also developed a technical ability that enabled them to quarry huge limestone blocks without metal tools and transfer them to construction sites without the help of the wheel. The settlers of the Mirador Basin learned to collect rainwater runoff from the roofs of buildings and store it in large reservoirs and cisterns. They also kept accurate calendars and preserved the records of their civilization. El Mirador is considered one of the most fascinating cities created by the Preclassic Maya.

The archaeological site was mapped, studied and underwent conservation projects; much of the work is support, to keep the ancient buildings from collapsing. Specialists learned not clear all of the trees from the temples; they now leave some shade trees to minimize damage to the delicate stucco carving from the intense sunlight of the tropics. Descriptions of the wonders found as El Mirador inspire a strong desire to explore the archaeological site. El Mirador was finally opened to the public in 2010 but unfortunately has very few visitors, as there is no road to the ruins. The spectacular site is often avoided by tourists because to reach the city from the nearest town that has an accessible road requires a two or three-day arduous and dangerous trek through the jungle. Another means to reach the ruins is an extremely expensive helicopter ride from the nearest town with an airport. El Mirador's inaccessibility is an asset in that it has been protected from looters and damage that have afflicted so many other Maya settlements. But unfortunately visitors are prevented from easy access to admire and experience the remarkable ancient city. Hopefully the government will tackle the problem to provide easier access in the future.

The Maya City of El Mirador

The enormous archaeological site of El Mirador is located in the northern Peten region of Guatemala, north of Tikal within the Mirador-Calakmul Karst Basin. El Mirador is part of the Mirador-Rio Azul National Park within the Maya Biosphere Reserve, an 8100-mi^2 area of rain forest in northern Guatemala. The archeological ruins were first noticed in 1926 and photographed by air in 1930 but only mapped in the 1960s. The Maya city had been totally lost to history and overgrown for many centuries. Due to the remoteness of the location, it was not surveyed until the 2000s.

El Mirador had been swallowed by the forest along with a huge number of other Preclassic cities, many of these are among the earliest and largest of the Maya world. The greatest concentration of Preclassic sites have been discovered deep within the rain forest of the Mirador-Calakmul Karst Basin. Studies of El Mirador changed the understanding and established history of the cultural and architectural importance of the Preclassic period. Amazing to Mayanists, they have now established that all of the components of advanced Maya society were already in place several centuries before the beginning of the Classic period. Dating of advanced Maya settlements continue to move earlier and earlier.

El Mirador was settled very early in Maya history, during the Middle and Late Preclassic period (1000 BCE–CE 250). Because of its monumental architecture, archaeologists expected that El Mirador would be comparable in age to other Classic Maya cities but studies revealed that many of the monumental structures there were not built at the same time as Tikal and Uaxactun. Instead, El Mirador thrived centuries earlier in the Preclassic period. El Mirador's peak period lasted from about the sixth century BCE to the third century CE. The city is especially relevant since it represents the birth, rise and fall of the first great Maya city of the lowlands. Some experts call the city, "The Cradle of Maya Civilization". El Mirador's early age, enormous size and advanced complexity was not realized until the 1960s.

Before El Mirador was thoroughly investigated, archaeologists believed that all Preclassic period settlements in the densely forested lowlands were primitive villages settled by hunters and maize farmers using slash-and-burn agriculture. They thought that their pottery would be primitive, unadorned or of elementary quality. When first discovered, it was assumed that because of the city's great size and elaborate layout, it had to be another impressive Classic period city but its architecture was not in good condition. But the site's poor condition was due to its much greater age and it building stone was of lesser quality. The surprising city had lain undisturbed for many more centuries than they had suspected. Researchers were amazed when they discovered that much of the monumental architecture at El Mirador did not date to the Classic period of CE 250 to 900 but reached back to 600 BCE.

It was shocking to realize that the huge city of El Mirador dated at least 800 years earlier than anyone had thought. For nearly 40 years the only Preclassic building known to scholars was a nearly 27-ft-tall, truncated pyramid excavated in the 1920s at Uaxactun, twelve miles north of Tikal. Scholars assumed that urban construction, and its political and economic systems would be smaller and less sophisticated at that early date. But the ancestors of the Classic period actually built much larger architecture than their successors and developed an even more complex society with powerful political and social organizations. The newly discovered Preclassic inhabitants were just as talented in their artistic creations; in pottery, in stucco friezes and in painting as those in the Classic cities.

After the archaeological site was cleared of its thick cover of jungle vegetation, the full extent of El Mirador's expanse was revealed. El Mirador quickly grew to become the largest site of its time, the earliest capital city of the Maya civilization. The ruins covered four times the area of the huge and powerful city of Tikal, called the "Crown Jewel" of Maya architecture. Experts believe that at its peak, El Mirador likely had a population of at least 200,000. If the surrounding area was also included, the total could have reached a million, while Tikal had only 90,000 at its peak.

During the late Middle Preclassic, El Mirador rose to become the foremost city in the Peten region; it replaced the smaller site, Nakbe.

El Mirador had all the components found in the great Classic era cities, monumental architecture, large platforms, majestic pyramids, agricultural terraces, water collection systems, raised fields, E-Group complexes, triadic architecture, extensive causeways, and an enormous population. When including its extended residential compounds and terrace systems, the city covered 132 km^2, which made it the largest in the Mirador Basin, much more than the impressive site of Tikal. Its monumental architecture rose to soaring heights; the site holds a series of architectural complexes and individual constructions that dwarf anything built by the Maya in later times.

El Mirador's architectural program included large numbers of E-Groups, a great number of raised triadic pyramids and carved monuments illustrating their rulers. The scale of architecture at El Mirador reveals the development of a level of social and political complexity that is equal to the giant city of Mesoamerica, Teotihuacan. Another striking feature at El Mirador was its expressive architectural embellishments. Its building facades furnished artistic display areas for panels depicting the highly honored Maya deities. The artworks were mainly carved stucco masks that were colorfully painted. The elaborately decorated structures created the most impressive display of ceremonial architecture known, even considering all Maya sites, even to the end of the Classic period.

Farming at El Mirador was successful for many centuries; their skillful management allowed their fields to produce enough to feed their inhabitants. When the Maya farmers discovered that the surrounding land, even when cleared of the jungle, was not fertile enough for sufficient agricultural production. They found an ingenious solution to solve the problem by relocating nutrient-rich mud from the many seasonal swamps, now called *bajos*, to the barren land. The inhabitants created raised agricultural fields, large reservoirs and many canals, which are essential elements in farming and water management systems. They also created terraced fields to grow maize, beans, squash, cocoa and many other crops. These improvements enabled them to turn the formerly barren jungle grounds into productive farmland to sustain their constantly growing population. Through its rich farmlands, El Mirador increased in size and importance, becoming the center of power in the Maya lowlands during the latter part of the Middle and Late Preclassic periods.

Another essential feature of El Mirador was the number and length of their causeways (*sacbeob*) linking important architectural compounds with the city center. Externally the roadways linked the site to the other major cities within the Mirador Basin El Mirador sat at the center of a series of radiating causeways that connected it to an extensive network. The *sacbeob* within the city stretched for 13 km (8 miles). Very labor intensive, their impressive roadways were raised 2–6 m (6.5–19.6 ft) above the level of the surrounding landscape and they measured from 20 to 50 m (65.6–164 ft) wide. A lengthy *sacbe* crossed the swampy bogs from El Mirador to the nearby site of Nakbe that is almost 12 km (7.5 miles) away; another roadway linked El Mirador to El Tintal, 40 km (25 miles) away, the longest *sacbe* in

Mesoamerica. The *sacbeob* functioned efficiently moving people, trade goods and ideas; they served as accessible pilgrimage and ceremonial routes.

Through their excavations, researchers found that El Mirador had a perimeter wall 3–8 m (c.10 × 26 ft) high in some sections of the city; the wall possibly reached 60 m (197 ft) in length. Built in the Late Preclassic, the second century CE, the stone wall surrounded the northern, eastern and southern borders of the city and has been dated around the same time that the city was initially abandoned. The wall was built just before the total desertion of the city, suggesting that the residents were aware of a possible military threat at that time.

Architecture and Art of El Mirador

An ancient city's architecture is the most essential evidence of its material culture and tells the story of its era of history. Much can be learned about the city's origins and ethnic heritage though its architecture. The changes apparent in the architectural program reveal the changing values, ideals and religious practices. Through the years the original design of the architectural programs change with modification of choices and more importantly with the dynasties of the rulers in power. The architectural assembly that survives at El Mirador is the accumulated results of centuries of change (Fig. 8.6).

The vast civic center of El Mirador covers more than 10 mi^2 (26 km^2) and includes several thousand structures; its monumental architecture reaches from 10 to 30 m (33–98 ft) tall. Most of the buildings were originally faced with cut stone and then adorned with large stucco masks portraying the esteemed gods of Maya mythology. The city appears to have been designed from its very foundation, since extraordinary astronomical alignments have been found between the architectural groups and the main temples. The major structures have solar alignments which shows the importance of urban planning and sacred spaces since the first settlers. From the time of the initial inhabitants, urban planning and sacred places were an essential part of the site development.

The architecture of El Mirador includes around 35 structures that belong to triadic groups; they are placed atop large artificial platforms. Although they are composed of a group of buildings, they have a different arrangement than the E-Groups. Average cities have only one or two E-Groups, the city center of El Mirador has a remarkable total of eight E-Groups complexes. The triadic group is composed of one large temple pyramid with two inward-facing, smaller pyramids on either side. This common triadic grouping is repeated in many of the Preclassic cities found in the Mirador Basin. Many of the largest architectural structures throughout the Mirador Basin are the triadic configurations, a dominant structure flanked by two, smaller mounds. No triadic structures are yet known from the Middle Preclassic period; yet the format appears suddenly throughout the entirety of the Mirador Basin by the early Late Preclassic period (c. 350–300 BCE). Mayanists believe that the triadic or three-pyramid groupings represents early cosmology of a celestial

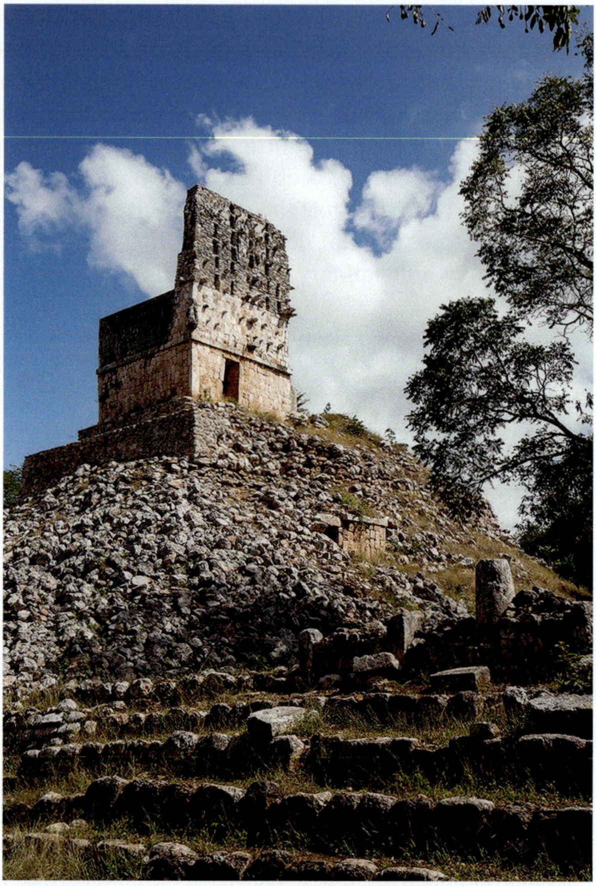

Fig. 8.6 El Mirador Temple atop tall eroded pyramid (Credit Juergen Bochynek)

hearth, "the three hearth stones of Maya creation". The Maya believed that three of the stars in the constellation Orion (Alnitak, Saiph and Rigel) represent the hearth-stones surrounding the original fire, which is a fuzzy nebula called M42, visible just below Orion's belt. The celestial hearth contains the essential fire of creation as written in Maya cosmology.

From the earliest investigations of Maya archaeology, the towering temple-pyramids of Tikal had been considered the tallest structures ever built by the Maya. Their highest reaching temple-pyramid was 65 m (213 ft); the tallest pyramids at El Mirador were expected to be only 8 m (26 ft) high, but they now know pyramids were much higher. Scholars were surprised to discover that one of the largest temple-pyramid at El Mirador stands even taller at 72 m (236 ft). The most massive of the many step-pyramids at El Mirador is La Danta, whose peak rises above the jungle canopy and can be seen for miles. Archaeologists estimate that perhaps 15 million days of labor were spent constructing La Danta, needing twelve men to carry each block, which weighed about a thousand pounds. Researchers have found nine quarries where the stone blocks were cut and shaped, located about 600 m, less than ½ mile away. The construction of such massive architectural projects at El Mirador would have required a high level of organization not thought possible at that early date.

The huge La Danta Complex is located on the east side of the main plaza and Central Acropolis. La Danta was built upon a highly forested elevated platform made of cut stone and rock fill that measures 299 m (980 ft) wide and 609 m (2000 ft) long that covered nearly 45 acres. Even though not quite as tall as the pyramid-temple El Tigre, its elevated location on a hillside makes it the largest building of the Maya world. The base of La Danta is strikingly adorned with jaguar and vulture heads built into the sides of the smaller temples of the complex.

From the top of the pyramid other nearby cities, such as Nakbe and Calakmul, can be sighted. Its total volume (2,800,000 m^3),including its platform, is considered to be the largest pyramid ever built by the Maya, in fact the largest structure in the world. (The volume of the Great Pyramid of Egypt is 2,600,000 m^3) La Danta's central staircase was flanked by the remains of monumental stucco masks picturing the gods like those found at Cerros in Belize and Tikal. There is evidence of overlaid construction done during the reoccupation of the site during the Classic period.

The other Late Preclassic pyramid-temple group at El Mirador is the massive El Tigre complex found at the west side of the city center; El Tigre reached 55 m (180 ft) in height and has astronomical alignments. The Tigre and La Danta complexes face each other on an east-west axis. The buildings are decorated with large stucco masks and panels depicting common deities of mythology. Part of the West Group, the El Tigre pyramid was estimated, based on replication experiments, to have required a minimum of 2,803,595–5,000,000 person days of labor for its construction. Atop the massive central pyramid is a triadic arrangement of three smaller temple pyramids, echoing the triadic arrangement of the central pyramid and the two that face each other across the front plaza. This double pyramidal complex covers a surface area six times larger than Temple IV at Tikal, the largest building at that Classic period site. The Tigre complex is thought to be the temple where rulers performed human and animal sacrifices. The lower pyramids on either side of El Tigre display gigantic stucco jaguar masks.

El Mirador's Great Central Acropolis was constructed upon a massive platform base measuring 400 m (1304 ft) and rising 10 m (32 ft) above the plaza at ground level. The Acropolis consists of nineteen major buildings and many small structures, seated on elevated levels. The Central Acropolis is formed into a narrow plaza bordered on one side by a series of small structures. Most of the structure facades were covered with cut stone that were decorated with large stucco masks. The Great Acropolis has evidence of palace architecture as well, but the residences of the commoners were situated at the periphery of the civic-ceremonial center. Their house groups, like typical Mesoamerican homes, were composed of three or four one-room structures, each on a low platform, surrounding a small central plaza that served as a patio.

Structure 34, the Temple of the Jaguar Paw, is found on the south side of the Acropolis platform; this temple rises to a height of 17 m (55 ft). A wide central staircase leads up four staggered levels to a broad terrace which has a triadic arrangement. A secondary staircase, flanked by huge deity masks set on sloped panels, rises to the central structure, the Jaguar Paw Temple. The temple overrides an earlier sub-structure that has a beautiful mask that still has its original, red-painted surface. This early temple has been dated to an incredibly early date, 600 BCE, contemporary with the Olmec culture. Its unexpected dating and the findings at El Mirador has forced a new interpretation of this time period because of its advanced sophistication.

The Acropolis area also had a large number of carved monuments illustrating rulers who already had power in the Maya lowlands during the Preclassic period. Additional artwork found in the central area depicted portraits of main Maya deities, such as the Feathered Serpent, the God Chaak, and the God Itzamna. The extraordinary discovery of excellent modeled stucco art that has been preserved, provides an excellent example of the Preclassic worldview of the ancient Maya. The reliefs of the Central Acropolis provide important information in their depictions. They present early examples of royal figures in the sculptural program visible in public spaces. The ruler images illustrate that they are adopting certain characteristics of the gods for themselves. Although there is a scarcity of text and glyphic symbols compared to later sites, the images are capable of transmitting information on the interests and social functions of the rulers. Their historical events and specific dates are still enigmatic as the stelae with images and glyphs have not yet been deciphered.

The most significant artwork found at El Mirador is called the Central Acropolis Frieze. In 1979 archaeologists were excavating a room of Structure 34, the Jaguar Paw Temple, part of the Tigre complex in the western part of the city and made a most significant discovery. This temple was one of the most thoroughly studied of all the buildings at El Mirador; its sophisticated decor was thought to belong to the later Classic period. But pottery evidence from the site proved that the temple dated much, much earlier, back to 600 BCE.

Researchers uncovered two 26-foor long carved stucco panels that depicted the Hero Twins of Maya cosmology. The Popol Vuh is the sacred book of myths, and history, the Maya story of how the world was created. The main figures in the book are the twins, Hunahpu and Xbalanque. The carved frieze depicts Hunahpu wearing a jaguar headdress and swimming while holding the head of his father, an exact depiction from the myth. Finding images derived from this sacred mythological tale at a Preclassic site was quite incredible, since it was believed that the Popol Vuh creation story had been corrupted by the Spanish priests who translated it. Scholars thought that after the Conquest the Maya had been greatly influenced by Christianity. This extraordinary artwork proves that this Maya account of their creation had been established for more than 2000 years before the Spanish arrived. The only copy of the Popol Vuh available, one of the few surviving Maya narrative texts, was dated 1550. Since then, brilliantly painted frescoes that also depict events from the Popol Vuh have been uncovered at the Preclassic site of San Bartolo, near Tikal. These very early artworks are still being analyzed by scholars to decode the information they retain.

Astronomy at El Mirador

Recent studies undertaken by archaeoastronomers have demonstrated that many important buildings of Maya cities were aligned astronomically, mainly to sunrises and sunsets on specific days. These orientations aid in the use of observational calendars that allow useful scheduling of agricultural and ritual ceremonies throughout the yearly cycle. The astronomical orientations in the center of El Mirador provide the earliest evidence of the utilization of these observational designs of urban centers. They demonstrate the importance of astronomical and calendrical concepts even in the earliest architectural design and urban planning. Also important is the 17° family of orientations in city planning which is the most widespread alignment group in Mesoamerica. These studies show that this common alignment of city planning did not originate in the Mexican city of Teotihuacan, (which is aligned 17° east of north), as formerly believed, but rather at the early Maya sites within the Mirador Basin.

In different areas of Mesoamerica, archaeoastronomical studies have proved that orientations of monumental architecture have a clear non-random distribution. This signifies that in general civil and ceremonial buildings were aligned mainly on an astronomical basis. The other important early urban centers of the Mirador Basin have not been studied for their alignments as yet. Recent research at El Mirador has discovered patterns in architectural alignments and their urban design. This indicates that complex rules connected to astronomy and calendrics, like those of later cities in other areas of Mesoamerica, were already established in the Maya lowlands by the Late Preclassic (300 BCE–CE 200). Most architecture at El Mirador is now degraded to earthen and stone mounds covered by thick tropical vegetation; these conditions make accurate orientations of the buildings difficult to determine.

Astronomers have determined that the architectural orientations at El Mirador show a consistent clockwise skew from the cardinal directions, prevalent throughout Mesoamerica. In addition several groups of orientations dominate different sections of the city layout. Mesoamerican architectural alignments focus around certain azimuthal values; the only conceivable way to create this chosen directional focus is by observing the position of astronomical bodies at the horizon. In addition, the geometrical patterns in the placement of some structure are aligned in a favored east and west direction. Most Mesoamerican architectural orientations are related to the Sun's positions on the horizon and most of the east-west orientation azimuths lie within the annual motion of the Sun along the horizon as it follows its yearly cycle north and south. At El Mirador, all of the east-west building orientations are positioned within this angle as well, indicating that they specify the sunrises and sunsets on precise dates.

Astronomical studies undertaken in several Maya areas have shown that the sunrise and sunset dates marked by orientations at a site are usually separated by multiples of 13 and 20 days, these are elementary periods of the Mesoamerican calendrical system. The close connection between the most frequently marked date of the architecture and the important dates of the maize cultivation cycle suggest that the buildings were aligned for that specific purpose, for appropriate scheduling of agricultural activities and the related rituals. The astronomical orientations at El Mirador reveal the application of observational calendars that were similar to those well known in other parts of Mesoamerica during later historic periods. At El Mirador most alignments of buildings and building groups were to the east, only two were oriented to the west or sunset positions and these structures were notably different. The eastern building alignments appear to mark the dates separated by intervals of 143 days (from October 11 to March 3) and 221 days (from March 3 to October 10), both of these are multiples of 13 days (143 = 11 × 13; 221 = 17 × 13).

Archaeoastronomers have established that Group-E type architectural complexes seem to have a concentration of alignment dates around February 19 and October 22. The Group-E complexes at El Mirador seem to fit this usual alignment. Because of the difficulties of ascertaining exact orientations with no advanced instruments, the dates marked by architectural alignments must have had utmost importance to the Maya. They must have marked the required times for performing obligatory rituals; the exact scheduling of agricultural tasks depended on both practical and religious factors. Modern ethnographic evidence shows that exact times of planting and harvesting vary from year to year, depending not only on certain climate conditions and maize varieties but also on the cyclical phases of the Moon. If the astronomer-priest could depend on various alignments to confirm their calendar, they had a better chance of predicting precise times of the agricultural cycle, even if adverse weather conditions impeded direct observations of the Sun on the important dates. The exact dates for enacting particular agricultural tasks and related rituals had to be determined with accuracy.

The alignment patterns at El Mirador indicate another astronomical practice that has been little studies. Astronomers have noticed that buildings were not only aligned to the horizon but also located with respect to each other. By making observations from one building, another structure situated far away could have been placed as an exact marker of the Sun's position on important dates. In mountainous areas, prominent peaks along the horizon were used in a similar way. The astronomer-priests of El Mirador, where few natural features offer possible markers, apparently found a means through the placement of buildings to form an artificial horizon.

The alignments embedded in Mesoamerican architecture and urban layouts must be explained as having not only practical purposes but significant symbolic importance. The goal of measuring time through the observation of celestial bodies might have been accomplished without building such monumental structures and then orienting them with precision. The aligned buildings should not be considered observatories, in the modern sense; but they were used for monitoring particular events in the heavens. These buildings were more than devices for observation. Considering their civic and ritual significance, the structure must have had a central position in the Maya world view and in their cosmology and ideology.

The Maya regarded the celestial order as absolute, perfect and superior to the natural order of Earth. They interpreted the cyclical motion of celestial bodies and cyclical changes in the natural environment as cause and effect. The resulting beliefs determined that events in the sky influenced and determined the seasonal transformations in nature. Architectural alignments to significant rising and setting points of celestial bodies verified the most important moments of cyclical time. The celestial phenomena replicated the ideal cosmic order that assured the survival of humans; the rulers, as semi-gods, claimed to be responsible. Building astronomical alignments into the temples, palaces and monumental buildings was an effort of the rulers to reconstruct and preserve the divine order seen in the heavens in their earthly realm. When the heavenly cycles were viewed at predicted times, they fortified the beliefs of the ruling class and legitimized their power which helped to sustain the current political order.

The alignment relationships in the urban layout of El Mirador display the remains of observational calendars used to schedule the dates for agricultural and ritual activities. They show that solstitial sunrises and sunsets were embedded into Mesoamerican architecture from the earliest eras. The consistencies uncovered in the urban patterns of El Mirador establish the earliest evidence known for observational patterns. The aligned buildings would have been of foremost importance to schedule events ahead of time, especially important in times of bad weather. The role of astronomy was embedded in ritual and intimately linked with social life, religion, and political affairs of Mesoamerican societies.

El Mirador's Final Decline

The weakening of El Mirador, between CE 100 and 200, was perhaps caused by the runoff of clay into the nutrient-rich mud which was caused by massive deforestation of the surrounding area. To make lime plaster that they needed to cover the roads, buildings, platforms and house floors, they needed enormous amounts of firewood. When the trees disappear, the soil becomes loose and will easily wash away; it could have covered the nutrient-rich muck that was needed for survival of the community. When the fields were not replenished the crops failed, which led to starvation and the collapse of El Mirador society. Archaeologists believe some inhabitants may have moved to Calakmul in the Yucatan Peninsula, which emerged as a powerful city-state and rival to Tikal in the sixth and seventh centuries. Between 300 BCE and the first century CE, nearly all the major sites in the Basin area were abandoned.

The site remained deserted for almost 500 years; then around 700 CE certain regions of El Mirador became reoccupied but on a modest scale. The new residents built only small structures that were fitted in amidst the ruins of the formerly grand monumental architecture. The largest buildings from this later occupation period were only about 8 m (26 ft) high. The new settlers stripped many of the Preclassic structures for stone blocks to build new structures and to grind for lime making, needed to paint them. Among the Late Classic inhabitants, however, were talented scribes and artists. They created fine polychrome ceramics, but their occupation of the site was brief. By about the end of the ninth century, 200 years later, the city was abandoned again and remained so to the present. Unfortunately, many of the Preclassic Maya cities are being threatened again by massive deforestation carried out by logging and road construction for logging and by looting. Even though El Mirador is located in a remote jungle, many people make the grueling trek to visit the incredible architecture at the site.

References and Further Reading

Hansen, R. D. (2002). *The architectural development of an early Maya structure at Nakbe, Peten, Guatemala research year: 1996*. FAMSI @ 2002.

Sprajc, I., et al. (2009). Early Maya astronomy and urban planning at El Mirador, Peten, Guatemala. *Anthropological Notebooks, 15*(3), 79–101.

Chapter 9
Early Classic Cities of the Maya

Maya City of Cerros, Belize

Located on a peninsula in northern Belize, Cerros was a small Maya city that thrived in the Late Preclassic (c.400 BCE) to the Early Classic period (c.CE 400). Cerros' ideal position on Chetumal Bay at the mouth of the New River was perfect for fishing and enabled merchants to move trade goods up and down the east coast of the Yucatan. Beginning as a small fishing village, the city became part of an early coastal trading network that became far-reaching, not only along the seacoast but through the inland river systems. Researchers have found evidence of Maya canoe trade along the Caribbean coast that dates even before Cerros was well established. The Preclassic Maya could have used astronomical observations for navigation on the Caribbean coast of Yucatan very early by marking the horizon where the land meets the water. This could have developed from an early interest in using astronomy to align their calendar rituals. Evidence that early calendars were available can be found at Preclassic sites with E-Groups.

As a successful trade city, Cerros then developed into an important economic and ritual center. Cerros is especially important because it demonstrates one of the earliest examples of the initiation of kingship rituals connected with astronomy in Mesoamerica. Cerros remained a provincial city throughout its existence but had monumental ritual architecture, temples, shrines, and residences; its population reached only 1089 at its peak. The city was also successful in farming as it developed an extensive canal system and raised-field agriculture. Eventually it developed into one of the largest influential trading centers of its area with 108 structures. Although large for its time, Cerros was never close to the size of the immense Maya cities of the Classic period.

Sited directly on the coast, the local landscape of the peninsula was dominated by water which played an important part in the creation of architecture at the site. The central core of the city sits next to the bay and is comprised of several relatively

M. Dolan, *Decoding Astronomy in Maya Art and Architecture*,
Springer Praxis Books, https://doi.org/10.1007/978-3-031-61092-9_9

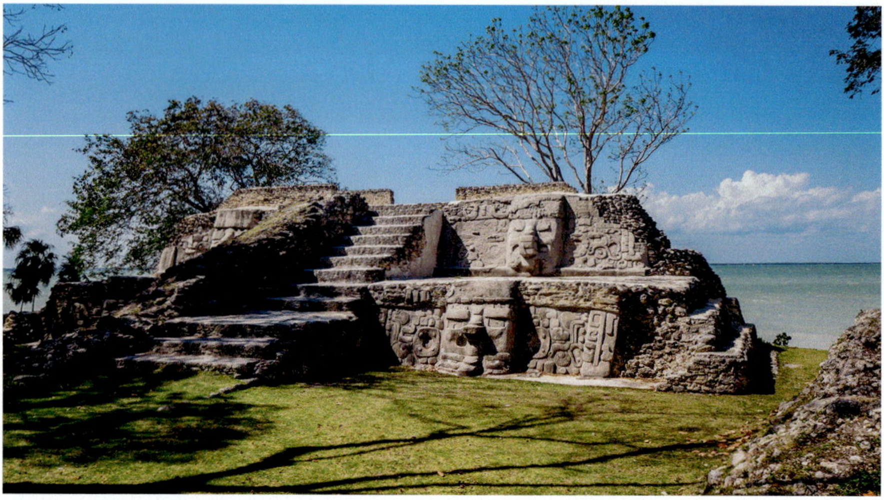

Fig. 9.1 Cerros, Belize, small pyramid with masks of the Sun and Venus (Credit Duarte Dellarole)

large buildings and step pyramids, an acropolis complex, large plazas and two ball courts. Most of the buildings were clustered on a series of elevated platforms at the northern tip of the peninsula. About 50 BCE the people of Cerros began to experiment with the system of kingship and also started an urban renewal program, destroying and burying their homes to make space for new temples and plazas. All earlier buildings were leveled as well and then the city's first major construction project was begun.

Cerros' new civic center was constructed over the early village and lies in the northern section of the site; the largest buildings and the highest density of architecture are found in this area. The central plaza has four large masonry pyramids with many platforms and additional plazas surround it. Most of the temples and elite buildings were clustered on a series of elevated platforms. A few of the monumental structures were adorned with detailed stucco masks; some are well preserved, some badly eroded. The construction of the new architectural program of Cerros was not done in one segment but the work spread over a period of 200 years. The astronomical alignments and symbolism of the iconographic décor preferences changed over that extended time period. Cerros was one of the few Maya sites that was not built over by a later era of the city's life (Fig. 9.1).

Astronomy at Cerros

The first major temple pyramid constructed at Cerros was built between 50 BCE and CE150; it sits on the northernmost edge of a small promontory with its back at the edge of the bay and its front faced the entire community toward the south. The pyramid has numerous astronomical associations. Archaeologists call this architectural complex unpoetically, Structure 5C-2nd, the most important feature at the site; it may have been built atop an earlier building. The temple marked the northern point of the sacred north-south axis that ran through the axial center of the new site plan, ending at the ballcourt located at its southern border. The pyramid was composed of two-tiers with south facing platforms; originally the pyramid had a perishable structure or temple above it which is now missing. In Maya cosmology, the most important direction of the sky is to the north where the celestial gods reside, while the south is the direction of the underworld. This primary temple was ideally positioned both physically and symbolically within the celestial realm, on the primary axis, between the Upperworld and the Underworld.

The missing temple atop the two-tiered platform functioned as a setting for private rituals involving the ruler and his connection with the cosmological forces. The king would use the rear area of the temple to perform private ceremonies, such as fasting, bloodletting and genital perforation. Experts suggest that the blood offerings were associated symbolically with creation events. The shedding of blood stands as a symbol of "a metaphoric death" allowed the ruler to reach the heavens and communicate with the ancestors. The two wide landings enabled the community to witness the public rituals performed by the ruler and priests. Since the entire settlement was facing south, the entire community could observe the rituals that took place on the upper platform or on its staircase. Structure 5C-2nd was greatly admired for its four remarkable stucco masks decorating the facade. The original masks are now protected by exact copies that cover them. These masks are the best preserved artworks at Cerros and would have been the focal point for ceremonial activities that acted out the ideology of the rulers and inhabitants of the city.

Experts argue over the symbolism of the four masks; emblematic masks became common pyramid adornments at many early Maya settlements. The masks on the two tiers represent the Sun and Venus in terms of their horizontal positions above the horizon (the Morning Star precedes sunrise, but the Sun sets before the Evening Star). On the lower level of the temple, scholars believe that the two masks represent the rising Sun (right side) and setting Sun (left side); on the upper terrace the masks symbolize Venus as the Morning Star (right side) and Evening Star (left side) with the apex aligned to the solar zenith. The masks are seen with the Sun rising from and then setting into the sea; Venus repeats the same actions. The masks at the upper level have long snouts and wear the headband of the king, while the lower masks have the *kin* or Sun sign on their cheeks. The faces of the masks all wear ear-flares that designate the royal attire that the king would have worn as well while performing sacred rituals. The masks also wears the "Jester God head band"; all the masks, except for the upper western one, have a sky band which symbolically marks a celestial location for the masks. The fourth mask is wearing an Earth band.

An additional meaning given to the masks are animal symbols related to jaguars in the lower two masks and symbols related to birds are seen in the upper masks. Scholars associate the jaguar with the zodiac sign we call Capricorn and the bird symbols to the constellation Cassiopeia. Some scholars also suggest that the pyramid also holds imagery of the Milky Way. Since the pyramid was painted white, the white stairway would have symbolically denoted a "white path". The temple would be aligned to the Milky Way when it reached a north-south alignment; it would appear above the stairs between the linked panels of the temple. Nighttime ceremonies could be attuned to the apparent connection with Milky Way and with the stars of the heavens which would have been most impressive.

Therefore, the placements of the masks represent the path that the Sun and Venus follow through the sky. The Sun and Venus were also identified as the Hero Twins of the Popol Vuh, central to the Maya myths; although some experts say the Hero Twins represent the Sun and the Moon. Some scholars suggest that the masks refer to the rule of the Hero Twins in Maya creation mythology and allude to the creation of the world by the divine brothers but others disagree. An alternate interpretation of the masks offered by scholars is that the lower panels represent the bundled bones and funerary masks of the Maize God and his twin brother. The upper masks are meant to represent to gods Itzamná, the God of Creation and Chaak, the God of Human Sacrifice. Together they destroyed and killed the Maize God. Perhaps the king was impersonating these gods when he would perform certain rites atop the structure.

At Cerros the large stucco masks on the temple facade played a much different role than the written texts that were engraved on architectural facades. The main purpose of the masks was to create an appropriate environment for a sacred ceremonial event. That along with the correct positioning of architecture would provide a more dramatic background, a cosmic stage for the rituals that took place there. The masks, in connection with the astronomical alignments and the surrounding landscape, are an important factor for understanding the function of the traditional rituals performed there. The building also contains four large postholes, once used for the placement of sacred trees, called World Trees. Together, the form, design, placement and iconographic décor of this special temple created a holy sanctuary embodying the entire Maya cosmology.

When ascending the stairs on ceremonial occasions, the king or *ajaw* was positioned at the center of the cosmic forces and the daily cycle of life and death; he would move through time and space to cosmic levels. From the coastal city it is possible to see both the Sun and Venus rising from the sea in the east and setting into the sea on the west but not throughout the year. The Sun appears to rise from the sea beginning in May and ending in August; this would be the most dramatic time period to enact certain rituals. Even though Structure 5C-2nd faces south, a skywatcher at the apex would have had a direct sightline to the zenith sunrise aligned with the eastern coastline on May 15, marking the onset of the rainy season. The rituals could imply that it is the power of the king, he alone interacts with the universal forces to assure the daily rising and setting of the Sun and the most important planet. The sacred activities integrate the symbolic interaction of the role of the king and his 'control' of the cosmos and the movements of the celestial sphere. Solar phenomena played an important role in the lives of the early inhabitants of Cerros.

Through the activities performed on symbolic architecture, most importantly bloodletting, the king became an institution rather than an ordinary individual. The ruler needed to establish his mastery over the powers of nature through communication with the supernatural. This demonstration of his divine power and his interaction with the cosmic forces reinforced his position as leader and affirmed his right to rule as divine king. The sacred architecture was a symbolic system of public art and rites of accession. These symbolic rituals assured the community of the king's approval from the heavenly gods who would care for their needs. Cerros had powerful kings who controlled their subjects through the enactment of ritual ceremonies, enacting creation myths, sacrificing his own blood and invoking celestial movements, all important to Preclassic Maya ideology.

All of these symbolic connections would provide the ruler with an ideal performance stage to enhance his ability to bring the blessings and approval of the heavenly gods to establish divine order to the local Maya world. The special dates of August 13 and February 5 were observed as "creation nights" or "creation cosmology" because of the astronomical alignments of the Milky Way, Ursa Major, Cassiopeia and Capricorn on these dates. The compelling ceremonies performed, interacting with the heavens and the gods, would produce a lasting effect on the community. Structure 5C-2nd was not only a symbolic stage or seat of power but a symbol of time, a calendric marker, a powerful site for communication with the spiritual forces. After it was built, other structures were constructed around it and the plaza area was enlarged. As the civic center changed over time, the ritual landscape shifted to incorporate more astronomical orientations (Fig. 9.2).

Astronomical studies of the early Caribbean settlements, such as Laminai, Belize, have shown the importance of observations of the Sun and stars in the seasonal cycle of the tropics. Native astronomical systems utilize the horizon as a

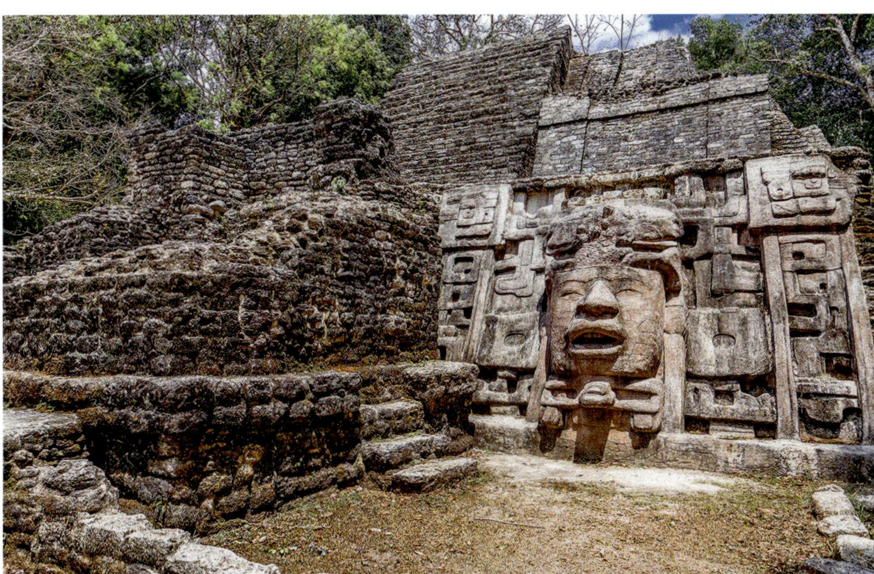

Fig. 9.2 Pyramid with large mask at Lamanai, Belize, preclassic (Credit Vadim_Na)

reference circle to the zenith overhead. The zenith can be marked by the Sun's position on the horizon at sunrise or sunset. The stars most useful for navigation have declinations that fall within the solstice extremes; their positions are directly connected with the solstices. The early Island Caribs had many names for the position of the solar zenith, indicating the seasonal passage of the Sun directly overhead at noon. Among the Maya, architectural structures are used as indicators of alignments that mark the Sun's seasonal or zenith position with precision. Cerros has a building that is oriented to the solar zenith.

An E-Group complex has also been uncovered at Cerros but its orientation has not yet been studied to determine if it has a specific alignment, but other buildings at Cerros show particular astronomical orientations that could have been useful in creating their early calendar. It has been suggested that Cerros' very early aligned architecture was probably used to create an early horizon calendar by observing the Sun as it crossed from the land to the lagoon. At Cerros the monumental landscape changed significantly over a relatively short period; many of the new buildings were astronomically aligned. Structure 4 stands as Cerros' largest and tallest building, 21 m (69 ft) high, 68 m (223 ft) long and 58 m (190 ft) wide at the base. It was built later than most of those in the civic center, so the redesigned site plan was almost completed at that time.

Most buildings at Cerros are aligned north to south, but Structure 4 differs by facing east, on the newly established east-west axis. Originally the building had a narrower pyramidal structure on the top with a vaulted room; a stairway lead to the upper structure. This building too was adorned with large masks which indicated that it was a site designed for ritual ceremonies along with Structure 5C-2nd. The greater size of Structure 4 could have provided a grander stage to conduct rites and a more open space for the audience to gather in the plaza below. The change to east in orientation may have signaled an association between the temple and the ruler, as the reborn rising Sun. It may have been the starting place for a ritual procession on particular occasions. The structure was also meant to be the king's funerary shrine; however it was never used for reasons unknown, as the chamber was empty upon excavation."

This building at Cerros also had important astronomical orientations; Structure 4 was an observation point for marking the Sun's movement from land to water at the day of the first solar zenith in May. Since the building rises above the tree line, a viewer standing on top of Structure 4 would have a direct view of sunrise over the flat landscape along the seacoast. Astronomers have determined that the building is aligned to sunrise on March 27, 5 days after the actual equinox (March 21–22), only an approximate orientation. At sunrise on the day it reached its zenith, the Sun seemed to be bathed in water when it announced the onset of the rainy season. At the latitude of Cerros the solar zenith falls on May 14 or 15; May has the largest increase in rainfall when compared with other months. The second zenith passage in the fall marks the date for harvesting. Even though the plaza in front of Structure 4 was designed for large public gatherings, the equinox could be better viewed from the top of the building, but space there was quite limited. Perhaps a viewing structure existed previously designed for ideal skywatching. Elites used these occasions to demonstrate their special knowledge and connection with the gods. The cosmological stories and myths that helped organize the community and inspire them to build these large temples was realized and transmitted through their shared experience and connections with the heavens.

Another important temple at Cerros is Structure 29C which faces westward and was built to the south of the original northern portion of the city. It is a triadic structure with three separate temple platforms on top of a greater pyramid and considered one of the clearest expressions of the Preclassic triadic plan at Cerros. The small, central platform faces west and is adorned with carvings of jaguar heads, while the sides of the two surviving temples display long-snouted masks. The temple was probably meant to be a war monument and was associated with the north and south ballcourts. Much of the site has not been excavated. The last large structure at the site called Structure 3A-1st was built around CE 100, but many of the other buildings must have been abandoned by then. Structure 5C-2nd represents a prime position where the solar zenith would have been observed along with the rising Sun and the coastline. During later construction periods, Structure 4 was erected with an alignment toward the solar equinox and zenith sunrise, this suggests a new interest in calendrical precision, and a formation of hierarchy at Cerros. The rebuilding program at Cerros demonstrates that the creation of massive ceremonial architecture was beginning to stratify. The rulers of the early city were able to direct the labor of the farmers and artisans to construct large public projects, including civic and ceremonial architecture. The ruler is becoming an institution rather than just an individual. Not only at Cerros, but in many parts of Mesoamerica, elites were distancing themselves from their subjects by ordering them to build remarkable temples and pyramid complexes where the leaders would live and commune with the gods. They took control of trade at the local markets and then the long-distance trading markets. They enjoyed finer homes and finely decorate pottery and trade goods.

The king takes his place alone at the highest point of his symbolic interactions with the heavens and declares himself to be the force that preserves the cosmic order. He becomes a major force of nature, expressed even in the architecture. He controls not only the city and the population but the movements of the heavens above. For the Maya the stars and planets were considered alive and animated by the gods for their various purposes. The king's public displays of power and domination became a method to re-enforce social cohesion. The ruling class had become the natural order appointed by the gods. This ideal form of government became embedded in Maya life and belief systems and described its social reality for the next thousand years. At Cerros the institution of kingship was in its initial stage; it did not become fully developed until the Early Classic period more than 300 years later. The foundations of sanctifying an elite ruler began at this early time.

The architecture at Cerros reflects the transformation of the ideas and ideals of the society and utilized the motions of the heavens to legitimize the divine right to rule. There were four of these aligned ritual temples, possibly each new ruler or dynasty built his own version. During the late Preclassic, Cerros began to decline, with changes in its economic and political fortunes. When the elite class abandoned

Cerros, termination rituals were performed, which included burning and smashing of ritual articles at the royal temples. By the Classic period, Cerros no longer functioned as a key trade city but continued as a residence where domestic activities were carried out. New construction was very limited and the population slowly began to dwindle. For a time, Cerros returned to its days as a simple fishing community. The city's slight habitation continued until it was totally abandoned before CE 400 and left to deteriorate. Cerros remained unknown until 1900 when it was finally rediscovered.

Maya City of Tikal

History of Tikal

The vast city of Tikal, one of the major population centers of the Maya World, laid hidden deep within a sultry rainforest of northern Guatemala, was not discovered until the nineteenth century. The forested area is so impenetrable that the ancient city sat silently secluded for 1000 years amidst the brilliant, multicolored birds and exotic creatures of the tropical jungle. Only those living in local settlements were aware of the extensive ruins. Tikal is the city's modern name; the Maya's name for the powerful kingdom is thought to be Yax Mutal. The initial settlements at the site of Tikal date back to 1000 BCE, during the Middle Preclassic period. Tikal became settled with small villages from about 600 BCE, the Late Preclassic; while some of the city's earliest monumental architecture survives from the fourth century BCE. By the Early Classic period around CE 200, the city had grown into a metropolis; it attained its peak around 550 when Tikal became the most powerful city-state in Mesoamerica. But by the mid-ninth century the great city's fame had dwindled and its huge population had greatly declined (Fig. 9.3).

Tikal's population may have reached as high as 90,000 residents, larger even than European cities at that time. Although Tikal grew into the largest, most impressive Maya city, it was independent; there was never a Maya empire with one central ruler controlling the other cities. Even though they shared the same language, religion and cultural behaviors, each city-state was clearly independent. During the Classic era, the numerous city-states were continually at war with each other. A ruler achieved prestige by capturing a rival king; he might capture him, inflict torture or behead him but he did not expand his city's borders. The more powerful cities, like Tikal, would command weaker sites in a client relationship, which Tikal did with Copan.

Archaeologists have excavated thousands of Maya structures found within a 6 mi^2 radius but Tikal probably extended even beyond that expanse. Archaeologists think that Tikal extended as much as 48 mi^2 since a large percentage of the site has not been cleared or mapped. In spite of two decades of archaeological work on the ruins, much of Tikal's history is still a mystery, much still lies buried, waiting to be uncovered.

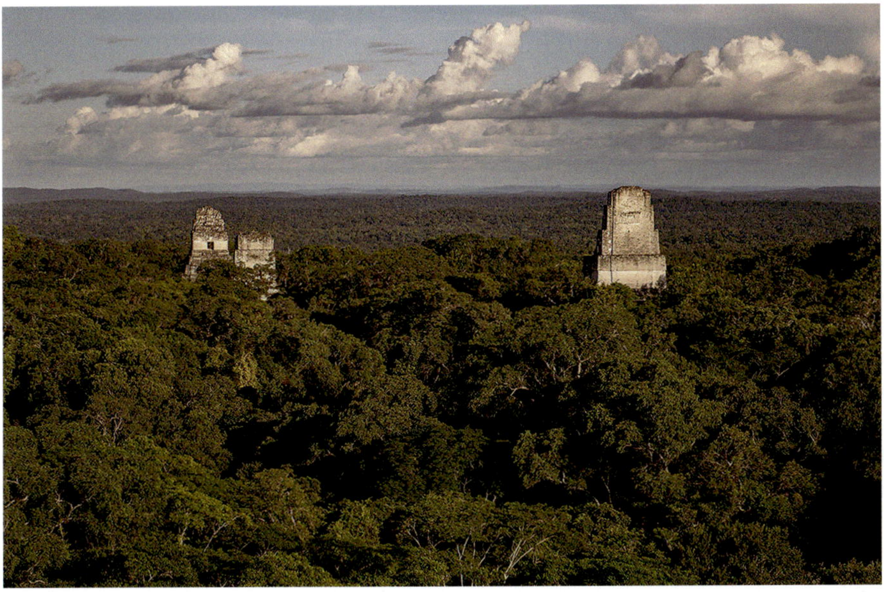

Fig. 9.3 Aerial view of Tikal lofty pyramids peeking through forest canopy (Credit Wirestock Creators)

Many of its 200 pyramids, its numerous temples and decorated monuments survive in a remarkably good state of preservation. Because of its remote location far from other settlements, Tikal was not rediscovered until 1848 much later than most Maya cities. Stephens and Catherwood had heard of the site but it was too thickly covered in jungle overgrowth and inaccessible at that time. The site had few investigators until the early twentieth century; the majority of the serious archaeological work at the site was done by the University of Pennsylvania during the 1960s. The core area around the Central Plaza has now been completely mapped; more than 3000 structures have been uncovered at Tikal. Although uncovered, only a small number of buildings have been fully excavated or studied by archaeologists.

A number of engraved monuments and lintels have been excavated; these are covered with important texts that revealed insight into the history of Tikal. It is now one of the best understood of the large Mayan cities. Epigraphers have established the names and dates of the long reign of dynasties holding power during the kingdom's Classic period, along with its alliances and its bloody military encounters. The carved inscriptions are so detailed that specialists have verified their chronology of thirty-nine contiguous rulers of Tikal over 800 years. The earliest known king lived during the third century CE; while the last known ruler came to power in 768. The palaces and tombs of many Tikal kings have been discovered, revealing substantial information concerning their years in command. Wealthy tombs with elaborate grave goods have been excavated dating back to the first century CE.

The rulers of Tikal appear to have been more ambitious and powerful than the rulers of other Maya cities; this is confirmed by their enlarging ruling class and numerous privileged members of their society. These powerful rulers of Tikal dominated the surrounding area economically, politically and militarily; they also controlled several smaller nearby cities, such as Uaxactun. Tikal maintained its prestige and dominant status throughout the Maya world during the entire Classic period.

Although isolated within a tropical jungle today, Tikal did was part of an active trading network that linked the city with other important Mayan and Mexican trade centers. Inscriptions tell that the city rulers were interested in expansion and reached south to Copan, which they founded. Tikal's leaders founded the city of Quiriquá as well. Copan was first ruled by a leader from Tikal, who dressed in some of the stylistic costumes worn in Teotihuacan. Perhaps Tikal was imposing its authority on the whole southwest area of the Maya, for it soon dominated the territory in trade, warfare and wealth. Evidence of this collaboration and the many interactions between Tikal and Copan were repeatedly found in their hieroglyphs.. Even though 630 miles apart, Tikal also communicated with the most important Mexican city, Teotihuacan; their distinctive ceramic designs have been discovered at Tikal along with depictions of their storm god, Tlaloc, who was also honored at Tikal.

Architecture at Tikal

The earliest monumental stone structures at Tikal appeared about 400 BCE around the North Acropolis. The more elaborate pyramids, temples and buildings that survive today were constructed during the Classic and Late Classic periods. Fortunately, Tikal's outstanding architecture was well built and well preserved. In addition to the hundreds of pyramids and temples, the site contains royal palaces, administrative buildings, residences, altars, seven ball courts and a multitude of inscribed standing stelae. The many temple-pyramids were constructed using local limestone blocks over a core of rubble and covered with a thick layer of stucco. Like other Maya cities, the most honored architecture was built on areas of higher ground and on raised platforms (Fig. 9.4).

Similar to most temples, Tikal's were built atop the soaring pyramids; they usually had a room or two within but due to their corbelled arch vaulting, they were quite narrow and would allow only small a number of people inside. Tikal structures were often built atop earlier buildings; a practice that was common with all Mesoamerican major architecture, enhancing the power of the building by encasing a structure previously built by their honored ancestors. The various buildings of the huge city were joined by carefully built roads raised over the swampy areas. Maya scholars think that the active building program must have supported a full-time workforce of stonemasons, sculptors, painters and laborers to carry out their constant building and rebuilding.

The city's spacious Great Plaza fills the center of the site and is surrounded by monumental buildings. Two tall temple-pyramids sit on the east and west sides and

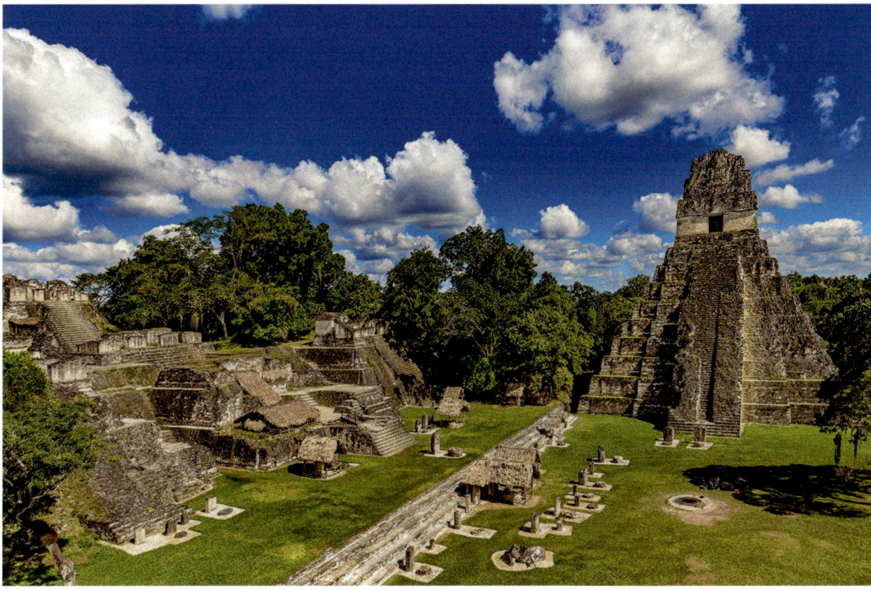

Fig. 9.4 Tikal Pyramid 1, Great Jaguar Temple**,** one of twin pyramids (Credit Witr)

on the south lies the Central Acropolis. The various plazas of the city were paved with stucco and graded with deep channels so that rainwater would run off to fill their reservoirs that were plastered and waterproofed for preservation. At the north of the Great Plaza sits the monumental North Acropolis which was the most honored place of the city. The North Acropolis was an architectural compound that functioned for over 1300 years as a royal necropolis and a center for funerary activities. Excavations of the North Acropolis revealed a series of tombs rich with treasured grave goods, proving that this section of Tikal held royal and elite burials. At the center of the great city the North Acropolis sits atop stepped terraces that are decorated with grotesque giant masks made of plaster and masonry.

The North Acropolis contains the most impressive and admired pyramid-temples of the Maya World. The six very large pyramid-temples filling the Acropolis, called Temples I–VI, were named during the earliest surveys of the ruins. They all face each other and are all aligned to the cardinal directions. The tiered buildings each reach over 120 ft tall and are among the most studied, most complicated and best understood of all Maya architecture. The most impressive of the six, Temple I, sits to the east and its twin, Temple II, sits on the west side of the complex, facing it; both were built during the Late Classic period. In addition to the pyramids, the North Acropolis also includes thirty altars and forty-three stelae, all completed by the ninth century. A row of tall carved stelae stand at the bottom of the North Acropolis; each stelae is adorned with the image of a highly adorned king. The huge number of stelae decorating the entire city are carefully inscribed with hieroglyphic text that eventually revealed the lost history of the people of Tikal through its long existence (Fig. 9.5).

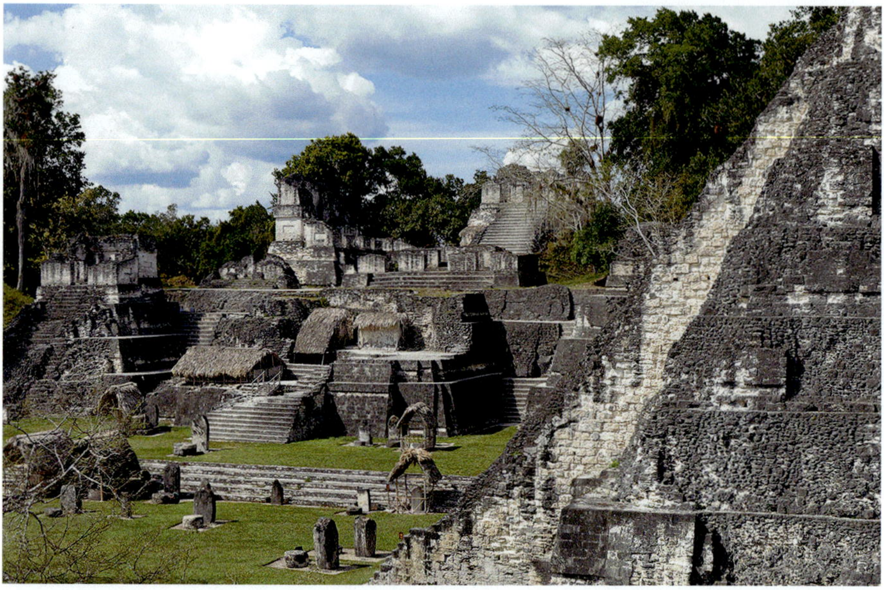

Fig. 9.5 North Acropolis at Tikal, built over hundreds of years (Credit Svetlana Bykova)

Tikal's Temple I is a very steep pyramid with an even steeper stairway on its western face. The stairs display huge masks on each side that symbolized the institution of kingship. This pyramid rises 47 m (154 ft) to the funerary temple at the apex. Amazingly, a substructure was discovered inside the pyramid dating back to 600 BCE. Temple I, also called Temple of the Great Jaguar or The Great Pyramid, includes ten sloping levels with rounded edges; it slants at a very steep angle of more than 70°. At the upper level sits a three-room temple which is topped by the majestic roof comb towering above. A portrait of the ruler Ah Cacau, or Jasaw Chan K'awil, (who reigned 682–734 CE) appears on the roof comb, sitting on a jaguar throne. In 682 Jasaw Chan K'awil also had a stela carved and erected calling himself *kaloomte'* which means "High King", indicating that he was the ruler of a kingdom not just a city. The tomb of Ah Cacau was discovered and excavated beneath Temple I; it contained his skeletal remains wearing a massive jadeite necklace. Sixteen pounds of jadeite were discovered within his tomb, along with decorated pottery and other valuables. His mortuary shrine is dated CE 734, but he died much earlier. The builder of Temple I was not King Jasaw Chan K'awil but his eldest son, Yik'in Chan K'awil, who commissioned it to honor his father.

Temple I's twin, called Temple II, stands across the plaza to the east; it too wears a two-level roof comb that towers almost as high as Temple I at 38 m (125 ft) above the plaza. Built around 700 CE, Temple II is stylistically much different than its twin. This pyramid was designed having only three tall tiers with well-defined edges, stockier than Temple I. A steep two-level stairway extends out into the plaza that lies before it. At the top of Temple II sits a shrine with three small rooms that have wooden lintels spanning the doorways. On a lintel inside the temple, an engraved portrait of a seated woman was discovered; the image is thought to be the wife of Jasaw Chan K'awil, Lady Kalajuun Une' Mo', who died in 704. Temple II served as Lady Twelve Macaw Tails' mortuary temple but her burial site has not been discovered as yet.

Temple I and Temple II are not the only double pyramid complexes. Within Tikal there were a total of nine Twin-Pyramid Complexes, not all are still standing; they differ in size but all have two pyramids that face each other on an east-west axis. These double pyramids are all flat at the apex with four stairways climbing the sides. The first set of monumental pyramids was constructed in the sixth century. Within the rooms in the upper temples were found depressions carved in the walls that amplify sound and project it in every direction. This feature allows the voice of a speaker within the temple of one pyramid to be heard by those within the upper rooms a great distance away.

Five of the key pyramids were not only ingenuously designed and perfectly constructed, they also incorporated accurate astronomical alignments with such precision that they continue to function as a giant astronomical instrument today. The sightline between Temple I and Temple IV marks the sunset position on August 13. The sunrise position on the winter solstice is marked by a sightline between Temple IV and Temple III. Because Temple I and Temple II are situated exactly east-west of each other, they mark sunrise and sunset alignments at the spring and fall equinoxes.

Temple III (or the Temple of the Jaguar Priest) was the last of the great pyramid/temples built at Tikal and the largest one built in the Maya lands during the eighth century. The temple was commissioned by a little known king and was probably his funerary temple but it has not been investigated by archaeologists as yet. This king, sometimes called the king of the Dark Sun, was second to the last king of Tikal. Temple III dates to 810 during the Classic period, which proves the economy was still stable at that time. It rises about 55 m (180 ft) tall and has nine stepped levels with an east-facing stairway. The summit was decorated with a tall roof comb, but it has been seriously damaged. Temple III has only been partly restored and is not open to the public. The temple at the apex contains only two rooms rather than the usual three. One of the rooms has a sculpted lintel showing an obese figure wearing a jaguar skin which is still in its original placement. The figure has not been firmly identified.

Although Temple III is narrower than most, it is majestic in scale. Temple IV stands as the tallest of Tikal's buildings, reaching 70 m (230 ft) to the tip of its large roof comb which measured 12 m (39 ft). The blade-like structure contains 250,000 cubic yards of construction material. Glyphic texts cover the back and sides of the roof comb. Temple IV honors Yik'in Chan Kawil, son of Jasaw Chan K'awiil. The roof comb portrays an enthroned Yik'in facing his father on Temple I, a quarter of a mile away. Like the other temples, it contains three inner rooms; the date 741was carved into two wooden lintels over the doorway. The view from this pyramid would have been unmatched; priests could overlook the vast city with colorfully painted buildings as far as the eye could see.

Temple V stands south of the Central Acropolis and was also built as a mortuary pyramid but the ruler has not been identified as yet. This temple/pyramid is the second tallest building at Tikal reaching 57 m (187 ft) and was dated with radiocarbon testing to about CE 700. The dates of ceramics found near the temple associate the construction to the reign of Nun Bak Chak who ruled during the second half of the seventh century.

The Lost World complex was the first area of Tikal to be developed and the last to be abandoned during the decline of the Terminal Classic period,. This special complex was regarded as sacred from its beginning and its temples were used continually for ceremonial occasions and for astronomical viewing throughout Tikal's lengthy history. The Lost World also held the greatest grouping of structures at Tikal, consisting of thirty-eight different buildings in an organized arrangement. Some of the buildings were designed for specific celestial observations, including the reoccurring cycles of Venus and the Sun and for marking the solstices and equinoxes.

Generations of masons at Tikal constructed five distinct versions of the Lost World complex, beginning around 600 BCE; each layer overlaid the previous structures, growing larger and larger over the centuries. The complex site gradually expanded over an area of 60,000 m (196,850 ft), reaching its peak at the fifth renovation around CE 250. This enclosed group of buildings endured in its final version and remained unchanged while many other areas of the city continued to be improved. During the Classic period, the North Acropolis and the Lost World competed as the chief ritual center of Tikal; both honored complexes were used as royal burial grounds.

The Lost World has two separate areas named the High Plaza and the Low Plaza; they are centered on the large central plaza. The Great Pyramid sits on the High Plaza and rises only about 31 m (102 ft) high; it has four tiered side with stairways adorned with large masks that are now too eroded to identify. Surprisingly, there was never a temple at the apex of the Great Pyramid unless it had been built with biodegradable materials that left no trace. The grand structure shows evidence that it was originally painted red and blue.

Eight adjoining buildings divide the High Plaza from the Plaza of the Seven Temples to the east. The organization of a group of large temples on the east side of the complex adheres to the layout of the E-Group complexes. These particular temples are identified as solar observatories or as an astronomical center. The E-Group in the Lost World contains a pyramid with a platform which supports three east-facing temples. The orientation of the grouping is aligned to a central axis which was maintained by other buildings in the complex. Structure 5D-46 in the North Zone faces west and is aligned to sunset on March 31 and September 12; the E-Group buildings marked sunrise on March 11 and October 2. Therefore both pairs of solar dates may have comprised a single observational calendar. Alignment to these particular dates is probably intentional with a specific purpose and perhaps associated with patron of the nearby residential palace and tomb; the ruler who died in CE 376. Further research will help clarify the E-Group and the other Lost World buildings' exact astronomical usages.

Tikal's Final Decline

By the end of the Late Classic period the breakdown of the Classic Maya civilization was spreading throughout the entire Maya region. City after city saw dwindling populations that moved closer to city centers; the numerous societal stresses that were occurring had initiated an increase in warfare over possession of scarce supplies. At that time the ancient and long-lived city of Tikal began its population decline as well. Just as the Tikal area reached its highest population, the site began to experience erosion, deforestation, soil deterioration, contaminated water sources and perhaps drought and famine. The land was overworked and crops began to fail; as a result, resources for the necessities of life dwindled. The failing conditions caused people to believe that their gods had abandoned them. Much of the population abandoned Tikal between 830 and 950, causing its central ruling system to crumple.

Major construction projects had ceased by the end of the eighth century and only one pyramid was built after that date, Temple III. After that last pyramid project, just a few small monuments were erected; the last around 869. At Tikal the archaeological record of decline seemed particularly harsh, pointing to sudden chaos and an abrupt breakdown of the political and social functions. The society became unstable and careful excavations of the ruins has shown that elite palaces were burned about that time. The collapse of Tikal is not a mystery, although Mayanists still debate the exact circumstances. An additional cause has been detected in its ancient reservoirs; many buildings were found to be covered with mercury-bearing cinnabar which would be washed off by rainfall and pollute some of their reservoirs. Numerous causes including environmental depletion, overpopulation, weather conditions and societal collapse, all contributed to the end to the 1500 year culture at Tikal. The city was totally deserted by the end of the tenth century or early eleventh century. There has been very little evidence found of occupation during the Postclassic period (Fig. 9.6).

The towering acropolises were left unattended to be reclaimed by the jungle, only their roof combs peak above the forest canopy. The ruins were left undisturbed by human contact until the ancient city was rediscovered in the mid-nineteenth century. For over a thousand years Tikal had led the Maya civilization with its successful royal life, its thriving trade network, its unsurpassed art and architecture. There is still much to be learned about this most important kingdom. Recently, researchers using lidar technology have discovered that Tikal was even larger than previously suspected. In 2018 aerial analyses detected 60,000 previously unknown structures around Tikal and in other parts of Guatemala; this includes pyramids, houses, and defensive fortifications. Future investigations will reveal more wonders from this remarkable culture. Tikal National Park is a favorite site for visitors who come to

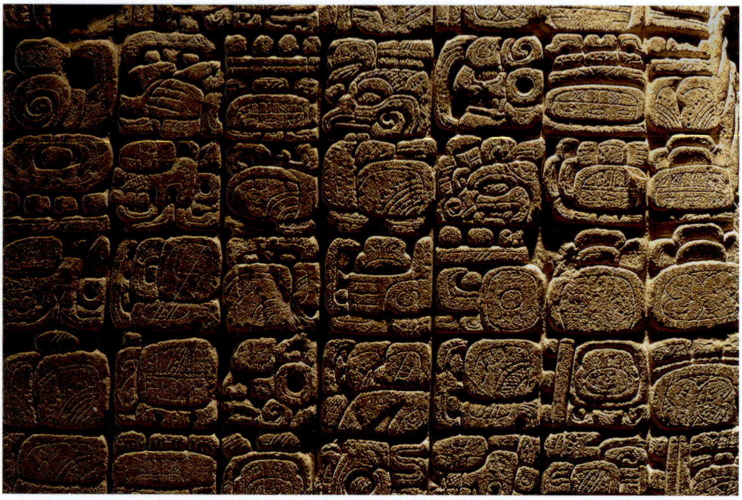

Fig. 9.6 Section of inscribed text that tell the history of Tikal (Credit T Photography)

admire the numerous towering pyramids; more than 300,000 visit each year from all over the world.

Maya City of Calakmul

Located deep in the Calakmul Biosphere Reserve about 35 km (22 miles) from the Guatemalan border, in the south-central area of the Yucatan Peninsula, Southern Mexico, Calakmul was one of the largest and most powerful city-states in the Maya world. Calakmul was not the name given to the city by its people; they called it Ox Te Tuun, which means "Three Stones" in the Mayan language. Calakmul translates to "City of Two Adjacent Pyramids" and was so named in 1931 by the man who discovered the ruins, Cyrus Lundell. The name signified the two large central pyramids that were the most impressive sights.

Currently, scholars estimate that the population reached over 50,000 people. Until the beginning of the twenty-first century, most lands around Calakmul remained enveloped within the dense tropical jungle, the second largest tropical forest in the Americas, only surpassed by the Amazon jungle in South America. This huge city was set deep in the heart of it. Until new studies were undertaken, the thick forest had obscured the immense size of the lowland city.

Previously, scholars believed that the legendary city of Tikal was the most densely occupied of all the Maya cities. Remarkably, recent LiDAR surveys have proven that its neighboring rival, Calakmul was even more densely populated. Researchers conducted an aerial survey using LiDAR technology; they mapped Calakmul's ruins to collect precise data on the surface features of an area covering

95 km^2 (37 mi^2). Calakmul was a huge urban city with extensive architecture, not just an elite center that was surrounded by houses of commoners. The aerial study not only uncovered extensive areas of Calakmul, but it also revealed dozens of other ancient settlements hidden in the inaccessible forests, including a few major urban centers with monumental architectural complexes and inscribed monuments. This exciting new tool has been unearthing the secrets of Mexico's long-abandoned ancient settlements.

The scope of what scientists discovered through LiDAR studies is truly breathtaking. By stripping away the forest cover, LiDAR mapping has exposed a treasure trove of archaeological remains. Their findings were made in ultra-dense areas that researchers had thought were too rugged and inhospitable for people to thrive, too harsh to sustain any sizeable settlements. The survey revealed the presence of hundreds of structures that had never been detected previously. Their amazing findings at Calakmul included numerous unseen buildings, huge apartment-style housing complexes, some included sixty individual living spaces. These residential complexes were positioned around large groupings of shrines, temples and marketplaces. None of this extensive housing had been identified before the LiDAR studies, which proved much higher populations than suspected. Analyses of maps produced through LiDAR reveal the active and greatly populated city at the peak of its prosperity, most dominant during the last three centuries of the Classic period. The site includes roadways, quarries, agricultural terraces, reservoirs and water ponds supplied by a series of canals. LiDAR mapping also revealed that the city had a formidable defensive system and was protected by a major defensive wall, indicating evidence of warfare (Fig. 9.7).

Archaeological excavations have revealed that Calakmul had a lengthy history beginning in the Middle Preclassic and surviving until the Postclassic. Archaeologists have worked for more than 20 years at the site and have uncovered much valuable information about the city's construction and how such a large number of inhabitants could survive there. The great city of Calakmul is slowly revealing its ancient treasures. In addition to the enormous city, Calakmul rulers oversaw an extensive kingdom comprised of twenty secondary centers, some of which were large cities. The total number of inhabitants of these satellite cities has been estimated at 200,000 people. Calakmul's entire kingdom included political control of over 1000,000 inhabitants including the subsidiary settlements.

The Calakmul area is unique since the site preserves intact remains of the rapid advancement of a Maya kingdom in the hostile environment of the tropical forest. The archaeological site contains the remnants of at least 1500 years (500 BCE to CE 1000) of intensive population growth and the development of a complex society. The city's remains demonstrate the successful adaptation to its adverse setting. During the Late Classic period scholars estimate that the city covered an area of over 70 km^2 (27 mi^2). The city's ruins show technological achievements and cultural development evident in their architecture, inscribed texts, sculpted monuments and painted murals.

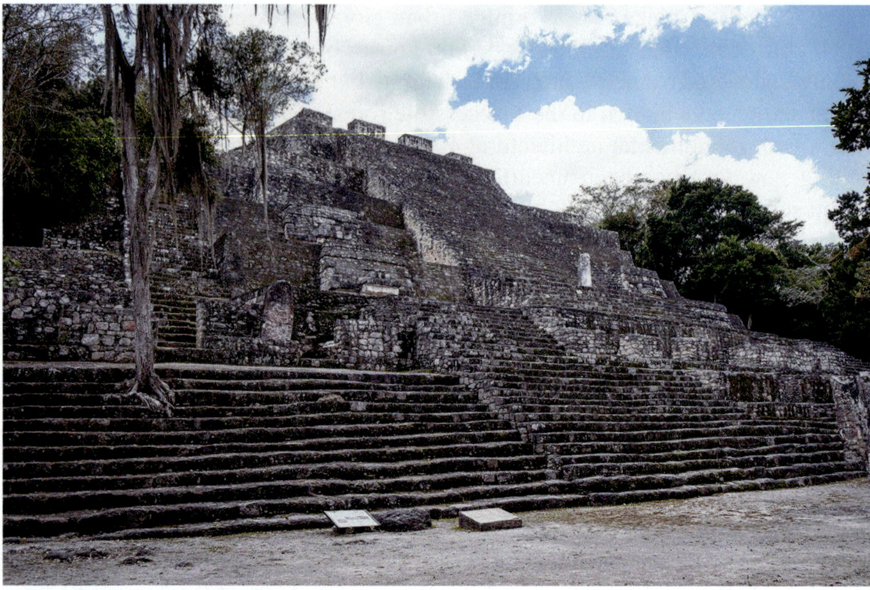

Fig. 9.7 Large tiered structure at Calakmul partially hidden by trees (Credit Guillame.B)

Architecture at Calakmul

The twelve centuries of history and construction in Calakmul saw the evolution of construction of architectural complexes such as acropolises, connected by open squares and courtyards. Calakmul is known for its huge number of monuments, most were erected surrounding the city's core. Archaeologists have noted 6750 ancient structures within the city so far. A version of the Peten style is evident in the towers and stone mosaic decoration of the building facades, including zoomorphic entrances. At the center of Calakmul, the main plaza measures 250 m (250 ft) by 150 m (450 ft) and has six structures standing around it. This is the site where the most important ceremonies and political events were held. This central section of the city is the oldest and had the longest lasting occupation.

North of the Great Plaza is a large temple-pyramid called Chii'k Naab; it is one of four buildings filling the central plaza. It reaches about 24 m (c.79 ft) in height; it faces the pyramid known as Structure I. There are five stelae in front of the pyramid and chambers on top. An eighth century tomb with jade masks and other goods was uncovered inside the building as well. In 2005 the substructure of Structure I was discovered; the inner building was built between CE 620 and 700. The base measures 12 m (39 ft) in width and 5 m (16 ft) in height. Beneath the temple are some of the most stunning murals ever found in the Maya world. Amazingly, the walls of the substructure were completely covered in vivid murals; thirty individual scenes have been exposed; archaeologists believe there are as many as sixteen more, not yet exposed.

The painted scenes were quite different from the usual depictions of royals, nobles and warriors in elaborate costumes. The extraordinary images had no religious, ceremonial or political scenes of the ruling elite, but instead portrayed images of everyday tasks carried out by commoners. The occupants were shown giving, receiving and consuming various foods, and transporting products in jars and on their backs. The murals preservation ranges from good to very poor. The colors are blue, green, yellow, red, and brown painted upon a gray-white stucco. Conserving the valuable images has been a major challenge for scientists since the tropical climate tends to destroy the paintings. Northeast of most important Structure I is Structure II.

The largest and most significant building discovered at Calakmul is the enormous pyramid named Structure II that is over 45 m (148 ft) high, the tallest of all the Maya pyramids. The incredible Structure II is a triadic pyramid and it sits on a 140-m square platform base. The main structure has four buildings over top of it. To the north are the palace buildings and to the rear is another temple which is the highest point of the huge pyramid, so tall that it cannot be seen from the ground. As with most grand pyramids of Mesoamerica, this pyramid compounded in size through the centuries when covered over with newer, larger structures (Fig. 9.8).

Archaeologists have uncovered nine tombs inside Structure II, including that of King Jaguar Claw, who ruled from CE 686 to 695. They also discovered an impressive carved frieze showing different characters; the pyramid was beautified with large stucco masks. The massive pyramid is assumed to symbolize the mythical mountain *Witz*, most sacred to the Maya; the interior of the pyramid was thought to provide access to the Underworld. At the base of the great pyramid of Calakmul are several stelae; one carries the date CE 431, the era of the earliest known ruler. The

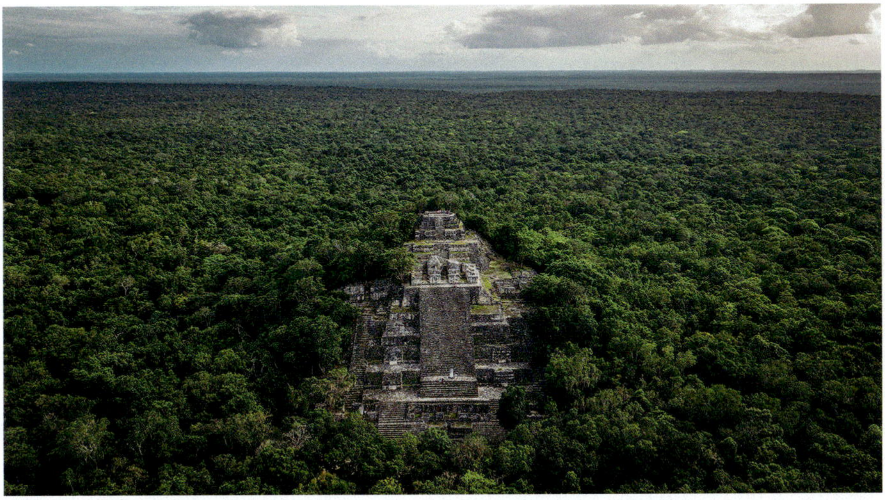

Fig. 9.8 Enormous pyramid at Calakmul, Structure II, largest of all Maya pyramids, with nine tombs (Credit Alfredo Mantus)

most prominent stele is believed to represent King Jaguar Claw's wife or mother performing a ceremonial ritual. Over the ages, this great pyramid-temple complex has had numerous uses: a residence for rulers and nobles, a burial site and most importantly, a place for political, ceremonial and religious ceremonies performed before gigantic crowds. Most significant buildings held burials of rulers and nobles.

Important stucco friezes and mural paintings have been uncovered at Calakmul in its many massive temple-pyramids and grand palaces. Calakmul is also known for its great quantity of stelae; the site contains at least 120, the greatest number of stelae in the Maya world. Most of the stelae are found in the West Group and the Northwest Group of architecture. Scholars learn a great deal from the inscriptions and carved illustrations that appear on stelae and public monuments. They relate the significant events and the opulent lifestyles of the elite class, but nothing of the lives of the lower classes of the society. The majority of the Calakmul stelae have been engraved with historic information but unfortunately, many have been badly eroded and are unreadable.

The Calakmul Kingdom had a vast network of interconnecting canals, reservoirs and raised causeways that linked the capital with other important cities, especially El Mirador, Nakbe, and El Tintal. There are at least eight white stone causeways, which suggests that there were strong political bonds between the four major cities. The roadways moved not only trade goods but ideas and people. The existence of extensive trade networks and exchange of ideas with neighboring regions is seen through their inscriptions, urban designs, architectural programs, pottery styles and funerary objects. The economic and trade connections continued through the entire Classic period when Calakmul grew to become the biggest and most powerful city-state of the Maya world. Equaled in size and power was Tikal. Calakmul was a political, religious, cultural and trading center, wealthy and prosperous (Fig. 9.9).

Kingship at Calakmul

By the Preclassic era Calakmul was already an important city; its early history is obscure but archaeologists have managed to establish a dynastic list that dates back to an ancestral past. Thanks to the inscribed building décor and painted ceramics, Mayanists have been able to establish a true but not a complete account of Calakmul history; twenty kings have been identified, covering a time span of over 400 years. Monuments relate epigraphic records and supply information not found elsewhere in the Maya lands. Numerous stelae, hieroglyphic stairways and ceramic vessels have recorded a long line of essential facts and events, such as royal births, accessions, deaths, marriages, alliances, conquests and defeats.

Early hieroglyphic texts inscribed on a stela, discovered in Structure II, recorded the investiture of a king of Calakmul in CE 411, the earliest glyphic date, and also describes a non-royal ruler of a settlement in CE 514. Five large stelae were erected in 741 but the name of the king at the time is illegible; he is called Ruler Y. After these markings there is a gap in the textual inscriptions that continues for over a

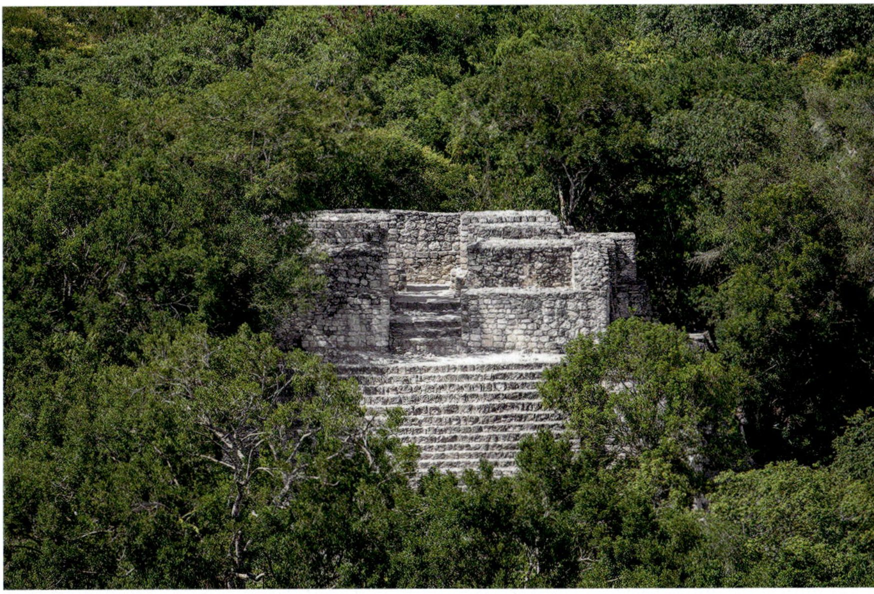

Fig. 9.9 Massive pyramid at Calakmul with very wide staircase and four temples on top (Credit LP Production)

century. The question of the lack of glyphic texts to document the important events of this period has not been answered, perhaps the Kaan Dynasty was living at some other location or that the monuments were later destroyed.

In later years Tikal's monuments and those in cities of its allies still emphasized single male rulers, while the buildings of Calakmul and of its subsidiaries give more prominence to the female line and appear to prefer the shared rule of king and queen together. Calakmul is unique for its matched sets of stelae with male-female portraits, depictions of kings and queens appear carved on many of the stelar. Excavations have revealed burials of kings and members of the nobility that contain rich body ornaments, elaborate jade masks, ear spools and colorful ceramic vessels. Altars and building décor are engraved with essential facts.

Calakmul was the capital of a large domain called the Kingdom of the Snake or Snake Kingdom that reigned for most of the Classic period. The city governed over areas as distant as 150 km (93 mi). he fierce Maya rulers from 635 to 850 were known as the Snake King (Kanu'l) dynasty. The lords of Calakmul also identified themselves as Divine Lords of the Snake. Through the entire Classic Period, Calakmul maintained a bitter rivalry and power struggle with the major city of Tikal to its south; the two Maya powers fought for dominance over the area. Calakmul's dynasty seems to be derived from the Preclassic city of El Mirador while the dynasty of Tikal appears to be derived from Teotihuacan. Both cities created extensive systems of allies and vassals. The vassals were smaller cities that joined with the powerful city to gain prestige and protection from their association. Tikal and Calakmul used their alliance networks against each other their relationship fluctuated throughout different years of the Classic period.

After gaining control of El Mirador, which was the dominant city in the Late Preclassic period, Tikal rose to become the leading Maya city-state, but later its supremacy was challenged by Calakmul when their warriors surrounded Tikal with a network of its allies. At that time they had greater power and a network of allies equal to Tikal. By the second half of the sixth century CE, Calakmul had gained control though it never totally blocked Tikal's influence. At a battle in CE 695, Tikal won a decisive battle and gained back the dominant position. This win eventually prompted the final collapse of Calakmul's once powerful alliance network, while Tikal secured a revival of its power.

In the Late Classic period the entire Kaan Kingdom centered at Calakmul may have totaled 1.5 million people and covered about 13,000 km^2 (5000 mi^2). The subordinate cities ruled by the Snake Kingdom were identified by archaeologists by the Emblem Glyph that was placed on monuments to indicate their rule. Some historians suggest that the Kan polity had ruled in the Mirador basin in the seventh century and after the collapse of El Mirador, the Kan traveled north to Calakmul and founded a new Kan dynasty; others say they came from a site further east. Many uncertainties remain and some eras of Calakmul's history are poorly understood; new studies must be undertaken to fill the gaps in understanding.

As yet few archaeoastronomical studies have been undertaken for the innumerable buildings at Calakmul. East and west of the Great Plaza is the traditional E-Group but the extent of its orientations has not been determined. The E-Group complex has the usual formation with a series of three temples constructed on a long narrow eastern platform and facing a western structure across a wide central plaza. The orientation of the E-Group's four components, Structures IVa, IVb, IVc, and IV, is slightly off the cardinal direction, 13° east of true north. Analysis of two doorways in Structure IVc have determined that are aligned with the setting Sun at the summer solstice. The stelae on top of one of the buildings are considered possible observation points. At the E-Group a stela depicting the birth of King Jaguar Claw and the tomb of a ruler were uncovered.

Calakmul's Decline

During the Terminal Classic the population of Calakmul declined dramatically bringing the population down to 10% of its former heights. Calakmul had played a prominent role in the history of the region for more than twelve centuries. The enormous amount of well-preserved architecture surviving reveals the size and power of the city and provides a vivid image of life in the ancient Maya capital. But by the end of the ninth century, neither Tikal nor Calakmul reigned in a meaningful way. A monument was erected at Calakmul in 790 but the name of the ruler responsible did not survive. The political authority of the city collapsed soon after that time. Two additional structures were raised in 800 and three in 810. A new stele was erected with a date of 899 or 909; later monuments appear but their style is crude, not the quality of the Classic Maya productions in the city.

The few inscriptions on these final monuments are meaningless attempts at writing. The remaining inhabitants were thought to be commoners who had occupied the elite architecture of the site. The presence of elite, indicated by finding of exotic materials such imported metal, obsidian, jade and shell, show that there was a continued occupation of royalty until the final desertion of the city. It was occupied until the mid-ninth century but was completely deserted after that time. Calakmul, like many Maya settlements, was abandoned suddenly for unknown reasons. Experts have determined that Calakmul became a pilgrimage site in the years between CE 1000 and 1500 due to devotional offerings uncovered inside some temples. The territory around the huge ancient city has now been reclaimed by the tropical forest and almost uninhabited.

References and Further Reading

Evans, S. T. (2008). *Ancient Mexico and Central America: Archaeology and cultural history.* Thames & Hudson.

Guenter, P. (2002). *Under a Falling Star: The Hiatus at Tikal,* Dissertation, La Trobe University, Bundoora, Victoria.

Montgomery, J. (2001). *Tikal: An illustrated history of the ancient Maya Capital.* Hippocrene Books. Inc.

O'Neil, M. E. (2009). Ancient Sculptures of Tikal, seen and unseen. *Res, Anthropology and Aesthetics, University of Chicago Journal* 55/56, Spring/Autumn 2009.

Vadala, J., & Milbrath, S. (2014). Astronomy, landscape, and ideological transmissions at the coastal Maya site of Cerros, Belize. *Journal of Caribbean Archaeology, 14*, 1.

Chapter 10
Maya City of Palenque

History

The architectural ruins of Palenque lie within the mist-shrouded jungles of Chiapas in eastern Mexico. The city-state was established amid the scenic foothills of the Chiapas mountains with waterfalls and shallow ponds and several rivers that merge creating one of the wettest and most lush areas of the Maya lowlands. Although much smaller than the colossal city of Tikal, Palenque can compete favorably with Tikal when comparing its remarkable buildings, lacy roof combs, and finely carved sculptural remains. Palenque was occupied from about 226 BCE, but the prime period of this remarkable city was much shorter than that of Tikal and other Classic cities, spanning only about 200 years, from about CE 600 to 800. Like most Classic cities its final decline began toward the end of the eighth century and was eventually abandoned to the ever-creeping tropical forest. Palenque's kings were among the most accomplished Mayan leaders. The Palenque kings were admired for commissioning grand temples including the Temple of Inscriptions, regarded as one of the important sources for ancient glyphic texts. The temple's inscriptions record nearly 200 years of Palenque's dynastic rulership, military campaigns and spiritual beliefs, rich and essential resources for Maya linguistic studies of today.

Palenque was visited initially by Spanish explorers seeking treasures in 1773 and again in 1784 but in spite of tearing down walls and digging for treasure-filled tombs, they found nothing they valued. The next visitors to track down the remote location were Stephens and Catherwood who investigated the ruins in 1834; they set up their working camp within the city's largest royal palace. They cleared many of the monuments, drew accurate illustrations and mapped the site but their stay was cut short by illness.

Archaeological excavations have cleared the dense overgrowth from the inner city but they have found that Palenque's ruins now extend over 2.5 km^2 (1.5 mi^2) and include hundreds of structures. Experts have restored many of the damaged

M. Dolan, *Decoding Astronomy in Maya Art and Architecture*,
Springer Praxis Books, https://doi.org/10.1007/978-3-031-61092-9_10

buildings. More extensive explorations have revealed that less than 10% of the total area of the great city has been unearthed; it is assumed that more than 1000 structures still lie hidden within the thick jungle. In addition to the many buildings, archaeologists have discovered three royal tombs, a burial of sacrificial objects, offerings to a royal person and the tomb of a noble.

Palenque is laid out with the same format as most Mayan cities; their Great Plaza spreads out in the center of the city and is lined with a series of small temples. Palenque's temple-pyramids have a delicate, serene style with open-work roof combs and are adorned with decorative stucco bas-relief panels. Surprisingly, the city has no free standing sculptures or stelae which is most unusual. Instead, portraits of their rulers and their gods were carved on stone panels or made from plaster and then attached to the buildings' walls. The kingly portraits were shown clad in elaborate dress, exhibiting various symbols of royalty. Like all Maya art and architecture, the decorative images were colorfully painted. Like all Maya rulers, Palenque kings claimed close associations with the gods and performed the required rituals to assure the continuation of a harmonious world order. The city's rulers fought wars, took captives and amassed great prestige and wealth that set them apart, high above the rest of society. Archaeoastronomers are pleased that Palenque's leaders displayed their strong interest and dedication to astronomy by encoding it into their history, their architecture, their artworks and on their innumerable glyph-inscribed walls.

Since Mayan hieroglyphic texts have now been mostly deciphered, epigraphers have been able to reconstruct much of Palenque's past through its quite detailed epigraphic records. Experts have established a long list of ruling dynasties from the fifth century CE and chronicled the history of the city's rivalries and competition with other states, in particular with the huge city of Calakmul. Carved texts at Palenque record an attack from the competitive, warring city of Calakmul in 599 and again in 611. After that second military assault, the greatest king of the Maya world was enthroned as its ruler, K'inich Janaab Pakal. Scholars suggest that the attacks on Palenque became the incentive for Pakal's vast building campaign.

King K'inich Janaab Pakal

The most well known and most carefully studied Classic king or *ajaw* of the city-state of Palenque was K'inich Janaab Pakal, or Pakal the Great (Sun-eyed Lord of the Shield). Pakal "touched the ground" on March 23, 603 (just after the equinox) and lived until 683. He ruled for an unbelievable 68 years, from 615 to 683 and his royal tomb was the richest and most spectacular of all the many Maya kings. Palenque's previous ruler had recently died when Pakal and his mother Sak Kuk arrived at the city, called from another area. Young Pakal was inaugurated as King of Palenque when only 12 years old. His mother Ix Ajaw Tz'ak (Lady Medicine) reigned as queen during his first 3 years and it is assumed that his mother played an essential role during his early years in power; they may have even reigned jointly.

Because of the attacks from the powerful state of Calakmul 5 years earlier and the death of their king, Palenque's position and power in the Maya world were in a precarious state when Pakal came to the throne (Fig. 10.1).

Assuming power during the city's vulnerable period of social upheaval, the young king was encumbered with many damaged or destroyed buildings. As an outsider his first need was to establish his right to rule and position himself as a powerful and unquestioned legitimate leader. In a recent study of Palenque during the seventh century, there appeared to be a community of scholars associated with the reigns of Pakal and his son Kan Bahlam and closely connected with the royal court. Evidence shows that a non-royal elite housing complex with inscriptions may name these scholars. These court astronomers designed a new mathematical system, the 819-day count, that they applied to combine important astronomical and calendric periods that linked historical events in the current reigns to the mythological predecessors thousands of years in the past. These astro-numerologically significant planned numbers legitimized the new dynasty of Pakal by joining it to the distant mythological past.

Fig. 10.1 Bas relief carving of bejeweled King Pakal (Credit FrentaN)

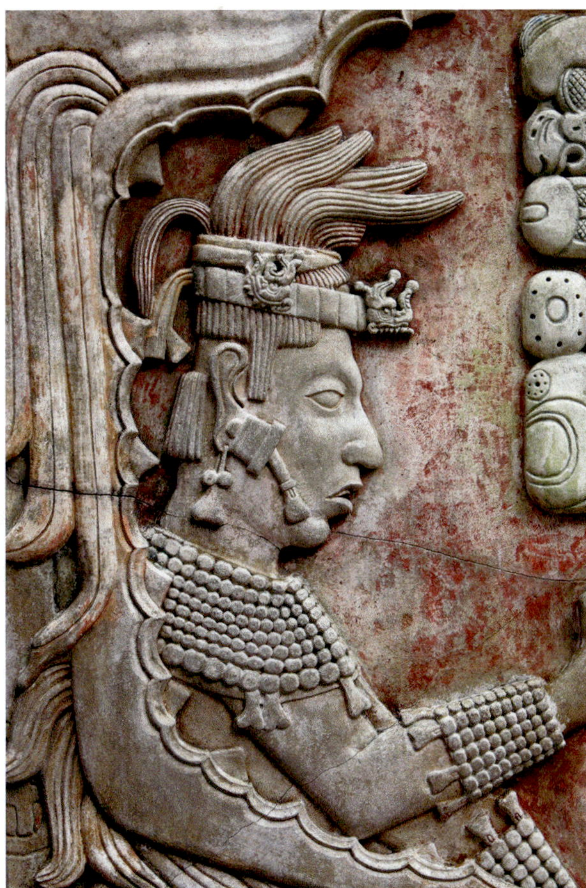

Quickly achieving great success, Pakal directed Palenque to major accomplishments in all areas, especially regarding their expressive art and architecture. He commissioned a great many of Palenque's beautiful buildings and temples; he succeeded in creating a powerful and prosperous capital city, rivaling Tikal in its splendor. Like King K'ahk' Uti' Witz' K'awiil of Copan, he became a master in controlling his public image by emphasizing his connection with the city's ancestors and patron gods, and through his visual depictions adorned with the iconography of the highly honored Maize God. In 624 Pakal married the princess of Oktán, a minor Maya city, called Lady Tzakbu Ajaw; they had at least three sons, perhaps four. Two of their sons became kings after the death of Pakal. Inscriptions on various temples tell that the queen was very involved in the activities of their kingdom and worked with Pakal in ruling. She died in 672, 11 years before Pakal.

Palenque's Royal Palace and Astronomy

One of Palenque's most exceptional creations was the Royal Palace. It was a long stone building with a thatched roof set upon a raised platform, the largest architectural complex in Palenque. Archaeological evidence indicates that parts of the Palace were used as the royal residence for Palenque kings as early as 250 BCE, but the surviving buildings seen today all date to the Late Classic (600–800). The Royal Palace was not only a noble residence; the site functioned as the administrative heart of the city as well. The palace complex was also used by the city's nobles and officials for civic functions, religious rituals, and entertainment.

The Palace complex included several adjacent buildings with a maze-like interior. It has an assortment of 12 rooms, called "houses", organized around two central courtyards, the typical Maya house pattern. The structures have the main characteristics of Maya architecture, high corbelled ceilings and narrow passages, but it also has subterranean rooms. The palace is unique in that it includes a four-level square tower that looks out in four directions, measuring 97 by 73 m (318 by 239 ft), providing spectacular views of the neighboring landscape. There are two entrances into the Palace, one on the north and another on the east side, both entrances are reached by climbing monumental stairways. Advanced in its design, the royal palace was provided with many large baths, latrines and saunas that were supplied with fresh water through their innovative water system (Fig. 10.2).

Like all Palenque's buildings, the palace walls are filled with informative glyphs and were beautifully decorated both inside and out; much of its interior decorative artwork has survived. The stucco sculptures adorning the walls, piers and roofs of the complex are among its most notable features. The images were painted with bright, meaningful colors; backgrounds and human bodies were always red. Blue was limited for royal, divine, and heavenly objects and persons; objects associated with the Underworld were painted vivid yellow.

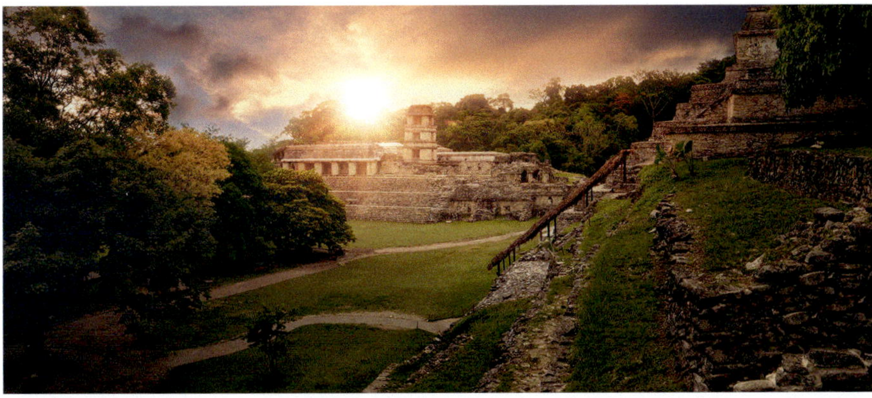

Fig. 10.2 Royal Palace at Palenque at sunrise (Credit Soft Light)

Between 654 and 668 Lord Pakal expanded the Palace, adding three additional chambers, Houses C, B and E, the most important. On November 9, 654 Pakal dedicated House E as his throne room used for royal receptions; it includes a stone bench, perhaps his royal throne. House E was a unique addition since it was one of the few Maya rooms painted white rather than the typical red seen on royal and ceremonial buildings. Behind the throne was the Oval Palace Tablet that describes the ascension of Pakal to rulership. A portrait of the king receiving visitors was painted over the throne. Pakal was crowned at the Sun's zenith passage, but he dedicated his House E at the nadir passage. Houses B and C recorded and celebrated Pakal's military achievements. Houses A and D were built by Pakal's son and contain a dedication date of August 10, 720 (Fig. 10.3).

Astronomical Alignments at the Palace

Archaeoastronomers have found evidence indicating that the palace incorporated numerous solar alignments; for example, at the zenith passage the Sun projected perpendicular shadows in the inner courtyards toward specific markers. It is noteworthy that House C within the palace was dedicated 5 days after a zenith passage on August 7, 659. On the west side of the Palace tower, a beam of light passes through a T-shaped window onto an oblique wall from April 30 to August 12. This light effect lasts for precisely 105 days; the light reaches its maximum width on the summer solstice. For the other 260 days of the years, the wall is in darkness.

In addition the central doorways of Houses C and A appear to be aligned with the rising Sun during nadir passages. Earlier archaeologists wondered why the walls of House A are slightly angled and staggered; this was explained by archaeoastronomers. As the Sun sets, there is an alignment of the central doorways of House A to the Sun at nadir passage sunrise, but that event is now obscured by trees. The zenith

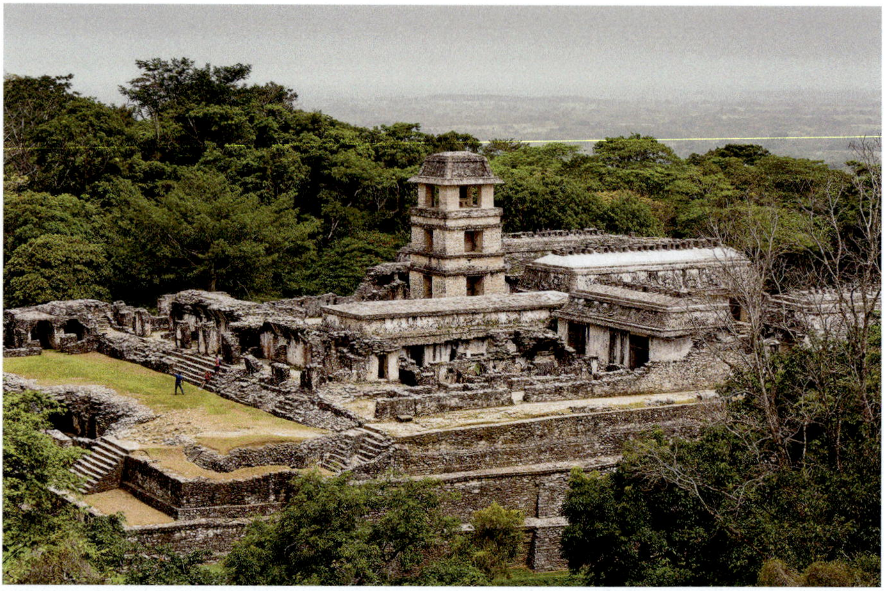

Fig. 10.3 Royal Palace with tower at Palenque, aerial view, begun ca. 250 BCE (Credit Boris Vetshev)

sunset would illuminate a ritual occurring in this entrance; further intensifying the occasion as seen from the public areas before it. Also, cultural astronomers believe that high priests and nobles would gather at the Royal Palace to view the setting Sun on the winter solstice; as the Sun receded, it would cast light beams directly into the central doorway of the Temple of the Inscriptions (Fig. 10.4).

The Temple of the Inscriptions

The monumental building most associated with Pakal is the Temple of the Inscriptions, the grandest step-pyramid in the city. This multi-purpose building was his memorial shrine, his pyramidal mortuary temple and contained his long-concealed tomb. His enormous sarcophagus had been buried deep inside, unknown for over 1300 years. Begun about 675, the mortuary temple was positioned upon a wide, rectangular terraced platform. The pyramid is comprised of nine levels, symbolic of the Underworld, with a wide, steep stairway climbing to the lofty temple above. The largest blocks on the exterior weigh up to 12 and 15 tons and these were positioned at the very top of the 27 m (90 ft) tall building. The small temple built atop the pyramid is unusual in that it has five doorways leading to the interior rooms rather than the standard three. The extensive glyphic text both inside and outside the building inspired the name of the temple.

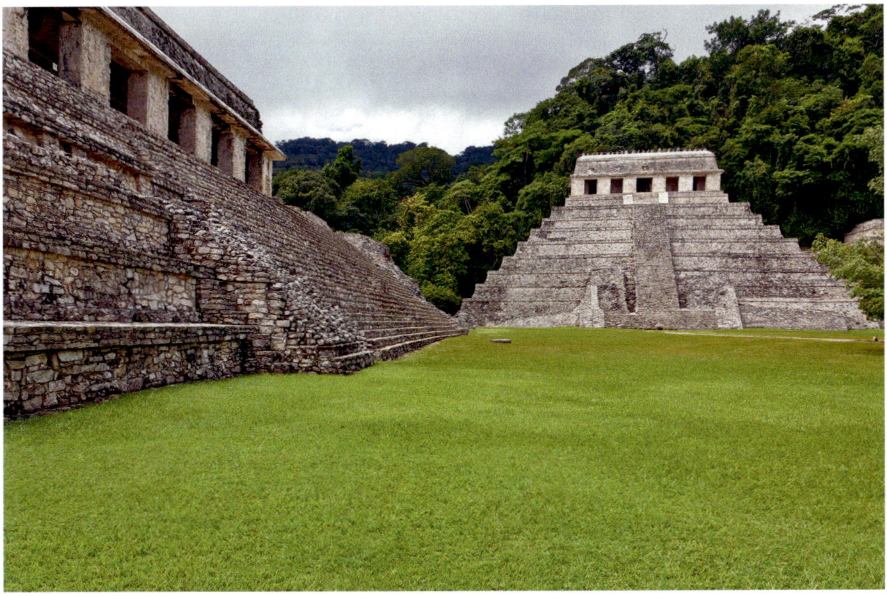

Fig. 10.4 Temple of the Inscriptions and the Royal Palace at Palenque (Credit Vadim Petrakov)

The exterior of the temple/pyramid was inscribed with a detailed and lengthy text; it is the second longest glyphic text found in the Maya world, after the famous Hieroglyphic Staircase at Copan. Fortunately the script survives and, like most of Palenque's engraved texts, remains well preserved and now can be easily deciphered. The galleries of the temple are supported by a series of corbel vaults that allow adequate open spaces. Three large stone tablets, covered with hieroglyphic text, were positioned inside the rooms. The innumerable glyphs on Pakal's temple document the city's history for almost two centuries, including the period-ending rituals. The inscriptions also describe Palenque's patron gods, called the Palenque Triad, known individually as GI, GII and GIII. The Palenque Triad were thought to have been created by an ancestral god in 2360 BCE in the mythical land of Matwiil.

Pakal's Sarcophagus

In 1952 archaeologist Alberto Ruz Lhuillier was investigating the inner room of the temple; he removed a stone slab that reveal a hidden passageway into the interior. The hidden passage had been packed and totally blocked with rubble before the city was totally abandoned. Removal of the impenetrable debris filling the secret stairwell required 3 years. Once revealed the long, deep staircase was shown to be corbel-vaulted; the walls of the stairwell were lined with large stucco figures representing the nine Lords of the Underworld. The first gruesome discovery at the bottom of the stairs was outside the doorway of a chamber; it was a stone box containing the bones of four sacrificial victims, three men and one woman.

An unusual triangular door opened into a secret room concealing a huge, mysterious stone block inside. The archaeological crew began cleaning and recording the contents and decor of the exposed chamber, but they were unable to go further since that working season had ended. The workers had to hold off their discoveries until the next season. But the news of the hidden chamber leaked to the press, creating headline news worldwide. The crew needed to raise enough funds to continue the excavation of the unknown mysterious chamber. Fortunately, they did raise the necessary funds; they discovered the richest and best preserved tomb of any excavated burial in Mesoamerica.

When excavation work began again, the team revealed that the colossal stone structure within the small chamber was a sarcophagus not the shrine or altar they expected. The sarcophagus filled the entire room, created from a single block of limestone. Its sides were engraved with portraits and hieroglyphics, probably coated in red cinnabar (mercuric sulfide) originally. Cinnabar is toxic to humans, so perhaps it was a deterrent to grave robbers. Around the edges of the sarcophagus lid were low-relief glyphic carvings that repeated important events from Pakal's life and those of his ancestors. The lid too consists of one huge slab almost 12 inches thick; a band with cosmological signs surrounds the lid, including glyphs for the Sun, Moon and stars. The research team required four car jacks and other props to raise the heavy lid which revealed a second "fish-shaped lid". This covering had four small openings in each corner to lift this lighter lid from the sarcophagus. Upon lifting the second lid, they discovered the skeletal remains of a single adult male. At that earlier time the scientific tools to determine the accurate age and physical conditions of the bones were not yet available. Archaeologists suggested that the occupant of the tomb had been about 40 or 50 years of age.

Inscriptions in the tomb identified the deceased as K'inich Janaab Pakal I and recorded the date of his death; his tomb was totally intact, exactly like it was buried in the seventh century. Later studies showed that he was near 80, not 50; the royal life was not as harsh as a peasant's. Pakal was honored with an astonishing amount of jade and other funerary treasures. He wore the costume of the Maize God, looking similar to his depictions shown on Palenque's iconographic sculptures and wearing a mosaic mask, made of the especially valued jade. In addition to the funerary mask, he was adorned with hair ornaments, earflares, a diadem with 41 jade discs, numerous bracelets, a jeweled collar, numerous beaded necklaces and rings on each finger. All were testament to the formidable power and prestige of their admired king. Maya rulers were revered as divine beings; therefore elaborate tributes to their power confirmed they were deities. Pakal's skeletal remains, his mask and regalia and the interior of his sarcophagus were all coated with red cinnabar, an element found in other Maya burials as well. The highly toxic powder was considered sacrosanct; red also signified blood, the sacred element of sacrifice to the gods. This hallowed but dangerous red pigment was used on Maya artworks and on walls as well.

The colossal sarcophagus must have been laid in place while the pyramid was under construction since it could not have been carried down the confined stairway. The lid alone measures 7 ft by 12 ft and weighs seven tons; Pakal may have supervised the design and start of his own mortuary shrine. The monumental temple also functioned as a place for remembrance and for communication with Pakal during ritual ceremonies in his honor as a celestial deity. An architectural structure was added to the temple to make this spiritual contact possible.

Within the pyramid is a unique structure, called a *psychoduct*; a similar duct was included in other Mesoamerican buildings for astronomical purposes. The duct is a small circular opening that leads from inside the coffin, up the stairwell and through a hole in the stone block that covers the entry to the tomb; it ends in front of pier C which pictures Lord Pakal. The duct or light tube in other Maya structures allowed light from a heavenly body to pass through the tube and fall upon a marker to establish a specific moment of time. It is thought that the duct in Pakal's tomb is aligned to the winter solstice and it directs a beam of sunlight to fall on Pakal's sarcophagus on that important date.

Others suggest the open duct alludes to the departure of the soul or psyche from the deceased and allows him to communicate with the living and with other realms. To the Maya, the open duct enabled the nobles or priests to speak through the shaft; it allowed communication between the priestly rituals performed in the temple above to access the burial chamber below. Perhaps it was also regarded as a channel for his rebirth as a god. The psychoduct could also provide access for a vision serpent to carry Pakal's spirit from his tomb to the temple above. The vision serpent grows from the foot of the Maize God; that same image was depicted on Pakal's sarcophagus and on panels in the upper chamber. Elite nobles entering the temple would have seen images of their beloved king being born into the otherworld as their patron god, Unen K'awil. Archaeologists are still studying the possible symbolic uses for this psychoduct within the pyramid.

The Sarcophagus Lid Artwork

The most incredible work of art of all those treasures found at Palenque is Pakal's sarcophagus lid; the entire stone covering was sculpted with symbolic iconography unique to Pakal, illustrating his apotheosis, meeting with the gods and attaining divinity in the Otherworld. The complicated scene carved onto the lid of the sarcophagus has had Mayanists arguing and altering their interpretations for years. The complex symbolism depicted on the lid reveals Maya beliefs concerning the events that occurred after the king's death. A king's journey to the Underworld along the World Tree was described in written texts but never visually portrayed in the past. These carved images make the post-death events visible to all; the illustration makes his journey real by materializing it. Pakal is depicted as the Maize God falling into and rising out of the Underworld and lifted to the celestial Otherworld. This complex scene combines the elements of all three Maya cosmic realms into one reality.

Behind Pakal's reclining figure, the central image represents the cruciform-shaped, jeweled World Tree or Cosmic Tree, which to the Maya was the symbolic *axis-mundi,* which connects the levels of the Underworld with the Upperworld of the Heavens. A great symbolic bird appears sitting atop the jeweled World Tree. The youthful Pakal appears adorned with jewels, an elaborate headdress and wearing only a skirt. Pakal's hairstyle is seen as both quetzal feathers, symbol of divinity and as maize, which is symbolic of the Maize God (Unen-K'awill – God K). Pakal appears in the center of the cosmos, as he begins the process of being reborn into eternal life. He wears the attributes and becomes enjoined with the Maize God, who is associated with food, fertility and abundance. Pakal wears the turtle ornament out of which the Maize God was born. The king emerges from a maize seed held by the Earth Monster. The Cosmic Tree behind Pakal will lift him to the heavens, where the Sky Dragon awaits him in the form of a bird, with two serpent heads on either side. The many symbolic icons show that after death Pakal entered the Milky Way near Scorpius and rose to heaven on the Southern Cross to reach his favored planet Jupiter.

The dramatic scene portrayed on the lid stands as an elaborate depiction of numerous cosmological events condensed into one image. The kingly figure, positioned upon the *kin* bowl atop the head of the Cosmic Monster, faces west. Facing west he rises up the World Tree like the zenith Sun. Pakal seems to be falling into the open jaws of a funerary serpent, an iconographic device that depicts an entrance into, or residence in the Otherworld. By falling into death, he ascends into a new life, reborn into eternal life. The World Tree is surrounded by the flowery otherworld, shown by the jewel-like emblems. The upper area of the lid shows a long sky band, with the heads of three figures above and below the main scene (Fig. 10.5).

The lid is one of the most priceless archaeological finds; the engraved image represents a forceful declaration of rebirth, vitality and continuity. After his death Pakal was worshipped as one of the patron gods of Palenque. He becomes the Sun and holds the ecliptic in his hand. The ecliptic was placed almost parallel to the Cosmic Monster while it sits perpendicular to the Gemini-Orion link. According to Mayanist David Stuart, "The sarcophagus lid, in its eloquent depiction of Classic Maya cosmology, is not a celebration of Pakal's moment of death but rather an apotheosis monument celebrating his rebirth." The images and date of the sarcophagus denote that the king's spirit ascended to heaven to become a god in line with astrological portents when the royal planet Jupiter stood motionless in its cycle to honor the deceased ruler. Pakal's apotheosis as God K coincided with Jupiter's first stationary point.

Since the opening of Pakal's sarcophagus, Palenque has become one of the most extensively studied archaeological sites of Mesoamerica. The death scene and glyphs on the lid have enabled scholars to understand dates, events and family relationships and how the Maya interpreted death and rebirth, as interconnected with the motions of the celestial bodies. It was a statement of dynastic vitality and continuity that was necessary after the death of their beloved and long-lived king. It is interesting to note that Pakal's magnificent sarcophagus was not intended to be placed in the plaza for the public to view and honor him. The detailed iconography

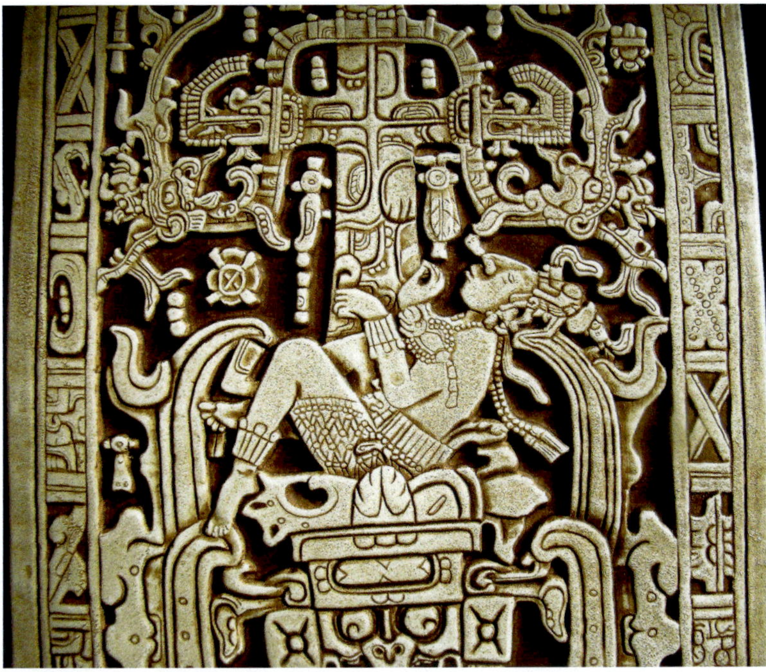

Fig. 10.5 Pakal's sarcophagus lid carving (Public Domain)

carved on the exterior was not designed to promote and legitimize his dynasty to the inhabitants; instead, it expressed his devotion to the gods and to his basic beliefs in Maya ideology. After scholars' comprehensive examination of the tomb contents, Pakal's remains were returned to his sarcophagus in 2009 and the massive lid was carefully replaced; the king is now enjoying his eternal rest.

Red Queen of Palenque and Temple XIII

Next to the Temple of the Inscriptions stands a similar but much smaller temple-pyramid, Temple XIII; it reaches about 36 ft high with a stairway on the façade leading to the shrine atop the pyramid. When archaeological investigations began there, the stairway of Temple XIII collapsed and was blocked with rubble. While working at the pyramid in 1994, researchers chanced upon another surprising discovery. A tunnel was excavated near the exterior stairway of the main façade of the Temple of the Inscriptions in order to access the interior of Temple XIII. The archaeologists found a corridor leading to three underground chambers; two rooms were open but totally empty. The third and largest chamber was sealed which seemed to indicate that something important lay behind the door (Fig. 10.6).

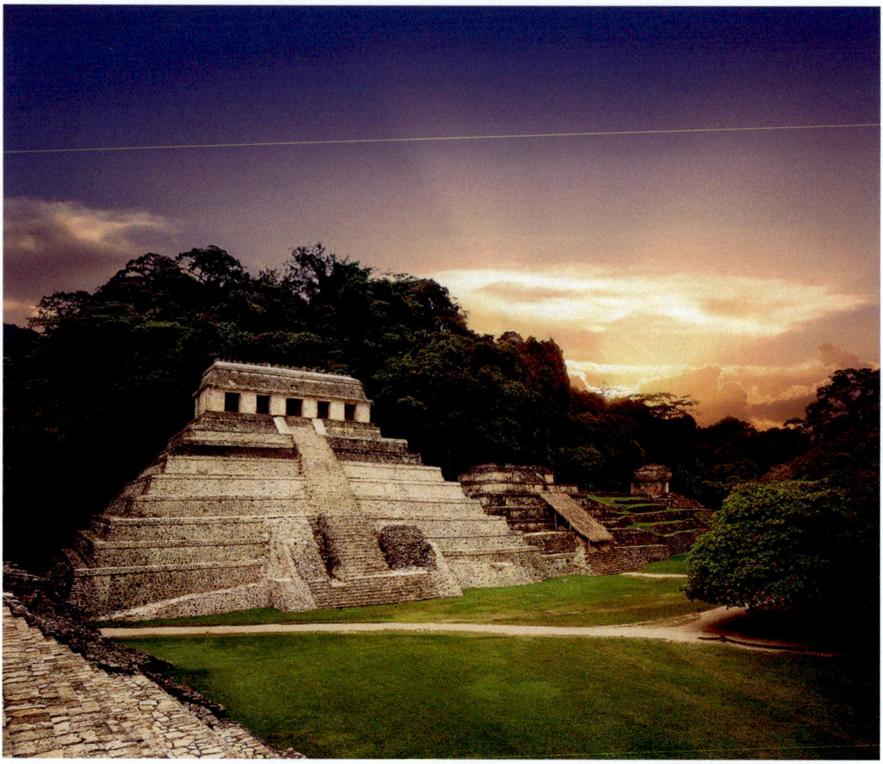

Fig. 10.6 Temple of the Inscriptions and the Red Temple at sunset (Credit Valery Shanin)

Proceeding cautiously, the first step was to create a tiny hole in the wall to observe the interior. A light was inserted through the peep hole which revealed a small, vaulted chamber about 14 by 8 ft. Like Pakal's burial chamber, the hidden room was almost totally filled with a monolithic limestone sarcophagus; grave goods and ceramic items were spread around the floor. This sarcophagus too was covered with bright red pigment. The team removed the stones blocking the entrance, opening the hidden chamber enclosed for more than 1300 years.

The first items that archaeologists discovered were skeletal remains of a young male about 12 years old with marks indicating cuts and blows to his rib cage, perhaps from a sacrificial ritual. The second burial, an adult female perhaps in her 30s, was lying outside the large sarcophagus; her bones also showed signs of fatal injuries. Perhaps she was sacrificed as well to assist the important person in the tomb in the afterlife.

When opening the large sarcophagus, the excavators were amazed to see that the occupant, like Lord Pakal, was dusted with bright red pigment, perhaps thought to be a preservative. The skeletal remains, identified as a female, were surrounded by tomb treasures, obviously a very prominent person of high status but not as grand as Pakal. Who was the mysterious woman? The size of her pyramidal tomb, her sarcophagus, her jewels, diadem and an abundance of precious grave goods suggest that she was a member of the aristocracy of Palenque and a contemporary of Pakal. The two burials were placed in adjoining temples and both contained human sacrificial victims. Based on the estimated age of this tomb, it is about the same as the Temple of the Inscriptions.

Wide-ranging studies of the bodily remains were undertaken by archaeologists; the carbon 14 test results have established that the bones belonged to a 60-year-old woman who was a bit over 5 ft tall and had osteoporosis. DNA studies have established that she is not a blood relative of Pakal; so it cannot be his mother. Tests on her teeth show that she was not a foreigner but from a local population of women. They have also done facial reconstruction. Because there were no inscriptions in the chamber, the identity of the Red Queen has not been established positively.

The magnificent funeral mask of the Red Queen is one of the greatest treasures found at Palenque but the ancient relic had crumpled into more than 100 malachite fragments. Through meticulous work, the myriad bits were reassembled. The intensive, staring eyes of the death mask were carved from obsidian and jade. Many indications in the search for the woman's identity point to Pakal's wife, Ix Tz'akb'u Ajaw, who came from another city to marry Pakal in 626. They had at least three sons; if their tombs could be found, DNA tests would determine if there was a match. Most Mayanists believe Lady Ajaw to be the most apparent occupant of the tomb, placed next to her husband for eternity.

K'inich Kan Bahlam II

After his death, Lord Pakal was succeeded by his son K'inich Kan Bahlam (635–702), (Sun-eyed Snake Jaguar) who became ajaw of Palenque in 684 and ruled until his death. K'inich Kan Bahlam II was named king of Palenque officially when only 6-years-old but his rule did not actually begin until age 49. Kan Bahlam was anxious to continue his father's ambitious agenda of construction to beautify Palenque with exceptional examples of art and architecture. Most of Kan Bahlam's architectural endeavors incorporated astronomical alignments and cosmic symbolism. His interest in astronomy was confirmed by his adoption of the planet Jupiter as his patron; most of the important dates in his life were scheduled close to the time that Jupiter makes its first observable departure after retrograde, when it returns from its period of reversal to the normal west-east celestial motion across the sky.

It is noteworthy that Kan Bahlam II was not inaugurated as the new ruler until Jupiter had reached a desired point, waiting for 4 months after Pakal's apotheosis event. It is likely that Palenque priests believed the period of Jupiter's retrograde motion was not an appropriate time to crown the new king. When particular astronomical events coincided with historical events, such as death of a ruler, they were at times adjusted to fit astronomical events or even timed to coincide with a significant astral event. Kan Bahlam waited 132 days after his father's death before scheduling his own inauguration ceremony. The delay had astronomical/astrological purposes; that date was linked to Jupiter's shift from retrograde motion; also the Moon was 1 day before full and conjunct Mars. Kan Bahlam's heir to the throne ceremonies, when he was young, were initiated on June 17, 641 and ended 5 days later exactly on the summer solstice.

Kan Bahlam must have been conscious of the enormous task ahead in fulfilling the rule of his god-like father. In satisfying that role, his first duty was to oversee the completion and decorative program of the Temple of the Inscriptions after Pakal was laid to his final rest. He, or his astrologers, scheduled the dedication of the Temple of the Inscriptions on December 23, 688, 2 days after the winter solstice. A stucco relief carving of Kan Bahlam appears on the exterior of the Temple of the Inscriptions; his image depicts the day he was presented as heir to the throne, shown as a child in the arms of his ancestors. His identifying feature was a deformity of six toes on 1 ft; this abnormality appears on his images when an adult as well. One of his legs ends in a serpent's head, which also denotes his divinity; he has a smoking ax in his forehead, both unusual traits are characteristics of God K or God GII. He is portrayed as the living incarnation of God K, the patron of rulers. Many sculpted portraits of Kan Bahlam survive in Palenque due to the large number of undertakings he achieved during his reign. He is easily recognized by his large nose and protruding lower lip.

The Temple of the Cross Complex

The most ambitious architectural works constructed during K'inich Kan Bahlam's reign are the three small temples atop stepped pyramids called the Group of the Crosses: the Temple of the Cross, the Temple of the Sun and the Temple of the Foliated Cross. The triple pyramids were built around 690 to honor K'inich Kan B'alam on his ascension to the throne. Situated at the base of their sacred mountain – the religious heart of the city, all three were built upon a large, elevated plaza with nine levels along the eastern side. Kan Bahlam commissioned the Cross Complex to provide homes for the gods to preside over if they so desired. The buildings are aligned to form a cross; the Temple of the Sun is directly opposite the Temple of the Foliated Cross; both temples are adjacent to the Temple of the Cross. Two smaller temples were built between the Temple of the Sun and the Temple of the Cross, the largest and most significant of the group. The facades of all three were fully decorated by artisans but only the exterior décor of the Temple of the Cross is still legible. The three temples of the Cross Group were considered symbolic mountains; each is astronomically oriented and embodies numerous levels of encoded astronomy.

The three temples' basic designs are quite similar; repeating the triple theme, each temple has three entranceways. The interiors are similar, separated into an antechamber and a larger rear room and a small room on each side; the rooms contained carved relief tablets. Within the central chamber of each temple, masons created a miniature roofed shrine, a room within a room. These holy portals were called "underground houses"; each carried its own proper name; they were thought to open into the Underworld. These special spaces were considered powerful centers of supernatural energy. Although the "houses" were only symbolically underground, they copied the real buried "underground buildings" of Palenque, the tombs of Pakal and other kings entombed in pyramids throughout the city, undiscovered as yet.

The inscribed text within the temples describes both real and mythical stories of the history of Palenque and of its many rulers. These carved glyphs narrate a continuous story similar to a Maya codex. The inner chambers of each shrine display an elaborately carved scene with two figures presenting ritual objects and effigies to a central icon. The iconography of these illustrations is closely related to the scene on Pakal's sarcophagus; they are all centered on the symbolic World Tree.

Kan Bahlam is pictured in each of the three temples but at different ages at each one. His image on the tablet of the Temple of the Cross is the oldest and heaviest, having a bulging stomach and deep facial lines. This illustration represents the accession of Kan Bahlam to the throne, when he receives a royal scepter from his father. At the Temple of the Sun he seems much younger; at the Temple of the Foliated Cross, his thinner body and his raised glance show him as the youngest of the three depictions. As their name suggests, the Cross and Foliated Cross Temples depict large crosses behind the king's images. They appear similar to Christian crosses but are actually depictions of the Maya branching World Tree, the symbol of the *axis mundi* and the cosmos. The World Tree stands in the center of the composition, symbolizing divinity and connection to the Underworld. This tree has attributes similar to the World Tree on Pakal's sarcophagus. These illustrations legitimize Kan Bahlam as a divine ruler, as he accepts heaven-sent sanction from his divine father.

The inscriptions on the stone tablets also describe the birth of the three gods, the Triad of Palenque, looking back to mythological times. This triad, which scholars call Gods I, II and III (their exact names are difficult to decode) were said to have been created by a progenitor god in 2360 B.C. in the mythical land of Matwiil. All three shrines were adorned with an array of cosmological symbols applicable to the patron god of each temple. The inner tablets also exhibit hieroglyphic texts explaining the purpose of each temple and relate information on the Palenque Triad of Gods, and their right to hold leadership. Previously it was thought that these three gods were universal throughout the Maya world, but now it appears that each Maya city-state honored its own individual triad of gods that had little relationship with Palenque's triad. The connection with their gods had its own deep cosmological meaning. Each of the three Cross temples is dedicated to a specific god of the Palenque triad and each signifies a path to the Otherworld.

Direct pathways to the Otherworld were essential for the reigning king; it was essential that he follow one of the sacred paths to bring back gifts of life and good fortune for his people. It was the ruler's responsibility to care for the Triad deities through periodic offerings and dressings, as was described in detail on the tablets of the Temple of the Inscriptions.

Temple of the Cross

Of the three temples, the foremost is the Temple of the Cross, which was dedicated to God GI, the first born of the triad, also known as Hunahpu. God GI's iconographic features were a shell earflare, a square eye and a fish fin on his cheek. He sometimes wears the quadripartite Monster as a headdress and is also linked with the Waterbird. The Cross temple seems to carry many levels of meaning. It is thought to represent a temple mentioned in the Creation myth that was founded in the north soon after creation day in 3231 BCE. This temple and its inner shrine recreate that ancient sacred place. A stucco molded *Witz* mask was placed on the exterior demonstrating that the temple was conceived as a sacred mountain. One of the many purposes for the Temple of the Cross was to display the inscribed panels honoring the triad of gods, and their right to be their patrons and to retain leadership. The Cross glyphic texts also document Kan Bahlam's ancestral history, his accession to the throne of Palenque and the divine origin of his ancestral history (Fig. 10.7).

The central sculpted image within the large chamber shows two bas-relief male figures who decorate the doorway. Rather than Kan Bahlam and his father, as originally believed, the latest interpretations and a clearer understanding of the glyphs reveal that both standing images depict Kan Bahlam, one as a young man at the ritual of passage, the other as an adult as he became King of Palenque. The World Tree stands between the two royal portraits. The jeweled World Tree emerges from a "*kin* or solar bowl", which is associated with offerings and blood sacrifice; these are important symbols for God GI. Here the World Tree motif serves to reference GI by highlighting his associations with the Sun, the sky and the dominant theme of ancestral resurrection. Atop the tree's jeweled branches sits the Principal Bird Deity, earlier thought to be the Pleiades, starry forecaster of the rainy season or perhaps the Big Dipper. The World Tree is iconographically similar to the one on Pakal's sarcophagus.

The Temple of the Cross and the Palace have important astronomical connections with the Sun's zenith and nadir positions which are revealed through the temple's artworks and hieroglyphic inscriptions, beginning with Creation and continuing through the succession of Palenque's rulers. Both the zenith and the nadir were essential to Maya concepts of spatial orientation, acting as the focus of ceremonial space and its position in the landscape. These solar alignments between major temples at zenith and nadir impart political and cosmological communications. The relationship between astronomy, myth, and history indicate that divine kingship was closely connected to the motions of the Sun, more than any other heavenly body. The position of the Palenque king was greatly strengthened when he timed his royal events to align with the key solar stations.

When archaeologists were excavating the terraces of the Temple of the Cross a few years ago, they discovered a religious offering of more than 100 incense burners that were ritually buried. The incense burners were carved with images of the Sun, as it climbed the cosmic World Tree, or the *axis mundi*, the ascent of the Sun at

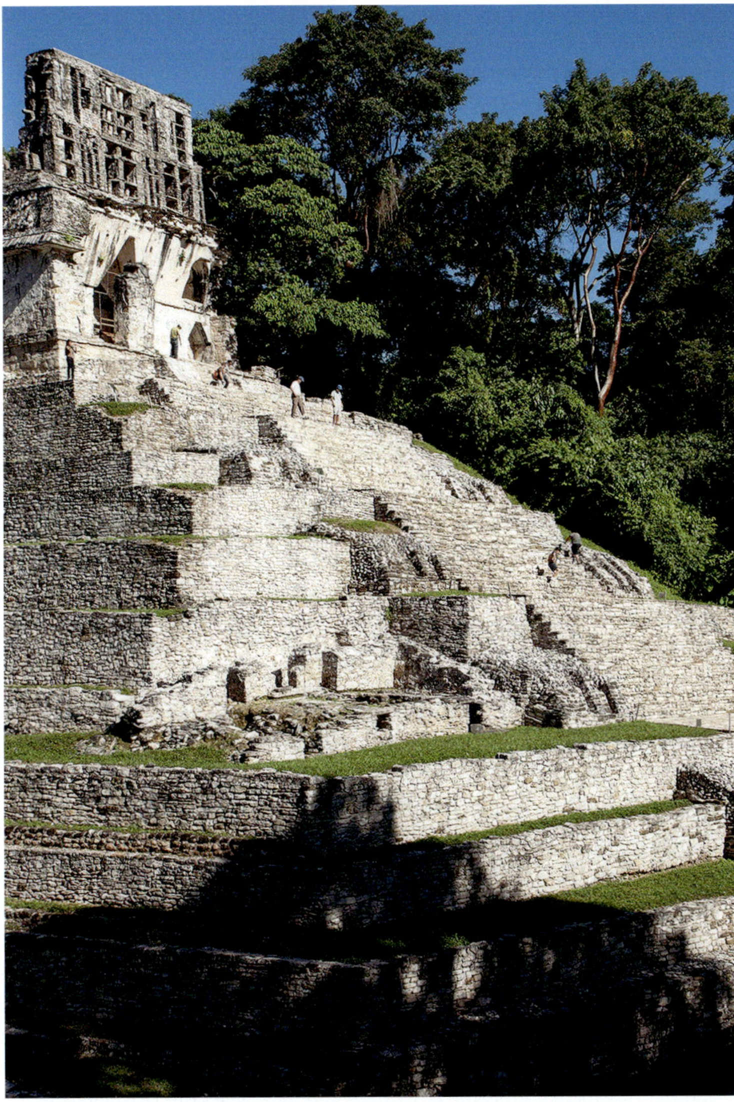

Fig. 10.7 Temple of the Sun with roof comb, late seventh century (Credit Gekkon)

zenith passage. Using incense burners as portable hearths during dedication ceremonies was appropriate because the ritual depicted and described in text on the Tablet of the Cross is intimately tied to the religious initiation of a "sky house". These holy incense burners were then carefully preserved, buried within the raised hillside of the Cross Temple.

The Temple of the Sun was aligned to allow the rising Sun to shine into the entrance at the winter solstice, but it only reached halfway to the rear wall. When the

Sun is at zenith, light again enters the doorway but its path is blocked by the corner of an internal wall. At the equinox and summer solstice, sunlight fills the entrance, extending all the way to the rear wall. These many solar alignments can track the path of the Sun through its annual cycle.

Astronomy at Palenque

Both Pakal and Kan Bahlam shared a strong interest in the celestial transits of the five planets as well as those of the Sun and Moon. As mentioned Pakal named his son as heir to the throne on the summer solstice; Kan Bahlam too linked important events to solsticial alignments. Multiple references to astronomical phenomena can be seen in the surviving artworks and literature from each of their reigns. By aligning not only their architecture but also their main observances, with key positions of the Sun, Moon and planets, they assured that the celestial gods were in harmony with their official ceremonies. For example, Pakal's katun-ending ceremonies were coordinated with several astronomical phenomena, especially positions of Venus. The religious rituals directed by Kan Bahlam were scheduled around the appearance of the Moon and the major planets, also at the day of Venus' maximum elongation. The dedication of the Cross Group was related to a major astronomical conjunction between Mars, Jupiter and Saturn which all set within 10° of longitude of the Sun and the Moon, entering the Underworld through the portal of the nadir. This special celestial conjunction was noted in both the Temples of the Cross and the Foliated Cross (Fig. 10.8).

Archaeoastronomers have discovered the accurate celestial orientations of many additional buildings at Palenque, for example, the Temple of the Count is aligned to Sirius; the Temple of the Foliated Cross to Capella, and House A and the east side of the Palace to the Moon at maximum elongation. Also Temple XIV is aligned with the winter solstice and the west side of the Palace is aligned with the setting Sun at zenith passage At Palenque's latitude the first zenith passage now takes place on May 7, the second on August 5. In addition a series of solar alignments occur in the tall Tower of the Palace. Specific light patterns and shadows within the Palace Tower enabled Maya astronomers and calendar keepers to track the Sun at the solstice and zenith passages. It appears that the Tower served as a working observatory for Maya skywatchers.

At Palenque we can see the process of birth, death and royal successions expressed in light. The last rays of the winter solstice Sun illuminate the Temple of the Cross to celebrate and commemorate the anniversary of Kan Bahlam's accession to the throne. The Temple of the Sun is oriented to face the rising Sun at winter solstice. He dedicated the Temple of the Sun on July 24, 690; that night Venus rose within 3° of the Sun's summer solstice's rising position. Venus would have been visible from the northeast doorway of the Sun Temple. After the temple's dedication, Kan Bahlam performed a bloodletting ritual to the gods on July 27, 690. The previous night the full Moon passed through the Sun's nadir position. It appears that

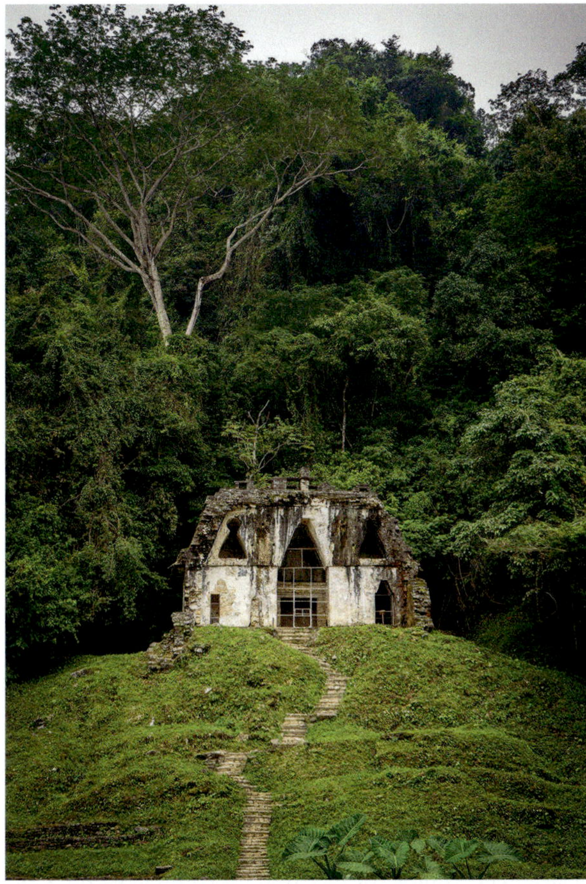

Fig. 10.8 Temple of the Foliated Cross, late seventh century (Credit THP Creative)

the king had waited for the Moon to reach its southern extreme and to begin its northern ascent before performing the bloodletting ceremony. On June 14, prior to the inauguration, a new Moon passed through the zenith (Fig. 10.9).

The Temple of the Sun was used to track major Sun stations as well as to mark important dates in the reign of Kan Bahlam. Recently four solar hierophanies have been identified within the temple. The morning light was observed entering the temple during the equinox and light entering the central doorway at nadir passage pointed to the transverse axis of the temple. From the Temple of the Sun, the morning Sun appears to rise out of the Temple of the Cross during zenith passage. That evening, looking west from the Temple of the Cross, the Sun sinks behind the Temple of the Inscriptions. The morning Sun, seen from the Temple of the Inscriptions, rises out of the roof comb of the Temple of the Cross during nadir passage on January 29. Also a bright beam of light pierces the eastern window of the Temple of the Inscriptions and marks the center of the corridor.

Many other rulers and patron gods of Palenque were equated with specific planets as well. The accumulated information about the heavens by astronomer priests

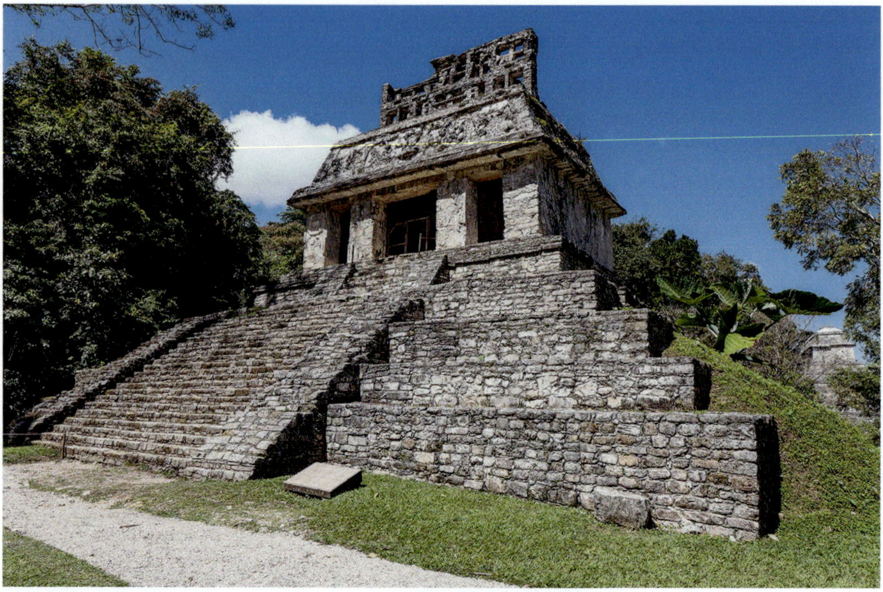

Fig. 10.9 Temple of the Cross, late seventh century (Credit Juan Ariza)

still holds many secrets not yet revealed concerning their design, function and interrelationships. Basic observations of the planets played a small role in the astronomical data encoded into the inscriptions, art and architecture of Palenque. But the zenith and nadir alignments were utilized to commemorate significant dates in the lives of Palenque rulers. Major alignments to the Sun at zenith and nadir passages can be viewed at the Temple of the Cross, the Temple of the Inscriptions, and the Palace. The Temple of the Sun, the Temple of the Count, and the eastern façade of House A in the Palace align to the nadir. The Maya myth of Creation can also be seen depicted in solar alignments at Palenque. Their texts and iconography indicate that the zenith and nadir passages were core themes in the Creation story. These motifs were duplicated in the king's efforts through art, history and the design of sacred space, to preserve their divine power.

These numerous targeted light rays documented at the equinox, summer solstice and zenith passage would have been witnessed by just a small group of astronomers-priests devoted to monitoring the passage of the Sun through its annual revolutions. The precision of the Sun Temple's design and subsequent modifications suggests that the Temple functioned as a commemorative structure for ritualized observations that reaffirmed the ruler's central place in the cosmic order. Public and private hierophanies depended on the harmony between scientific observations and mathematical knowledge.

These themes are repeated in the ruler's efforts, through art, written history and the design of sacred space, to establish and preserve their divine power by identifying themselves with the Sun and by aligning their temples to the zenith and nadir passages. Pakal and Kan Bahlam further assured their connection to the mythical past, as well as to the future, through cyclical astronomical events. The union of time and space presumable occurred on the date of Creation, an event that was eventually celebrated at major sites throughout Mesoamerica. That primary occasion is recreated in the alignment of the Cross Group to the local zenith and nadir transits. The Sun's nadir takes place on January 29 and November 9; at that time the setting Sun drops below Temple XXIV which stands behind the Temple of the Sun. On the anniversary of the day of Creation, the Sun rises from the Temple of the Cross, shrine of GI, god of the day Sun. The nadir evening Sun sets over the Temple of the Sun, shrine of GIII, god of the night sun. At noon during zenith and at sundown during nadir, the Sun strikes the World Tree at the center of the World inside the temple. This powerful, nocturnal occurrence was celebrated with inspiring rituals performed under the shimmering, visible, rotating celestial dome. The celestial and architectural unification bonded participants with one another, with their leader, and connected them on a physical and psychological level with the divine cosmos. The innumerable astronomical alignments at Palenque could not have occurred by chance and required an intimate knowledge of celestial phenomena as the stars and planets transited over the city.

Conclusion

Fifty-three days after his death, Kan Bahlam was succeeded as king of Palenque by his younger brother Kan Xul II at age 57. Maintaining the family tradition of commissioning outstanding architecture, Kan Xul added buildings on the north side of the Palace. Rulers that followed Kan Bahlam and Kan Xul carried on with their example of creating royal architecture oriented toward the three-fold Cross Group. After the Pakal's family dynasty, Palenque continued to enjoy a century of growth and grandeur. But by the eighth century the city was overwhelmed with mounting problems, as were most Classic city-states. Construction of major buildings, temples and monuments was abandoned; agriculture survived but only for a few more generations.

The last known inscribed glyphs at Palenque were carved on a vase dated November 17, 799. This inscription records the name of a noble called "Janab Pakal", perhaps the city's last ruler; his name honored the legendary king who had reigned more than 100 years earlier. Care of the great city was neglected and then finally by about 850 entirely abandoned; the glyph-covered pyramids and temples were slowly overgrown by the tropical forest and then hidden under a dense canopy for the next seven centuries; their myriad stories and royal history gradually eroded. The factors leading to the collapse of the wealthy Maya kingdoms are still hotly debated but drought caused by climate change is understood to be among the major causes. Unfortunately, no Maya king had the power to prevent the natural collapse and inevitable desertion of the city he ruled.

References and Further Reading

Geunter, S. (2007). The Tomb of K'inich Janaab Pakal: The Temple of the Inscriptions at Palenque. *Mesoweb*. https://www.mesoweb.com/articles/guenter/TI.html

Mendez, A., et al. (2015). *Astronomical observations from the Temple of the Sun*. https://www.academia.edu/82263955/

Schele, E. D. (2016). A history of the controversy over the age of the bones of K'inich Janaab Pakal. *Mesoweb*. https://www.mesoweb.com/articles/schele/Schele2016.pdf

Schele, L., & Villela, K. D. (1992). *Creation, cosmos, and the imagery of Palenque and Copan*. Texas meetings, University of Texas, Austin, TX.

Scherer, A. K. (2012). The Classic Maya sarcophagus: Veneration and renewal at Palenque and Tonina. *Res: Anthropology and Aesthetics, 61/62*(Spring/Autumn), 242–261.

Stuart, D. (2006). The Palenque mythology inscriptions from the Cross Group at Palenque. In *Sourcebook for the 2006 Maya Meetings*. Department of Art and Art History, University of Texas.

Tiesler, V., & Cucina, A. (Eds.). (2006). *K'inich Janaab Pakal: Reconstructing the life and death of a Maya ruler*. University of Arizona Press.

Chapter 11
Maya City of Copan

History of Copan

The ancient Maya city of Copan, in western Honduras, was founded alongside the Copan River; archaeological findings show that the site has been continually occupied for more than 2000 years, from Early Preclassic to the Postclassic period. The Copan River runs through the fertile, well-watered mountainous valley; the site sits near a river network that reaches far to the west and empties into to the Atlantic in the east. Therefore the city had abundant natural resources and was ideally positioned for controlling trade. Scholars propose that Copan was developed to increase and command the trade of the much-desired jade mined from nearby mountains, the Maya's only source of the mineral more precious than gold. The other highly valued commodity among the Maya was obsidian, volcanic glass that was fashioned into long, thin, razor-sharp blades. The black blades had many common uses but their most important purpose was for bloodletting since obsidian could make clean, quick incisions. The distribution of obsidian was not openly traded but controlled by the king.

The well-located river city developed into a thriving Maya territory abounding with stately pyramids, splendid temples, and tall stelae. All monuments were beautifully sculpted in the city's distinctive decorative style. Their buildings were completely covered with intricate hieroglyphics by craftsmen with minimal stone or wooden tools. Between the fifth and ninth centuries Copan functioned not only as a mighty royal city, but as one of the wealthiest, most powerful kingdoms of the Classic period.

A century of research by archaeologists and epigraphers has reconstructed much of Copan's history through its descriptive artworks and hieroglyphic texts. Archaeoastronomers have shown that Copan's astronomer-priests encoded astronomical information into many of their structures and into its decorative programs. Copan's history can be verified since its surviving monuments are dated and span

M. Dolan, *Decoding Astronomy in Maya Art and Architecture*,
Springer Praxis Books, https://doi.org/10.1007/978-3-031-61092-9_11

the entire Classic period. Even though only one quarter of the size of Tikal, Copan manifested the economic, social, political and ideological dimensions of a complex state organization. The city's buildings and sculpted monuments prove that it was much more sophisticated than a typical trading center. During Copan's peak the culture excelled in its achievements in mathematics, astronomy, architecture and in detailed hieroglyphic writing. Famous for its splendid architecture, Copan developed its own characteristic, high relief sculptural style (Fig. 11.1).

Copan today includes a main complex of ruins with several secondary complexes surrounding it. The primary complex contains the monument-filled Acropolis and several smaller plazas. The construction of the Great Plaza and the Acropolis demonstrates an enormous amount of effort in that a huge area was raised 30 m (98 ft) from the ground, then carefully leveled and covered with paving stones. The main temple-pyramid area sits atop an artificial terrace that was partly destroyed by an earthquake in 1934 and by the Copan River. The river's course has been altered so it is no longer a threat to the architectural remains. Within the Ceremonial Plaza are a large number of detailed stelae; a series of tall portrait stelae lined the processional pathways.

The Acropolis consisted of a collection of overlapping step-pyramids, connecting plazas, carved altars, lofty palaces and a ball court; all areas of the city were linked by paved roads, or *sacbeob*. Next to the temples stands the renowned Hieroglyphic Stairway, one of the most amazing monuments of the Maya civilization. The risers of the 66 stairs are carved with more than 1800 individual glyphs, the lengthiest Maya inscription known. Inscribed on a stone of the Hieroglyphic Stairway is a description of Ruler 13, Waxaklajuun Ub'aah K'awil; the text says that

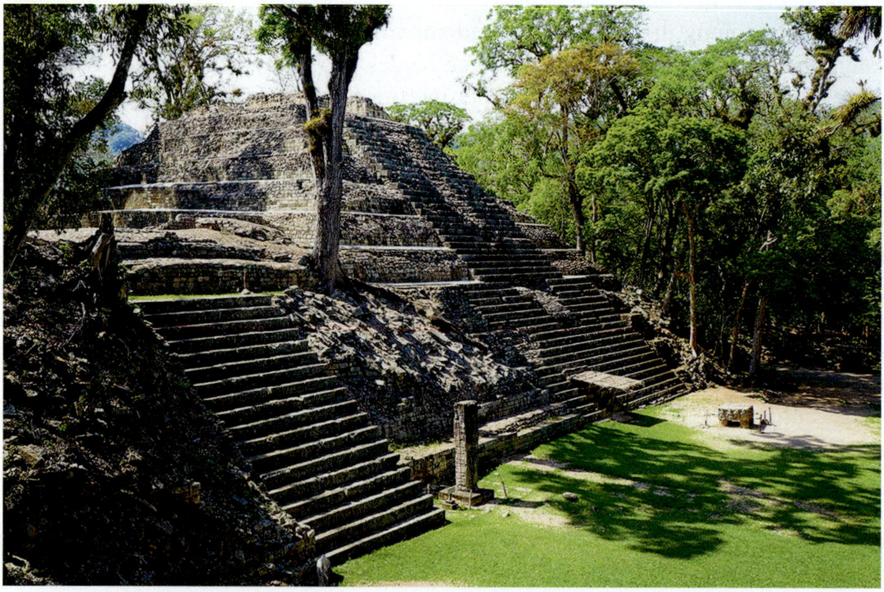

Fig. 11.1 Copan over-lapping step pyramids (Credit Inspired by Maps)

the king died on 13 May CE 738, "his breath expired in war". This decorated stairway embodies great significance since the texts document the long history of the city and its rulers including accurate details (Fig. 11.2).

The earliest inscription at Copan is dated to CE 427 when a new Maya ruler K'inich Yax K'uk' Mo' came to the city from Tikal. Once there he established his kingship by inscribing the momentous event in stone, overseen by the ruler of Tikal. All later kings of Copan's royal dynasty trace their ancestry to his fifth-century reign. A total of 16 succeeding kings have been recorded on the city's monuments. The nearby town of Quiriquá was settled at the same time for the same purposes and became a vassal of Copan. During the Early Classic period, scholars believe about 20,000 people had settled in the area, the highest level of Copan's occupation.

Copan is significant among Maya cities for many reasons; most important is its vast quantity of hieroglyphic texts. The huge number of inscribed buildings, statues and stelae indicates that the city's literate culture was dedicated to recording its history and its long centuries of kingship. Many epigraphic researchers have focused on Copan because its continual series of informative texts over a lengthy time period enabled scholars to compare and analyze changes in the Maya vocabulary. Deciphering Copan's innumerable glyphs played an essential role in finally breaking the Maya code. The translated texts have helped scholars piece together the lives and accomplishments of the city's 16 rulers during the five centuries of its peak. They have uncovered relevant facts to better understand the lives of the people of Copan. Archaeologists now are able to date minor changes in ceramics and architecture to within a few decades. The texts showed how the rulers of Copan expanded

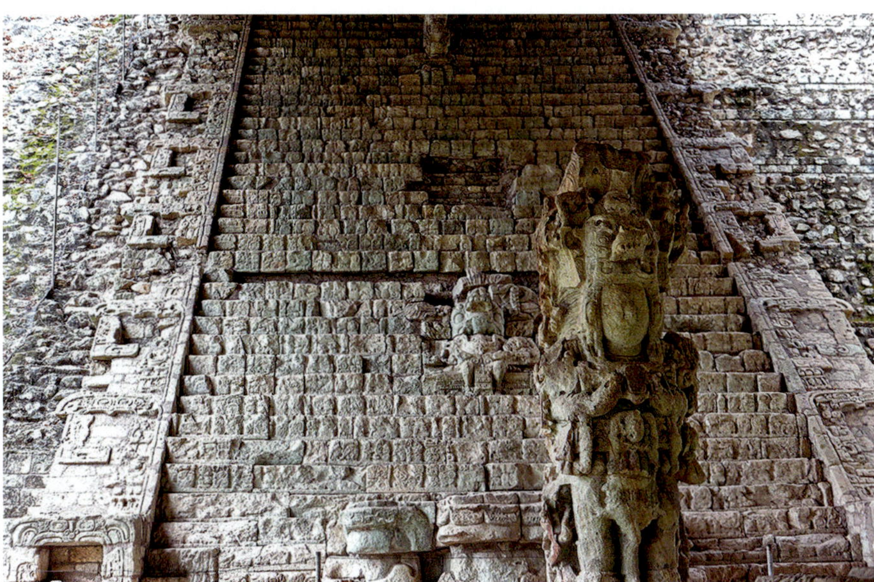

Fig. 11.2 Base of Hieroglyphic Staircase with royal stela at Copan, partial view (Credit Autumn Sky Photography)

the Mayan understanding of calendrics and the cosmos. Copan leaders introduced new forms of hieroglyphic writing and made significant advances in architecture; artistic expression flourished during their years in power.

King Waxakllajuun Ub'aah K'awiil

One of Copan's most capable and famous royal figures was Waxakllajuun Ub'aah K'awiil, known in earlier literature as 18 Rabbit. He was the 13th *ajaw* or ruler of the city and reigned from CE 695 to 738; he is considered the greatest supporter of the arts in Copan history. He directed huge improvements in architectural design and in innovative artworks. The gifted sculptors of Copan reached their highest achievements in the decorative artworks adorning the buildings commissioned by Waxaklajuun Ub'aah K'awiil. In addition to his many temples, palaces and stelae, he also created the Hieroglyphic Stairway to enshrine the glorious deeds of his ancestors. His final achievement was Ballcourt A-III, impressive for its sloping benches and six macaw markers, which are symbols of the Sun. It is thought that perhaps the ballplayers acted out symbolic battles between light and dark (Fig. 11.3).

The early temples and pyramids of the Maya were decorated with glyphs and depictions of their gods merely as monumental masks; only the name glyphs of the current rulers were engraved near the top or along the stairways. During the Classic period extraordinary carved images of the city's rulers, clad in their elaborate regalia, were displayed in public places not on pyramids but on large freestanding stelae. Between 711 and 736, Waxaklajuun Ub'aah K'awiil erected seven intricately carved stelae of himself that are considered masterpieces of Classic Maya sculpture (Fig. 11.4).

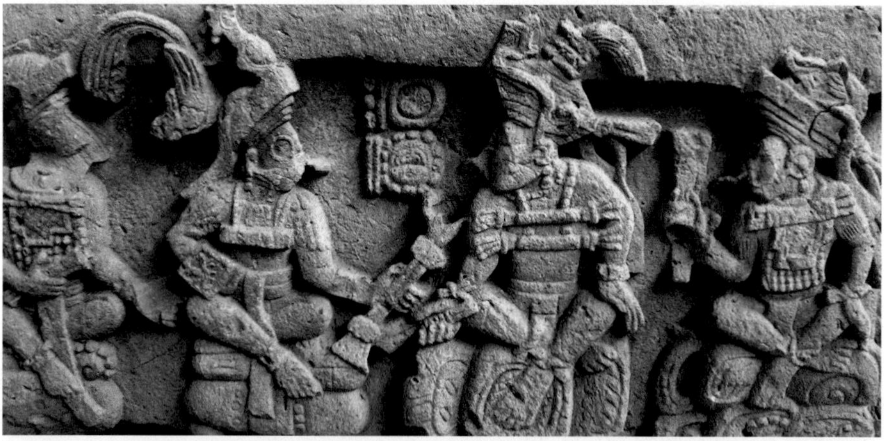

Fig. 11.3 Bas relief carving depicting K'inich Yax K'uk Mo' and three other Maya kings at Copan (Credit by Vladimir Korostyshevskiy)

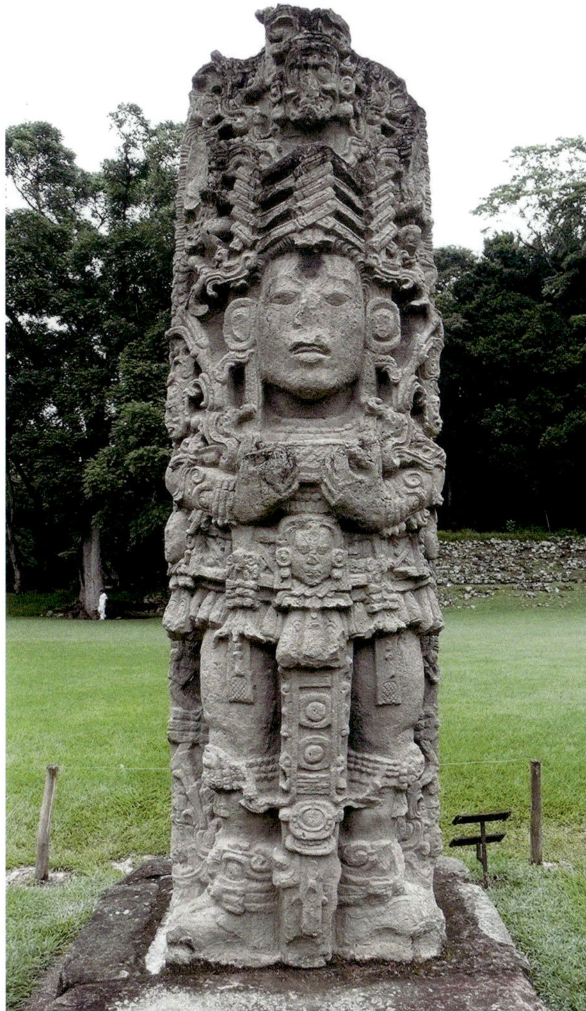

Fig. 11.4 Monolithic Stela of K'inich Yax K'uk Mo' with ceremonial headdress at Copan (Credit Oscar Espinosa)

These stelae reached the highest level of artistic achievement in the Maya kingdoms. They showed the king ritually posed; each stela included the iconographic characteristics of different Maya deities. Considering the numerous monuments commissioned during Waxaklajuun Ub'aah K'awiil's reign, scholars think he is the most important patron of arts in Copan's history. Traditionally, the temples at Copan honored the familiar Maya gods along with the special patron deities of Copan; these religious images are prominent on many temple facades and the Mayan glyphs include words of worship. Among the many buildings of Copan, Temple 22 was one of the first to display new developments in Maya's religion and artworks. Here the king was depicted on the temple not as an abstract god-like regal figure but as an actual historical being with a physical body, highly honored by the images carved on the exterior of the temple.

Temple 22 highlights the traditional roles of the divine king, projecting his central role in eighth-century Maya history, myth, and cosmology. The temple actually served as a physical and conceptual extension of his royal body. By commissioning life-size images of himself upon architectural sculpture at Copan, Waxaklajuun Ub'aah K'awiil asserted a different kind of kingship that focused on his human rule. He also wanted to assert Copan's leading role on the southern boundary of the Maya world because his standing as a powerful leader was beginning to weaken. At this time Waxaklajuun Ub'aah K'awiil was struggling to hold his large kingdom together; for that purpose he commissioned Temple 22 as a response to his challenging situation.

Temple 22 at Copan

Most experts agree that the remarkable Temple 22 ranks as the most important, most studied and most admired building at Copan. This temple was constructed by Waxaklajuun Ub'aah K'awiil and dedicated in 715 to commemorate his 20-year reign as king of Copan. The temple's orientation, its design, its symbolic representation, religious functions and artistic renderings all have clear astronomical significance. The architectural form of Temple 22 was not unique but was a recreation of a design that had been built by his royal ancestors many times in the past, incorporating the Maya's special vision of Creation. Waxaklajuun Ub'aah K'awiil commissioned this temple when Copan was at the peak of its authority and the pinnacle of his accomplishments. The temple played a role not only in the religious and cosmological viewpoints of the culture; it also performed as a vital symbol of power and competition for Waxaklajuun Ub'aah K'awiil. The new temple would raise his position among his enemies and rivals through its impressive design and innovative iconography. But soon after its completion, economic, environmental and political problems were becoming evident; this era marked Copan's peak and also the beginning of its decline.

Similar to most sacred buildings of the Maya, Temple 22 was created to function as a dwelling place of the gods and a sacred place where the God-King of Copan could communicate with the deities and with the esteemed ancestors. Encased within its physical form, Temple 22 was designed as a reconstruction of the cosmos in accord with the belief system of the Maya people. The temple served as a sacred cave within the Underworld, deep below the earth, as well as within the hallowed heavens rising all around it. The temple's copious iconographic decor encoded its meaning into the carved artworks. But even more, Temple 22 also acted as an astronomical observatory. Heavenly motions could be tracked through its carefully positioned window and from the building's orientation and alignment with other Copan monuments. Astrological portends too could be inferred from observations made and recorded there. Temple 22 functioned as a multi-purpose portal to strive for supernatural powers and to receive divine interactions. Additional research over the last few years has provided abundant evidence to interpret and decipher the deeper meanings of the iconographic program displayed on the temple exterior and interior walls.

Looking at the temple's sculptural program shows that the designers wanted to incorporate the entire Maya ideology into the building's form and its pictorial narrative. Temple 22 did not represent merely the usual symbolic mountain but instead embodied the complete Maya universe. Its mosaic façade is admired as the most sophisticated assemblage of representative icons. Its meaning can be interpreted on many levels, appropriate for each viewer's level of understanding. At the simplest interpretation it was the site of royal audiences and rituals required of the king and perhaps the queen. On a symbolic level the temple represented the mythological and cosmological beliefs of the Copan culture. On a more scientific level it incorporated the motions of the heavens, more particularly Venus and its cyclic pattern. The people of Copan could read the pictorial narrative played out on the temple walls, in the same way a Christian could reaffirm their history and religious belief system when reading the paintings, frescoes and stained glass windows of a medieval church.

By studying the temple's physical form, its hieroglyphic scripts, and earlier written surveys, archaeologists have worked to re-create the original structure of Temple 22. Essential themes in Maya art and literature are the sacred mountain and the carefully ordered universe. The images depict the temple as a fertile mountain decorated with flowers, maize and flowing water. These special factors, along with the temple's design, represented a three-dimensional diagram of the Maya universe including a reference to cosmic and human time. These main factors are repeated in the artwork found on the interior doorway of the temple but in more precise details.

Raising distinctive buildings to a higher elevation was a common practice in Maya state architecture. The Acropolis of pyramidal buildings overlooking the city had been constructed by the earliest rulers of Copan. While creating the many elaborate stelae of himself, Waxaklajuun Ub'aah K'awiil also reshaped this central area of Copan into an impressive pseudo-mountain range. At the peak of the Acropolis, Temple 22 was built upon its own E-shaped platform positioned on the 20-ft terrace that framed the entire court. A flight of seven wide steps raised the temple enough that it towered over the nearby buildings. The risers of the steps too had been sculpted with various glyphs but are now much eroded. When visiting Temple 22, people had to climb to the height of the Acropolis and walk through several corridors before reaching the East Court (Fig. 11.5).

Temple 22 faces south from its ideal position on the north side of the East Court. Since the temple sits at the highest elevation of the city, it can be easily seen from most areas. The courtyard in which it sat was designed for civic gatherings and for attending performances and celebrations. The temple was relatively small compared to many temples in the Maya world. Surprisingly, this royal temple was not build over an existing building which was common with monumental structures and pyramids. Doing so greatly enhanced the power and prestige of a new building for it held the mystique of an older, historic building hidden deep within.

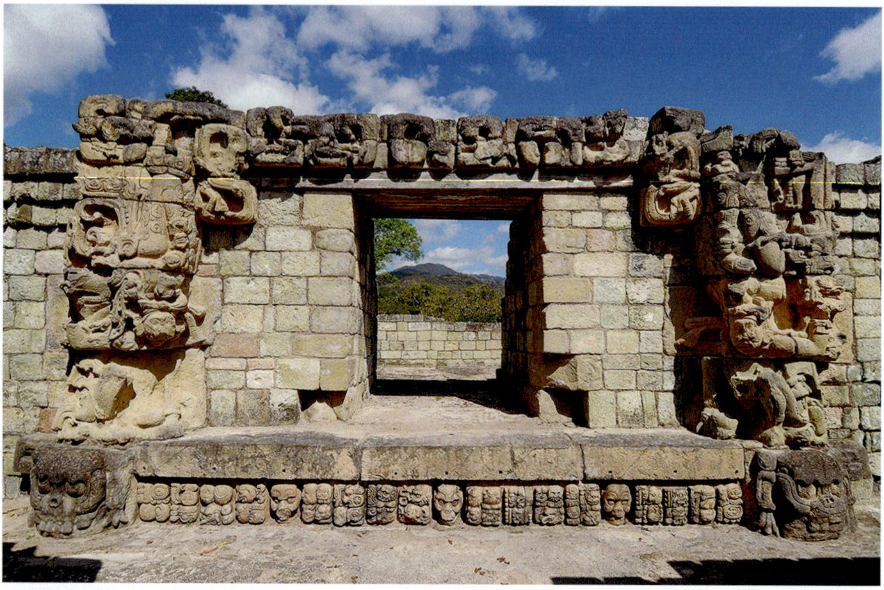

Fig. 11.5 Inner doorway of Temple 22 surrounded with carvings, Copan (Credit VojtechVlk)

The latest research projects restoring Copan have shown that Temple 22 most surely had three stories; its three-level façade must have been at least 17 m (c.55 ft) tall. The huge amount of sculpture attributed to the building would require three stories to encase all the relief carvings. Also the temple's thick load-bearing walls suggest they were intended to carry a great deal of weight. The building was not constructed with the usual stone or wood but with mud mortar, a material that required continual maintenance to prevent seepage that would damage the walls. Although only sections of the temple façade survive intact, this temple endures as the best preserved of the buildings erected during Waxaklajuun Ub'aah K'awiil's long reign.

Best known for its beautifully carved sculpture, Temple 22 is admired as the most well-executed of all extant Copan structures. The temple's overall sculptural program had the most advanced concepts, construction and artistic representation of the era, adopting a distinctive style in their creations. Today most of its sculptured images lie in jumbled piles, but enough remains affixed to the structure to make the identification secure. The sculpted figures that currently decorate the entrance and the walls of the temple comprise only 1% of the carvings that originally appeared on the building. At present over 4000 carved stones have been recovered from the surrounding grounds. The temple also shows a striking increase in the amount and complexity of human and quasi-human figures applied to both the inner and outer walls. Temple 22 was envisioned as a model of the primordial, maize-producing mountain prominent in Maya cosmology.

The temple's exterior facade held more than 36 sculptural motifs that represented a variety of human, animal, hybrid, floral and astronomical features. The known artworks include 20 ornate life-size torsos of the Maize God; this number may correspond to the 20 years that Waxaklajuun Ub'aah K'awiil presided over Copan. The numerology encoded in the temple's iconography seems to carry multiple levels of meanings, joining the maize plant and ruler to prosperity and the Creation myth. The maize-deity sculptures with their serene expressions are among the most beautiful examples of Copan artistry. Their down-turned eyes were fashioned in the Classic Maya style; some have bits of red and black pigment residue.

Adorning each of the four corners of Temple 22 were three large, sculpted mosaic images of long-nosed masks with cleft foreheads positioned one atop the other with maize sprouting from each. These stacked depictions are known as Chac masks since their elephant-like noses are similar to drawings of Chac in Mayan codices. Other adornments include, the Maize God, Venus symbols and images of Waxaklajuun Ub'aah K'awiil surrounded by other Maya deities, local gods and an arching sky-serpent.

In the belief system of the Classic Maya, all natural openings into the inner earth, meaning caves or cenotes, were actually portals to the Otherworld/Underworld. The design of Temple 22 incorporates this belief twice, outside and again inside. The central portal into the temple was decorated to replicate the mouth and gullet of the great Witz or Cosmic Monster. The *Witz* Monster in Maya cosmology stood as the symbol of the sacred, living mountain in their creation story. The *Witz* Monster's zoomorphic face with a huge gaping mouth and a cleft in the center of his forehead is showcased on the temple façade; his frightful open mouth granted entrance into the symbolic living cave, the core of the sacred mountain, the unknown, mysterious Underworld.

Beneath the entranceway and lining the riser of the step were carved glyphs and a long row of varied skulls. Two huge skulls, each carved from a single stone, were positioned at either end; referring to the place of death. The inscriptions along the bottom of the wall, between the smaller skulls, record the conclusion of the first katun of Waxaklajuun Ub'aah K'awiil and include the date, 29 November, 711, which agrees with the time Temple 22 was constructed. The entire iconographic program, both inside and out, was originally covered with a thin coating of plaster and carefully painted. Archaeologists are still struggling to piece together the complicated images and cosmic symbolism of Temple 22's artworks. This depiction of the Creation Mountain with an inner cave was presented first in Temple 22.

Temple 22 and Astronomy

After passing through the outer entranceway of the temple, one enters a narrow passage that approaches the gaping jaw of a second carved entryway, set back and hidden by the outside walls. Within the mountain-temple they created an inner sacred chamber; this cave within a cave represents the deeper Underworld, or Place of Fear, home of the ancestors. The entire second doorway was framed with carvings. The inner cave/room was raised about a foot above the large room of the temple.

The sculptural composition surrounding the inner doorway was one of the most remarkable artistic creations ever imagined in the Classic Maya World. The striking artworks that enclose the opening incorporate various levels of meaning.

Atop the doorway of the inner sanctum is a full image of the *Witz* Monster interpreted as the Maya Sky-Band. Spanning this inner wall the Monster's long snake body represents the Milky Way in its east-west position and it weaves through scroll-like clouds, or perhaps cloudy areas in the Milky Way. The Monster was sculpted in such high relief that it seems to be struggling to free itself from the wall. The *Witz* has two heads; the head on the left (west) is formed like a crocodile/reptilian with a Venus sign; the other end (east) holds a *kin* or Sun sign. The Sky-Arch does not only symbolize the sky during the day, but also the sky at night. The double-headed serpent's body represents stars and constellations. Facing west, the sky-bar images align with the ecliptic along which the Sun and planets travel. Facing the opposite end, the Sun glyph sits at the junction of the Milky Way and the ecliptic.

The snaky body of the *Witz* curves on either side of the door but over the center a series of S-shaped symbols appear with seven small figures intertwined. Susan Milbrath suggests they represent the Sun, Moon and five planets. Venus glyphs are carved at each side of the serpent's body. The sky-band is held in place by two crouching male figures at the base, similar to Atlas figures, two *Bakabs*, or *Pauahtun* (god N), the Maya creator gods who hold up the sky. Monster heads rest on the shoulder of each *Bakab* sky-bearer. The human-like figures with elaborate headdresses and jewelry appear to be grasped by the snake's spiraling body. Their arms and legs were interwoven as if struggling with the Monster. This Celestial Monster lying above the door embodies a form of Venus-Sun duality.

The Maya regularly tracked the rising and setting positions of Venus, always close to the Sun. These maximum points could be determined simply by visual observation of Venus's 8-year oscillation along the eastern and western horizon. Archaeoastronomers have found many Maya structures oriented to the Venus extreme positions, verifying their interest in making these stellar observations. Venus glyphs and Venus associations are abundant at Temple 22. Artists also adorned Temple 22 with many images of the young Maize God and carvings symbolic of the rain gods. Below the inner doorway the sculpted step contains high-relief glyphs carved with three large Venus symbols.

This special temple at Copan ranks among the few Maya structures in all of ancient Mesoamerica to include a window. This lone window appears on the west side of Temple 22 and looks toward the western horizon. The window is also parallel to two additional monuments, Stela 12 and Stela 10, which form a 6.5 km baseline leading to the east for astronomical purposes. These stelae were commissioned by ruler 12 called K'ahk' Uti' Witz' K'awiil or Smoke Imix during his period of architectural improvements. Researchers have assumed that the primary purpose of the temple window was to observe the setting Sun on April 12, significant for the agricultural calendar. On the eastern side of the Copan valley, astronomer-priests could view the Sun rising along the baseline between the window, Stela 12 and Stela 10.

A study by archaeoastronomer Michael J. Grofe investigated other possible alignments of the two standing markers by studying the textual inscriptions found on Stelae 10 and Stelae 12. His article presented a new interpretation for both the baseline orientation and the western window. Textual clues on the stelae suggest that these alignments are not solar but are stellar instead. Initially it was proposed that the eastern solar alignment was associated with the termination of the dry season which occurs around mid-April.. Through astronomical analysis Grofe proposed that baseline commemorates the astronomical observations made on the eastern horizon, involving sunrise on the very important K'atun ending 9.11.0.0.0. This event occurred at the same time as the rising of the brightest star in Orion's Sword called Na'ir al Saif. This nebulous area within the constellation of Orion is recognized by the Kiche' Maya as the smoke from the hearth fire in an asterism called the Three Hearthstones of Alnitak, Saiph and Rigel. As we have seen, these stars are the basis of the Maya Creation Story and crucial in their mythology. The baseline orientation applies to both the point of sunrise at the previous era ending and stellar events encoded in their cosmology. Grofe wrote: "Such solar, lunar, and sidereal observations as we find in the monuments of K'ahk' Uti' Witz' K'awiil suggest that Maya astronomers were capable of making precise astronomical measurements over time involving calculations of the tropical year, the sidereal year, and the eclipse year".

Temple 22 Window Alignments

Initially it was proposed that this western solar alignment from the window of Temple 22 was associated with the termination of the dry season which occurs around mid-April. But that was not the only possible observation from the window; there were important stellar as well as solar sighting positions. Astronomers have analyzed the possible alignments to the Venus cycle that might have been visible from the window and discovered that the potential sightings of Venus almost always occurred in April or May. They showed that Venus's northernmost setting or its northerly extreme was aligned with the window in a significant manner. In the year before Venus' great extreme, the first day of Venus sighting was always on or very close to the 8-day period from April 25 to May 3. During the year following a great extreme, the last day of visibility of Venus could be viewed from the window as well.

These alignments with Venus through the temple's window present the same type of calendar relationship as the great northerly extremes of Venus. The initial appearance of Venus in the window coincides with the start of the rains in the year preceding a great northerly extreme. The Venus sightings suggest that this entire area of Copan may have had a special relationship with the brightest planet. The discovery of maize and then providing it for the nourishment of the people is attributed to Venus, but the rain gods are considered responsible for its growth. To the Maya, Venus has a much higher status than the rain gods who are in a closer relationship with the people. This cosmic journey of Venus to the north relates closely with the production of life-giving maize.

The two types of Venus events can be shown to generate close observation of its seasonal pattern which would be connected with the planting of maize and the onset of the rainy season. If and when the Copan astronomer-priests actually applied these Venus observations cannot be proven but their compelling concern and charting of Venus appearances provide evidence of its probability. Perhaps this cosmological connection of earthly weather patterns with the cycles of Venus may add to the importance of that planet in the Mesoamerican world. Another reference to the brightest star appears on the building next to Temple 22, three large Venus glyphs are inscribed on Temple 21a as well.

Additional celestial references are the GI (fish god) sculpture from Temple 22 which has star symbols on it, suggesting that it relates to an astronomical object. Both the human-like heads on Temple 22 and those found under Structure 26 were accompanied by forms of falling water, perhaps representing a celestial body such as Venus that accompanies the onset of the rainy season. The symbolic motifs that represent the sky, celestial bodies, and patron deities are thought to have decorated the upper levels of the temple, so that the water imagery symbolized rain pouring down from the sky onto Flower Mountain.

As mentioned, the numerical glyphs on the exterior step located furthest to the right seem to record the Maya date, 29 November, 711 CE, near to the temple's inauguration. But this date also refers to an important astrological event. On that date Venus rose as the evening star while the Sun was aligned with the Milky Way in the west; while in the east, the Moon was conjunct the Milky Way. The Maya created their calendar applying these kinds of heavenly conjunctions. These important heavenly encounters were celebrated with festivals, rituals and religious sacrifices. This dual aspect in the heavens is also recorded on Stela C at Copan. The stela confirms two important events, the rising of Venus as the evening star, and the transit of Orion (the Cosmic Turtle) directly overhead. These celestial positions can be easily verified using astronomical software.

Waxaklajuun Ub'aah K'awiil restored the East Court to create a design that is a diagram of Maya directional cosmology. The East Court is oriented to the cardinal directions and we saw that the sculpture on the stair refers to the celestial symbolic phenomena. The dominant thematic iconography adorning Temple 22 affirms the Venus-Maize-Rain symbolism so essential in Maya ideology.

Temple 22 Inner Room

Mayanists currently suggest that the inner room of Temple 22 was multi-purpose; it was utilized as the royal throne room where Waxaklajuun Ub'aah K'awiil and later kings directed their rule over the vast networks of the Copan kingdom. The placement of powerful images and the extensive decorative program designed for the inner room combined to heighten the aura of divine kingship. Amidst this compelling decor the king would be venerated while surrounded with historic, mythic and cosmic symbolic images honored throughout Maya culture. The conspicuously decorated space would have also been used for religious rituals that reconfirmed the king's unique connection with the gods and affirmed his right to rule.

The religious rituals included customary sacrifices and most importantly genital auto-sacrificial bloodletting. These activities were performed by kings and queens, as well as elites of the society in order to maintain their social structure. Offering blood sacrifice to the deities demonstrated that the royals had supernatural powers and were in good standing with the gods who would then continue to protect their people. Through his painful blood sacrifices and by acting out traditional rituals, the ruler was mysteriously transformed into the God-King. The benevolence of the Maya gods was crucial for the onset of the annual rains and to assure plentiful crops of maize, all in association with the nightly appearances of Venus. Waxaklajuun Ub'aah K'awiil commissioned Temple 22 to assert his power by showing his knowledge and control of earthly and heavenly forces. He crafted images of himself adorned as a supreme ruler and sometimes in the guise of a god. He used art and architecture to codify and amplify his royal position as skillfully as the Tudors in England or the Hapsburgs in Europe.

Temple of Yax Pac at Copan

Another fascinating finding at Copan concerns the temple of the ruler Yax Pac and astronomy; inscriptions reveal a royal relationship with the planet Venus. Yax Pac ruled as the last king of Copan. Yax Pac's Temple 11 in the west court, built in the second half of the eighth century, is his most significant and final construction effort at Copan. Temple 11 is an effective three-dimensional map physically depicting what the Maya imagined was the appearance of the cosmos, which must have awed the commoners who came to the temple to worship their rulers. The south side of the temple carries symbols of the Underworld, the north has symbols of the afterworld; the middle world of the present was in between and contained the king's throne. The Maya Underworld was regarded as an exotic plane of existence through which rulers and diviners could travel, a portal to the Underworld, but only when in a trance. The Underworld was believed to be a real world with real, though supernatural, inhabitants. Each night after the Sun set, the Underworld flipped over and became the heavenly dome.

At the base of Temple 11 sits the unique Altar Q; the glyphic text written on the altar has been entirely deciphered. All 16 members of the Copan Dynasty are depicted on the altar, including Yax-Pac; to affirm their identity, all are seated on their name glyphs. Yax-Pac receives the scepter of rulership directly from the founder of the dynasty, Yax-Kuk-Mu.

An important text inscribed on one of Copan's altars, located in front of a sacred temple, tells of a first predawn appearance of Venus that was linked to the date of a king's inauguration whose name was New-Sun-at-Horizon, or Yax Pac. This king is especially noteworthy because he appears to have ruled at a time of the construction of large ceremonial centers that are believed to have signaled the institution of the *ahau* principle. This innovative idea maintained that the king himself, now endowed with a new royal title, was conceived supernaturally and that he derived his power directly from nature, directly from the gods of the heavens. Experts propose that a major change was taking place in Maya society, from egalitarian to a more strictly hierarchical social system. This particular phase of Maya civilization also produced the elaborate astronomical tables in their codices. Another date referring to Yax Pac's rule is accompanied by a Venus glyph that says, "It shined the Great Star – Lord *Ahau*". This inscription was inscribed in hieroglyphs on the eastern doorway of his temple. Modern calculations show that the date given coincides with the actual time Venus would have been sighted from the doorway.

It appears that Yax Pac connected most of the historical statements inscribed on the temple pertaining to his reign to the movements of the planet Venus, particularly its occurrence as evening star. His accession, the conduct of war, the dedication of stelae and his monumental architecture, almost all he accomplished, were recorded under the sign of his patron planet. Yax Pac manipulated history and managed his behavior and actions according to the motions of Venus as it moved through its heavenly cycle. He did this with profound consistency (Fig. 11.6).

Copan's Final Decline

During the later reign of Waxaklajuun Ub'aah K'awiil, the political situation began to change, the increasing population, and decreasing resources in the area added to the city's problems. In 738 Copan experienced a profound political upheaval, Waxaklajuun Ub'aah K'awiil was captured and executed by the king of the small city of Quiriquá. After the death of the great king Waxaklajuun Ub'aah K'awiil, a new king was installed; the ceremony took place 39 days later, very close to the maximum elongation of Venus as the morning star. The violent death of 18 Rabbit caused a sharp decline in the city's prosperity; no new monuments were created for the next 18 years. In spite of that weakening, Temple 22 at Copan was utilized for another 100 years and continued to symbolize the Maya ideals with its imposed mythology, history, astronomy and cosmology. The temple was utilized to seek benefits from the gods, to gain access to supernatural powers, and to honor their ancestors (Fig. 11.7).

Between 800 and 830 Copan's lengthy royal sovereignty abruptly collapsed. The population of approximately 20,000 fell to less than 5000. Although the city still had a small population that remained active for some years, by 900 the site was abandoned. Copan was never totally consumed amidst the tropical jungles like many Classic cities of the Maya; a few farming communities still occupied the area when the Spanish arrived in the sixteenth century. The ruins were known by the Spanish but they had little interest in them.

The elaborate sculptures that decorated every building, temple and monument of Copan slowly eroded; some had detached, openly accessible for theft or destruction. The vibrant shades of red, blue, green and yellow pigments that had once decorated

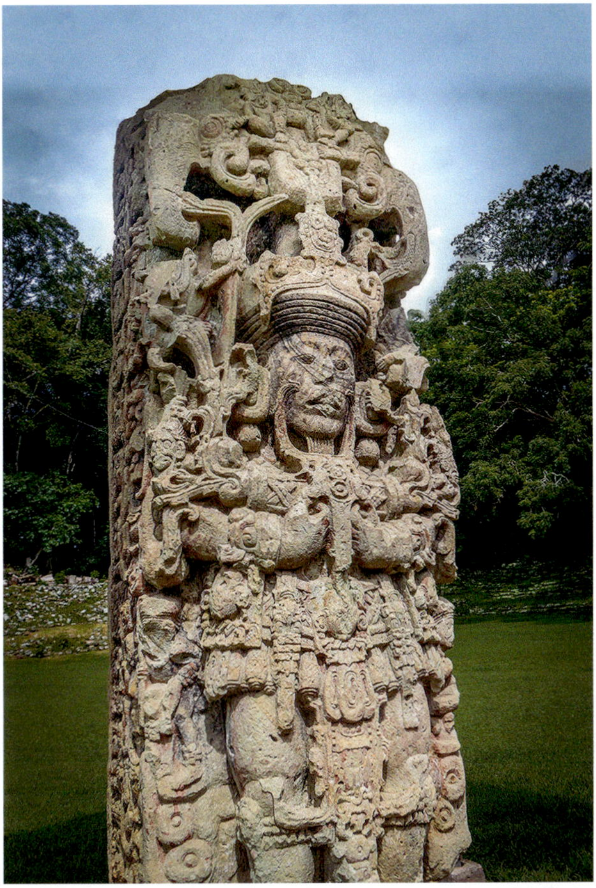

Fig. 11.6 Monolithic royal stela at Copan (Credit Diego Grande)

the temple walls and artworks began to fade to a dismal grey and become blackened with mold from the tropical climate. Thankfully the buildings of Copan are being maintained and restored as well as being studied. The city's glorious story is still being uncovered. Viewing the surviving sculpted stones and picturing the vivid colors, one can reimagine the original beauty, majesty and mystery of Copan and of Temple 22. Today scholars consider Temple 22 a masterpiece of Maya creativity.

Many theories have been introduced and innumerable papers have been published investigating the last decades of the Classic period, trying to explain the swift decline of the Maya cities. The numerous successful and prosperous city-states in different areas all appeared to collapse around the same time. Recent studies indicate that warfare in the Copan Valley may have played a more important role in both the development and decline of the Maya. The final century of the Classic period witnessed a huge number of monuments constructed at various major and minor settlements including pyramid and plaza groups and even grand palace complexes. Populations grew and the civilization prospered showing no sign it was all on the brink of collapse. Within a few generations, the great civilization faded into the past.

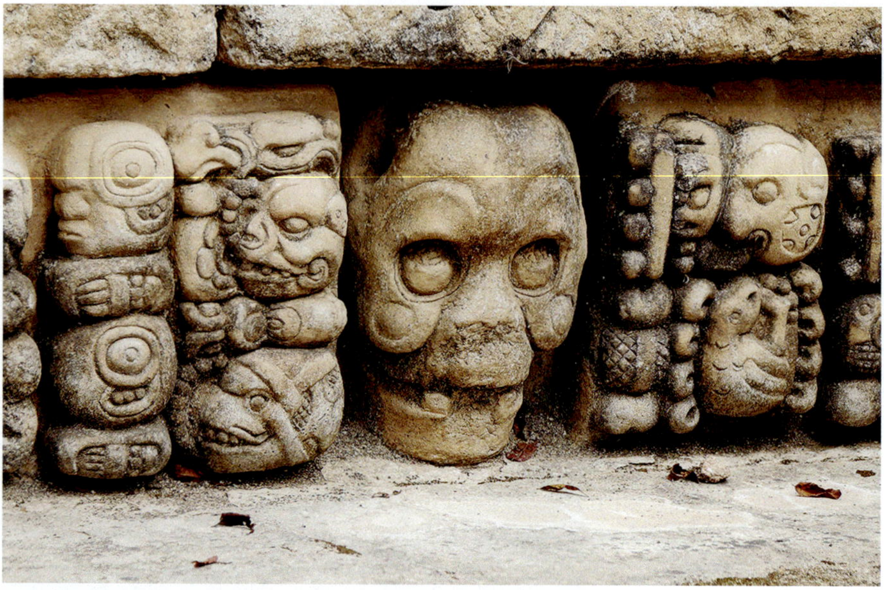

Fig. 11.7 Carved image with glyphic text (Credit VojteckVik)

Why did these well-established people desert their monumental structures that they struggled for years to build? The thick vegetation of the tropical jungle which the Maya farmers had overcome grew back and eventually buried the great temples, plazas and palaces where the kings and priests had celebrated their royal power.

References and Further Reading

Aveni, A. (1992). *Conversing with the planets: How science and myth invented the cosmos*. Times Books.

Closs, M. P., et al. (1981). The planet Venus and Temple 22 at Copán. *Indiana Journal, 9*, 221–247. University of Indiana, Indiana.

Fash, W. L. (1998). Dynastic architectural programs: Intention and design in Classic Maya buildings at Copan and other sites. In S. D. Houston (Ed.), *Function and meaning in Classic Maya architecture*. Dumbarton Oaks Trustees for Harvard University.

Freidel, A. L., et al. (1993). *Maya cosmos: Three thousand years on the shaman's path*. William Morrow.

Grofe, M. J. (2012–2013). *The Copan baseline: K'atun 9.11.0.0.0 and the Three Hearthstones in Orion* (Vol. XXV). University of Texas Press.

Milbrath, S. (2000). *Star gods of the Maya: Astronomy in art, folklore, and calendars*. University of Texas Press.

Sherer, R. J. (1994). *The ancient Maya*. Stanford University Press.

Sprajc, I. (1993). The Venus-rain-maize complex in the Mesoamerican world view: Part II. In *Archaeoastronomy* (JHA) (Vol. 18, p. xxiv). Science History Publications Ltd.

von Schwerin, J. (2011). The sacred mountain in social context. Symbolism and history in Maya architecture: Temple 22 at Copan, Honduras. *Ancient Mesoamerica, 22*, 271–300, Cambridge University Press.

Chapter 12
Late Classic Cities of the Maya

Maya City of Bonampak

Bonambak in Chiapas, southern Mexico, lies near to a tributary of the Usumacinta River and was a dependency of Yaxchilan that is about 30 km (19 mi) north. The small site was not near the size or population of other important Maya Classic cities but is valued as just as important. The monumental architecture of the city dates to the Late Classic period (CE 580–800) and flourished in the mid-eighth century. Four nearby cities, Bonampak, Yaxchilan, Piedras Negras and Palenque, all traded with settlements along the Usumacinta River at the same time; they competed for territory and dominance in commerce (Fig. 12.1).

The architectural center of Bonampak includes an enormous acropolis which is seated over a natural hill and overlooks the great plaza. The huge plaza of Bonampak is one of the largest of its region, stretching over 91 m (100 ft). Four stelae, inscribed with Maya glyphs, were raised in the central plaza; three of the stelae are carved with images of different rulers. The principal structure on the acropolis is a small building, Structure 1 called the Temple of the Murals. There are also two smaller platforms that support smaller buildings on either side. The central stairway of the acropolis leads down to the great plaza. The famed murals were found in the small three-room building on the first level of the acropolis. The most treasured discovery at Bonampak was the incredibly well preserved frescoes decorating the three rooms of Structure 1 in the Main Plaza. The only monumental painting to survive from the late Classic period; the finest example in the Maya world.

M. Dolan, *Decoding Astronomy in Maya Art and Architecture*,
Springer Praxis Books, https://doi.org/10.1007/978-3-031-61092-9_12

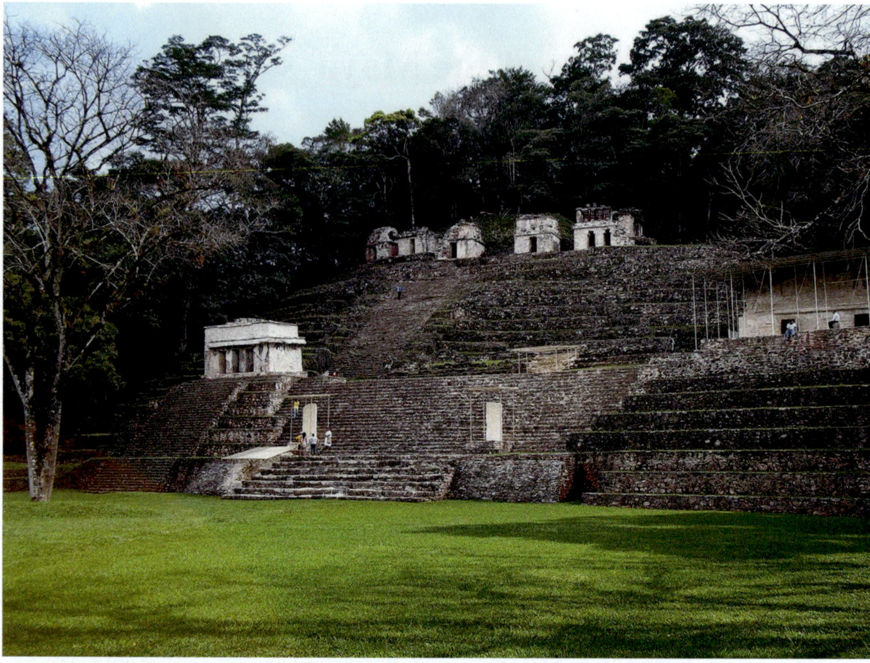

Fig. 12.1 Bonampak temples Structure 1, Temple of the Murals (Credit SL Photography)

Bonampak Murals

Bonampak means "painted wall" in modern Mayan languages and its vivid paintings gave the ancient city its name. The exterior of the temple was once colorfully painted as well with bright colors of Maya blue, red and green but its decor is now totally eroded. The exterior frieze contained more than 100-glyph text as well as paintings on the outer walls. The murals were first known to the outside world in 1946, after which they were photographed and accurate color copies were drawn. Efforts were made to preserve and protect the paintings and the architecture but both have deteriorated since that time. The Bonampak murals overturned the previously held theories that the Maya were primarily a peaceful culture and only sacrificed warriors in association with the ball game. The explicit murals depicted scenes that told a much bloodier tale of their war activities. Researchers previously thought that the Maya only began capturing and sacrificing their enemies after association with the more warlike Toltecs. The paintings at Bonampak were accomplished without foreign influence and demonstrate that the Maya were like most of the world populations, not placid and peace-loving.

The Temple of the Murals is relatively small with three doors that have inscribed lintels above them; each door leads into a separate room; each room contains a raised bench. The magnificent murals decorate both the walls and the vaulted ceilings of the three rooms. The narrative of the artworks follows the sequence beginning on the east wall, then south, west and north walls. The timing of the rooms begins with Room 1, then Room 2 (its caption gives the date August 6, 792 but perhaps describes an earlier event) and lastly Room 3, although some scholars dispute that temporal order. The images are divided into three tiers; the center tier often includes descriptive glyphic texts. The planning and execution of the extensive mural required a sizable team of specialists consisting of plasterers, pigment preparers, and calligraphers as well as painters and planners whose superior ability is apparent in the murals.

The informative murals cover the entire room, floor-to-ceiling. The bold colors and vivid actions illustrate key events in the life of the last ruling family of Bonampak, ruled by King Chan Muán and his wife Lady Rabbit. The historic information provided by the murals related not only major events in the reign of Chan Muán but also actions performed before the royal court. The three painted rooms portray 281 human figures and most have glyphic captions; 1/3 of the figures are given names but the rest bear no titles. The murals were not produced by a single painter but a team of expert artists who worked together to create the outstanding results. Each room relates a unique narrative story that is filled with minute details, most valuable to archaeologists.

The frescoes were created in the usual way by applying paint to wet plaster. There are no seams in the paintings which means that each room was painted in a single session since the plaster needs to be wet for the paint to adhere effectively. The paints were colored by using plant and mineral sources to create a great variety of shades, blue, red, yellow, mauve, purple and green. The narrative scenes in murals presented a new and clearer understanding of the official activities carried out by Maya culture not available in other archaeological sites. They provide archaeologist substantial information and incredible details about the life and beliefs of Maya royal courts. Each room tells a unique story.

The paintings found in the three rooms describe a number of regal events that occurred in 790–792, during the rule of Yajaw Chan Muán, including brutal warfare and acts of self-sacrifice. The upper levels in all three rooms depict images of supernatural beings. The scenes are read from left to right. The murals reveal the important role of music in Maya royal court life since the paintings depict a variety of musical instruments such as log drums, scrapers, gourd rattles and drums made from turtle shells, along with flutes and conch-shell trumpets. Images show that dancing was also an important activity in the ceremonial occasions, accompanying the various rituals (Fig. 12.2).

The first room depicts Yajaw Chan Muan celebrating the occasion of his son being named as heir to the throne, shown as a small child sitting in his hand. The scene shows three brothers, who are performing a ritual dance. The central dancing figure is the king's heir, although some historians have different interpretations. The king himself stands among a group of white-robed dignitaries, all adorned with elaborate headdresses. The artwork depicts the scene of tribute, formal dressing, dance, mummery and musical activities. Some experts suggest that the major scene of Room 1 may instead show an exhibition of wealth at either Bonampak or at Yaxchilan where dignitaries received necessary gifts. On the vault of Room 1 above

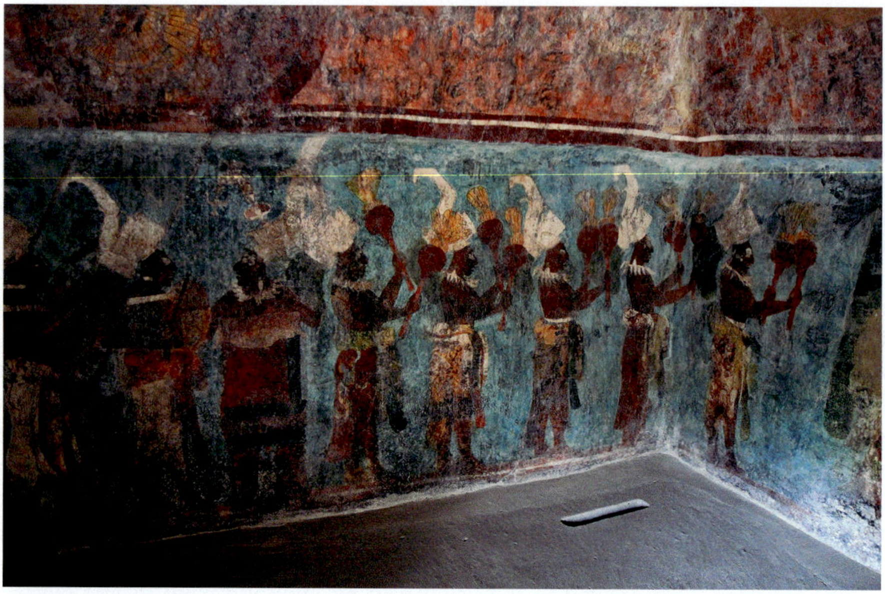

Fig. 12.2 Bonampak murals, musicians with instruments (Credit Anton Bkrd)

the scene, the different faces of the Sun God are depicted as it transits across the sky; while on the east wall a large crocodile, a symbol of the heavenly dome, carries the many avatars of the Sun God on its back.

Room 2 illustrates two separate scenes, a depiction of the actual battle, a scene of violent conflict, and then the exhibition of the captives. Called the most important battle scene of all Maya art, Room 2 is largest of the three and pictures 139 human figures, many more than the other rooms. The south wall illustrates warriors fighting with intertwined bodies on the lower and upper tiers. The scene celebrating the battle is about 15 ft long and 9 ft tall; the figures are clad in full regalia. Victors strip the losers bare and drag them by the hair. The next murals illustrate the aftermath of the warriors raid on a nearby center (Fig. 12.3).

The ruler Yajaw Chan Muan is dressed in full battle regalia and is addressed with the longest glyphic caption above, while nobles of the royal court stand in attendance. The victorious king is pictured standing with captured enemy prisoners weary and prostrate at his feet. The victors carry their weapons and wear clothing with feline elements, especially the jaguar. The ruler holds a jaguar pelt-covered spear in his right hand and wears an elaborate jaguar headdress, a jaguar jerkin and jaguar boots. The tortured prisoners are exhausted, nude and depicted with blood running from their fingernails that have just been removed. They await further punishment; their ultimate fate would be sacrifice to the gods The narrative scene is purposely slanted to honor the victors and humiliate the losers.

At the upper level of the sloping vault of the north wall of Room 2, above the prisoners, are four large cartouches holding animal and human-like figures. They

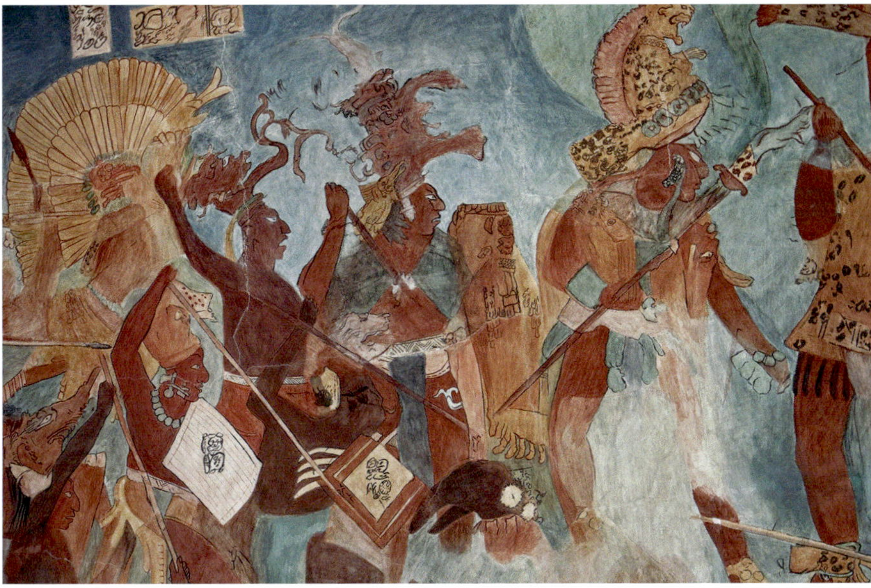

Fig. 12.3 Bonampak murals reproduced, battle scene with warriors (Credit Leon Raphael)

are placed above a sky band; this celestial connection indicates that they are symbolizing heavenly objects. The leftmost objects are two peccaries with star glyphs around them which may be linked with Jupiter. The two peccaries have also been associated with the two bright stars of Gemini, Castor and Pollux.

The rightmost image represents the turtle constellation; it carries three stars on its back and is clearly illustrated in the cartouche. Experts say the turtle is actually the double-headed turtle that cracks open for the rebirth of the Maize God. The turtle has been identified as part of the constellation of Orion. Because the three stars on the turtle's back are similar to the belt of Orion. Further evidence affirms that both the peccaries and the turtle were depicted as constellations.

The two images between the astral images show human-like figures with nearby star glyphs. One figure holds a star in his left hand and a different object in his right. The other figure seems to supporting himself with his left hand and throwing a dart or weapon with his right. These glyphs and signs appear to imply the influence of the zodiac and astrology. The date given marks the time when Venus was in inferior conjunction and the date of the second solar zenith event at the site. Another date written marks the midpoint of Venus' disappearance interval around its inferior conjunction, several days before the solar zenith. Perhaps these dates refer to Maya astrology that may have been considered when timing the battle that was being celebrated.

Room 3 depicts the ritual celebration scene, showing bloodletting by Maya nobles; it consists of 65 human figures. The most striking feature is the depiction in the upper level of three elegantly attired individuals, the three royal brothers also seen in Room 1. They are decorated with tall, green quetzal-feathered headdress and dancers' wings. The central figure, the heir to the throne, holds the heart of the sacrificial victim or war captive below who is dragged down the pyramid steps by two assistants. The mural depicts seven more dramatically dressed participants with similar headdresses to the three brothers. The upper tier of the east wall depicts the elite women holding stingray spines to their tongues performing a bloodletting ritual. The scenes include musicians lifting their instruments upward as the form a procession. The scene demonstrates a ritual dance with observers and ritual bloodletting. Noble women sitting on a bench watch the male dancers. The victory celebration shows the events that takes place after their successful campaign. Art historians continue to study the amazing murals as more scientific tools become available.

Also at Bonampak, Stela 1 shows a date recorded when Jupiter had been in retrograde motions for 3 weeks. Visible above the eastern horizon at dawn, Jupiter was quite bright because it was in conjunction with Uranus, which was bright enough to be viewed with the naked eye at that time. At that same time, Saturn was invisible in conjunction with the Sun.

Bonampak's Final Decline

The Bonampak murals offers an in-depth view of Maya court life with intimate details. No other known artwork depicts such detailed characteristics nor provides a reliable resource for understanding ancient Maya society. The extensive ruins of Bonampak have not been totally excavated, much of the archaeological site is still concealed deep beneath the Chiapas rainforest. By 830 construction at Bonampak stopped but monuments built just before the collapse showed no iconographic or stylistic adjustments. Leaders tried to reorganize the political organization but there was soon a political collapse. Such lavish representation as the Bonampak murals, however, were not enough to save the dynasty from eventual political collapse, which was followed later by the departure of the overlords around 810. The city was completely abandoned soon after and the murals were never completed. The spectacular royal ceremony was left unfinished, as if the artists had laid their brushes down in the midst of their work and just walked away.

City of Tula, Hidalgo, Mexico

History of Tula

The ancient city of Tula is located southwest of the Mexican state of Hidalgo and northwest of Mexico City. Unlike the Yucatan which has no lakes or rivers, the area around Tula has three constantly flowing rivers, chiefly the Tula River. The earliest settlements excavated in the Tula area date to around 400 BCE and the initial settlers were probably people of various ethnic backgrounds, later residents were Toltecs. It is possible the site was under the political control of the most important Mesoamerican city of that time, Teotihuacan. The earliest urban settlement contained only small public architecture and was considered to be of only minor importance. After the fall of Teotihuacan around 550 CE, surrounding city-states began to grow and take control of the smaller communities in the area. Tula too began its rise in importance but never equaled Teotihuacan in its vast settled area or huge population, that had reached 100,000 at its peak; nor was it equal to its Mexican successor, Tenochtitlan, the capital city of the Aztec Empire which also reached a population of at least 100,000 people. Tula was thought to be the city Tollan that the Aztec wrote of and so greatly admired.

From CE 650 to 750 Tula covered a small area known as Tula Chico. The city expanded greatly during the next phase, from 900 to 1150 and saw construction of Tula Grande, the major urban epicenter. Tula's occupation site stretched to over 14 km (8.6 mi) with its maximum population reaching 60,000 with another 20,000–25,000 in the surrounding areas. During the Early Postclassic period, Tula was the only major city in Central Mexico; other large cities of southern Mesoamerica had become depopulated around the tenth century and most of the area surrounding Tula was rural. Tula was well situated for success, located on a vital trade route, near an obsidian mine, and within a fertile agricultural area.

Between CE 900 and 1150 the Toltecs attained the height of their power and influence; Tula served as the capital city of the Toltec area and dominated much of central Mexico as well as adjacent regions to the north and the west. During the earlier period their buildings were mainly small, closed spaces where rituals were enacted by only a few elite people. But later the architectural style at Tula changed; they began to developed large, open spaces for ceremonies to be viewed by many participants. The city also acquired defensive fortifications and its political influence extended over most of the present state of Hidalgo. Tula's economic and political importance in the larger Mesoamerican world system was substantial. Although its realm was never considered an empire, Tula played an essential role in Mesoamerican history with international prominence, influencing societies far beyond its local area and attaining great wealth. Because of their successful military conquests, part of the city's riches may have come from tribute and looted treasures. Producing desirable crafts and ceramics and fashioning obsidian tools was the occupation of a sizeable number of Tula's population; other sectors were the ruling class, the merchant class and farm workers. The people of Tula enjoyed the advantages and rewards of extensive trade routes; traders were able to transport valuable turquoise, mined in the American Southwest, all the way to Chichen Itza. Although far-ranging, the trade network of the Toltecs was not as strong; its trade goods were not as varied as the ancient Maya. Profits were not as plentiful as those earned by merchants of Chichen Itza, since the Toltec's trade items were limited to objects made from obsidian, ceramics and textiles.

The Early Postclassic eras have not been studied nor published as extensively as most other time periods of early Mesoamerican history. These periods are the most poorly understood of the entire Pre-Hispanic past. No detailed archaeological map of the site has been comprised as yet. Chronologies of those eras are not well established, ceramic sequences are only marginally researched, and thoroughly excavated sites in the region are rare. In addition, few radiocarbon dates have been firmly established and the results attained are not strongly supported with archeological confirmation. Even though Tula was the most important city of its era, archaeological research and study of Tula's ruins has not been as extensive as most other Mesoamerican capital cities. The city did have an orthogonal grid layout in its initial design but it may not have reached the entire extent of the city. It underwent major reorientations through time when a new grid orientation was established. Similar to Teotihuacan, Tula's city design was astronomically aligned at 17° east of true north, but its alignment was moved toward true north between 700 and 900 CE. Unfortunately, the few studies conducted at Tula do not include its archaeoastronomical alignments.

Architecture at Tula

Excavations at Tula have shown that although less significant and less socially complex than other cities, Tula had an impressive urban center with varied residential architecture and workshops; the plaza was large enough for 100,000 people. Tula's ceremonial center has steep slopes on three sides, providing a defensible position. This great central plaza was surrounded by a variety of monumental buildings. On three sides of the plaza are long meeting halls that had ceilings carried by columns that were directed toward the plaza. The area has over 1000 m (3280 ft) of stone benches. The benches are decorated with stone carvings of warriors and groups of people walking in long processions (Fig. 12.4).

The chief structures excavated at Tula are its two temple-pyramids, two main ballcourts and several large civic buildings facing the Tula Grande plaza. Building 1 or the Palace of Quetzalcoatl sits east of Pyramid B but survives in ruinous condition. Building 3 sits at the north side of the plaza and is one of the largest colonnaded halls in Mesoamerica, it is enclosed with an arrangement of columns throughout the inner space that would have supported a roof. The columns would divide the interior for processions or other ceremonies. Many buildings at Tula were constructed of mudbrick adobe; these structures do not survive as well as the durable stone ruins seen in most Mesoamerican cities. The most impressive of the remains at Tula is the Pyramid of Quetzalcoatl, also called Pyramid B, which is fronted by many rows of columns that face the central plaza. Building C, also known as the Burnt Palace, sits in front of Pyramid D. A fire transformed the original mudbricks of the building into baked bricks. Recent studies suggest that the building burned after its monumental sculptures had been removed and the Toltec civilization had fallen. The fire probably took place after the site was abandoned around 1150. A serpent wall, called Coatepantli, surrounded the pyramid (Fig. 12.5).

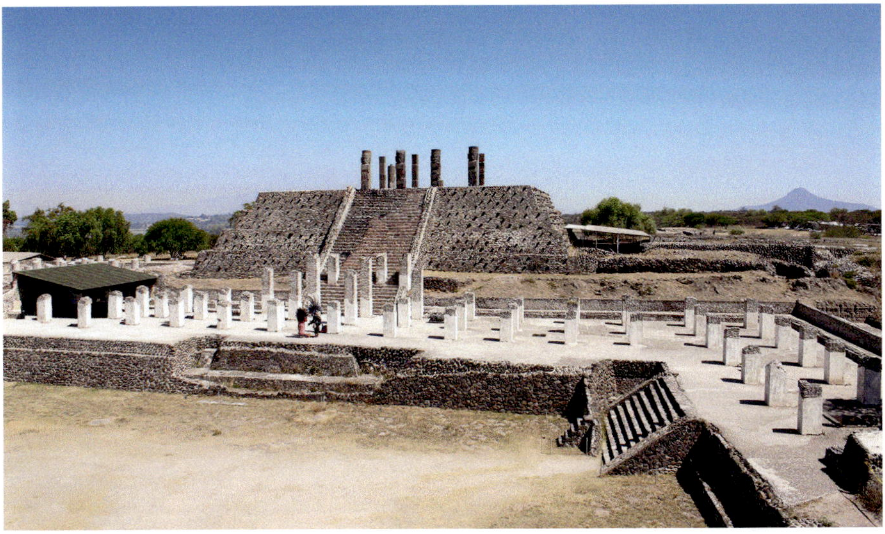

Fig. 12.4 Toltec city, Tula, Palace of Quetzalcoatl with Atlantean statues, tenth century (Credit FrentN)

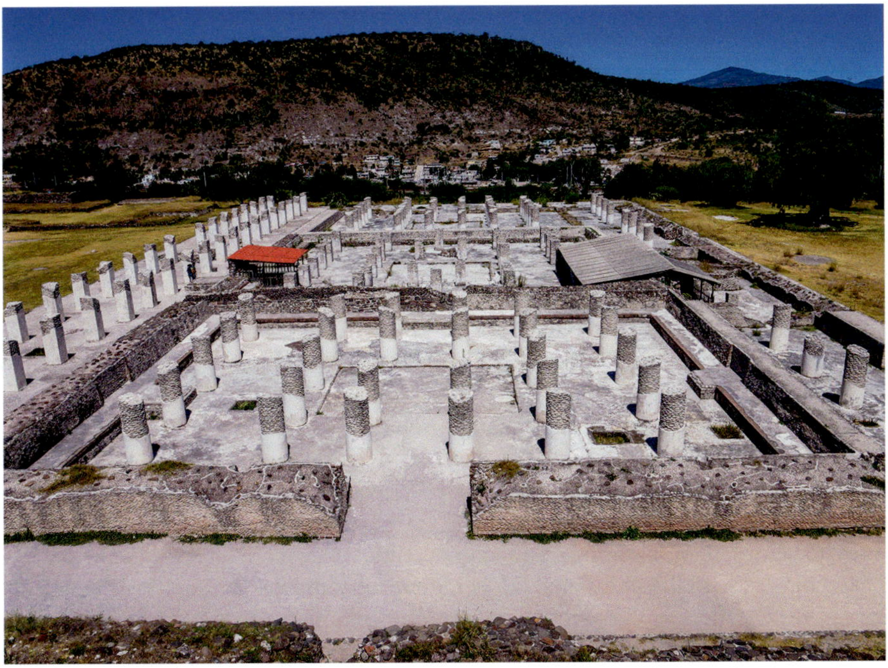

Fig. 12.5 Tula, Burnt Palace, burned around 1150 after abandonment (Credit Scott Yellox)

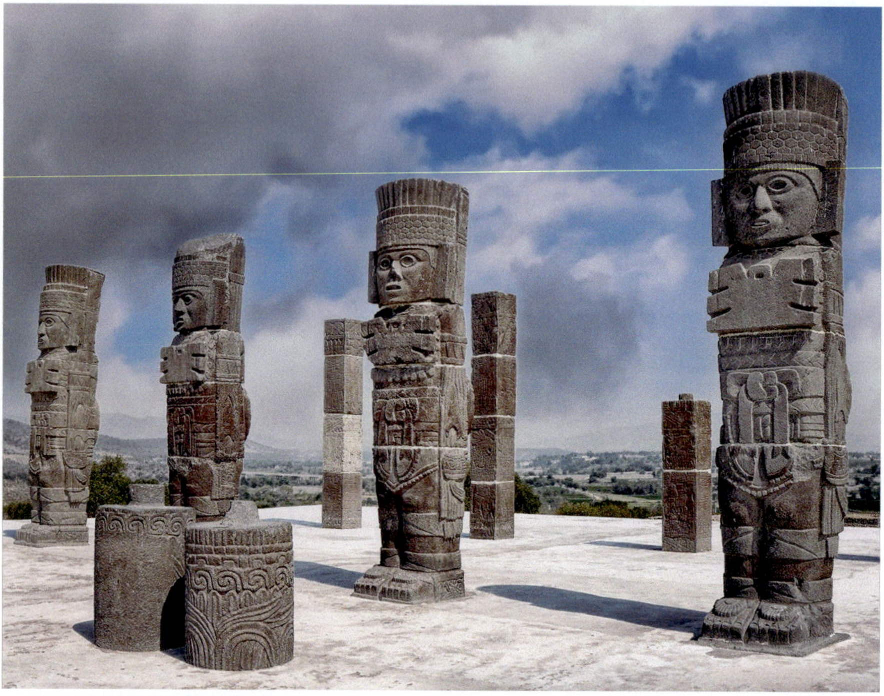

Fig. 12.6 Tula, Atlantean warrior statues, basalt (Credit Madrugada Verdeig)

A few archaeological excavations have uncovered some statues, stelae and art objects; including the atlantean statues that depict Toltec warriors dressed for war. Much more research must be done on the ancient city and it needs to be analyzed in context with other contemporary cultures, but these too have not been well studied. No written records survive explaining details of Tula's political or religious systems, nor its trade networks. The scribes of Tula probably had a hieroglyphic writing system, but there is no extensive evidence, only a few glyphs appear on structures. There are hopes that many more Toltec inscriptions and artifacts will be unearthed eventually. Tula's detailed history is yet to be written (Fig. 12.6).

The Toltecs were admired as excellent sculptors and artists and created numerous exemplary monuments and stone carvings. Ruler portraits carved on stone slabs that were so prominent during the Classic Period were attempted at Tula but were soon eliminated. In addition to the Atlantean statue, the Toltecs initiated other sculptures to adorn their buildings and monuments, such as standard-bearer statues, seated figures with flagpoles, and the enigmatic chacmools. Within the urban center carved stone reliefs with depictions of warfare, sacrifice and warriors were prominent subjects. The earliest interest in depicting decapitated skulls, bones and skeletons appears at Tula. Other gruesome pictorial images commonly seen at the site are serpents eating skeletal coyotes and eagles eating human hearts.

Toltec Warfare

Reputed to be fierce warriors, they were continually active in promoting warfare by their impressive attire. Toltec warriors were outfitted with imposing headdresses, chest plates, and padded armor; they wore a small shield on one arm. Warriors carried short swords and atlatls, weapons that could hurl darts at high velocity. Some held a curved blade that appeared to be a combination club and ax. They were also known as religious warriors who encouraged and spread the cult of Quetzalcoatl, their feathered serpent god, throughout the region. Their military was divided into units named for animals, such as jaguars, and for gods, including Quetzalcoatl and a dark deity named Tezcatlipoca.

Like most Mesoamerican cultures, the Toltecs also practiced human sacrifices to honor their gods; the presence of chacmool figures and the *tzompantli*, where skulls were displayed, provides evidence of this frequent activity. The earliest interest in the public exhibition of decapitated skulls and skeletons seems to appear among the Toltecs. The remains of a T-shaped platform, a *tzompantli* were discovered at Tula. The *tzompantli* had been carved with multiple rows of carved relief sculptures depicting skulls and bones. Recently decapitated human skulls would have been displayed atop the platform. The skull rack found at Tula was probably constructed during the twelfth century and may be contemporaneous with the one that survives intact at Chichen Itza. Like the skull rack at Chichen Itza, it was also adjacent to one of the city's ballcourts.

The Enigmatic Chacmool

The inscrutable stone sculpture called a Chacmool emerged late in the history of Mesoamerican artworks, probably during the ninth century. The unusual statue was created initially in the Toltec area of central Mexico and in the Yucatan at Chichen Itza. Later it spread through Guatemala, El Salvador and even Costa Rica where one was found dating to about CE 1000. Always male, the distinctive figure appeared fully formed during the Terminal Classic period; it did not evolve slowly from another or earlier sculptural prototype. Archaeologists have discovered 12 Chacmools in Tula and 16 in Chichen Itza, the largest number at any site. Their popularity or value continued through time since Chacmools were also used in the Aztec capital of Tenochtitlan until the sixteenth century, but they were unknown during the Classic period in the important cities of Teotihuacan or Tikal. When encountering one, whether atop a pyramid, sitting among the ruins, or in a museum, the Chacmool commands one's notice.

The figural image reclines uncomfortably on a raised platform while supported by his elbows holding his knees tightly bent and feet flat on the ground. His head turns at a 90° angle from the body, always looking away; his raised hands appear to hold a bowl or a flat disk that sits upon the figure's stomach. There are slight

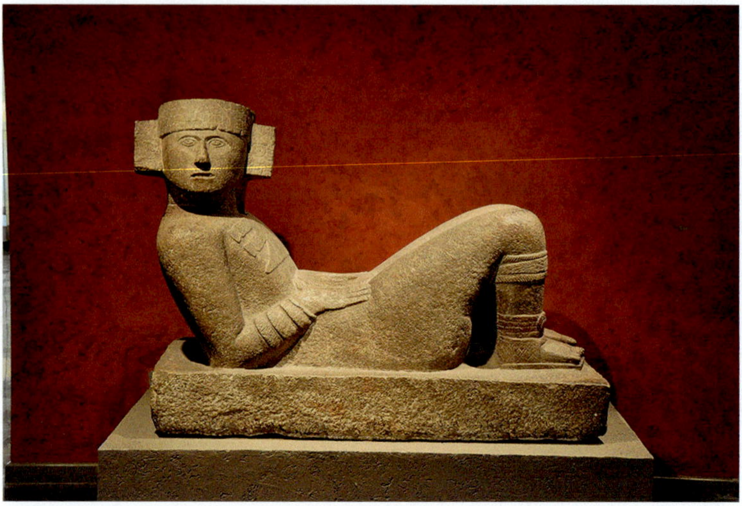

Fig. 12.7 Chacmool, Aztec, Museum of Anthropology, Mexico City (Credit jejim)

variations in a Chacmool's basic form; some heads face right and others face left. The statue's facial expression seems direct, totally blank; the large staring eyes of some look up rather than straight on, but they show no emotion. The Chacmools may be richly dressed in a colorful costume, wearing a small headdress or sometimes nearly naked. The earliest Chacmools appear to be young men with warrior attributes but the figures develop different characteristics in later depictions (Fig. 12.7).

The Chacmool was not an object of worship in itself but rather used by Maya priests during religious ceremonies. The plate or disk on the middle area of the figure may have served as an offering table to receive sacrificial gifts; or perhaps for receiving blood sacrifices or human hearts. Another suggested use of the Chacmool was as a sacrificial stone upon which victims were laid in a prone position so that their hearts could be removed. The reclining, defenseless depiction of the Chacmool is similar to the position of enemy captives seen in Maya sculpture and painting during the Classic period, but their true function, indigenous name and origin are still questioned by scholars and remain in dispute. Who were these reclining figures? What was their function? Do they represent generic males or a specific individual, an enemy, a noble or lord, a deity or supernatural spirit?

Most experts agree that Chacmools were connected with ritual sacrifices to war and storm deities, such as Tlaloc and Chac, and with water imagery, also with human heart sacrifices offered to these deities. The latest theories favor a Classic Maya origin for the form of the Chacmool which took place at Maya Chichen Itza instead of Toltec Tula. The preferred theory now accepted by many Mayanists is that Chacmools represent sacrificial victims who are captured in war. The odd, uncomfortable position that the statue holds imitates that of the bound captive who is often shown beneath the feet of a Maya ruler on stelae and in paintings. But other experts argue that there is no indication of binding cords, torture or humiliation on any Chacmool yet discovered. The sculpted figure shows no fright or shame but is calmly posed with an open, vague stare into the distance. Other scholars suggest the unusual pose is associated with the sacred ritual ball game and the sacrifice of "vanquished captive ballplayers" who, after the game, have their hearts and heads removed. Scholars will continue to investigate and hypothesize but will the true function and meaning of the Chacmool every be completely understood?

Tula and Chichen Itza

The Toltecs of Tula and Maya of Chichen Itza had a well-established relationship but its true nature is not clearly understood. Arguments associated with the aspects of their connections date back to the 1890s. When examining the long history of Chichen Itza and Tula, it appears that many pieces of this puzzle are missing. The exact connections that existed between the Toltecs and the Maya of Chichen has not been established and remains a hotly debated problem. The two societies, though not close geographically, show numerous similarities. Not only are their cities' architecture comparable, they shared their common devotion to Quetzalcoatl or Kukulkan and their trading routes. A close exchange of ideas is seen especially in the high degree of similarity in the design not only of their architecture, but in their sculptural programs and their choice of artistic décor. Their similar assembly of columns in colonnades, atop pyramids and inside halls and the presence of benches are other examples. Which culture developed these features first is an unanswered question. The discovery of colonnades, columned halls and skull racks at other sites to the north that may be even older has deepened the debate. But each culture needs to be better understood in itself before scholars can understand the relationship and interpret the influences that existed between them.

Through the years of research at Tula and Chichen Itza, many different theories have been raised, analyzed and accepted but then as new evidence and understanding develops, they are often rejected. They were the two largest cities of that era; although 800 mi apart, they seem to mirror each other in many ways and cannot be understood in isolation. Other cities that flourished during that key transitional period have not been thoroughly studied; these missing pieces are also needed to clarify their true connections. Scholars need more basic information on each of the archaeological sites and on other contemporary cities that interacted in that area.

Tula's Final Decline

During the twelfth century Tula began to have problems with the fertility of the soil, with drought and the changing climate. The tragic decline of the Toltecs occurred around 1150 CE when Tula was sacked and completely burned. Little has been discovered about who attacked the city or why it was totally destroyed. The Toltec's reputation for aggression and hostility may have incited an attack of retaliation or it may have been caused by civil war or internal struggles. Tula did not survive as long as most Mesoamerican cities but it produced a great quantity of sculpture and finely hand-crafted artifacts.

Even after Tula had been destroyed and looted of its treasures, the site was continually occupied throughout the Mesoamerican period, chiefly by artisans, stonecutters, masons, potters and scribes. Though greatly weakened, Tula continued to be an important influence on the Aztec Empire, for its history became part of Aztec mythology.

When the Aztec Empire came to power, long after the fall of the Toltec civilization, the Aztecs claimed to be descended from the elite rulers of the city called Tollan. The Aztecs described Tollan as a mythical paradise where its people were masters of learning, science and art. Aztec leaders adopted many of the traits attributed to the Toltec culture, attempting to imitate their success, mainly the worship of Quetzalcoatl and the practice of human sacrifice. They even copied Tula's central layout design for their own city. Because the Aztecs admired the Toltecs so highly, they removed many sculptures and artworks from Tula's ruins for their city and then copied and installed their artistic designs. Through the centuries local thieves too looted valuable artifacts from the ancient site. By borrowing their military techniques, artistic themes and political systems, the Aztecs hoped they would become as powerful and fearless as the famed Tollan warriors and legitimize their own rule for acceptance by the gods and the people. It has always been assumed that Tollan and Tula were one and the same.

Recently, researchers have examined whether Tula was actually the legendary city of Tollan that the Aztecs revered so highly. They have suggested that Tollan may have been a concept that referred to all of the great civilizations that preceded the Aztecs, not just one culture. Was Tollan the single city of Tula or an accumulation of past successful civilizations? The Aztecs attributed the spectacular Tollan with extraordinary accomplishments, the development of art, language, myth, militarism and rulership; they included the discovery of medicine and the calendar system. Of course these were long established before the age of the Toltecs. For legitimacy Aztec kings tried to trace their descent from Toltec kings and marry Toltec women. Archaeologists contend that the material remains of Tula are not equal to the splendor and majesty that the Aztecs ascribed to them in colonial documents.

The debates on the history and influence of the Toltecs of Tula will continue until more facts are established. Discussions of their true influence on the Aztecs and other Postclassic cities will persist. The long argued relationship and connections of Tula with its sister city Chichen Itza will also endure. Current opinions prefer an explanation of interregional interactions between the two prime centers with idea exchanges, innovative political systems and shared symbolism in their constantly evolving traditions.

The Worship of Quetzalcoatl – Kukulkan

Some scholars suggest that the name Quetzalcoatl may not refer only to the Feathered Serpent god but may have also been linked to a real person, called Acatl Quetzalcoatl, who became a legendary ruler of Tula. They propose that this person was later idealized gaining the attributes of a god. Myths connected to the fall of Tula tell that this legendary hero became involved with a dark deity named Tezcatlipoca and evil entities who caused the destruction of Tula. The disgraced ruler then wandered along the Gulf Coast, sacrificed himself, and after death he became metamorphosized into the planet Venus, glowing as the Morning and Evening Star. Legends say that Ce Acatl Quetzalcoatl, the fallen hero, predicted that he would return one day to resume rule over the people. Some believed that the death and resurrection of Quetzalcoatl re-occurred every time Venus disappeared in the western sky and returned to life in the predawn. The distinction between the historical figure named Quetzalcoatl, and the Feathered Serpent deity has become blurred through time. Legends also say that around the tenth century a man named Quetzalcoatl or Kukulkan, who was thought to be a leader or a priest, appeared at Chichen Itza as well. Since there are no evidential records or inscriptions to prove the stories, the identity or reality of the human, historic Quetzalcoatl heroic figure will remain a mystery.

The Toltecs felt great devotion to the Feathered Serpent and honored the god with temples, sculptures and religious ceremonies. The concept of devotion to Quetzalcoatl was not created by the Toltecs; they were not the first people in the Mesoamerican world to revere the god's power. The serpent god, whose neck is covered with the feathers of the sacred quetzal bird, held a long history in various societies of the Mesoamerican world. Images of the feathered serpent have been discovered even among the artworks of the ancient Olmec civilization; the earliest image was found at La Venta on Stela 19, dated to around 900 BCE, the Early Formative period. Manifestations of Feathered Serpent imagery continued into the Middle and Late Formative eras, but the attributes of the god may not have been the same as those of the initial eras. The Plumed Serpent was a common component in the iconography of Mesoamerican artworks but is most often found decorating the architecture of important cities (Fig. 12.8).

The first major city where the symbol of the Feathered Serpent became a vital religious and political icon was Teotihuacan. At Teotihuacan the Feathered Serpent ranked among the highest in its large pantheon of deities and seemed to act as both an earth and a water deity, closely aligned with the rain god, Tlaloc. The widespread appearance of Feathered Serpent imagery at Teotihuacan is well documented and reached its height in the numerous tenoned serpent heads decorating the Pyramid-Temple of Quetzalcoatl, begun at the end of the second century CE. The pyramid was a prominent construction on the central thoroughfare of the city, called the Avenue of the Dead by the Spanish. The three-dimensional stone heads of the Feathered Serpent project from the balustrades and side of the Teotihuacan Pyramid of Quetzalcoatl, adding more depth and power to the architectural design. Only one

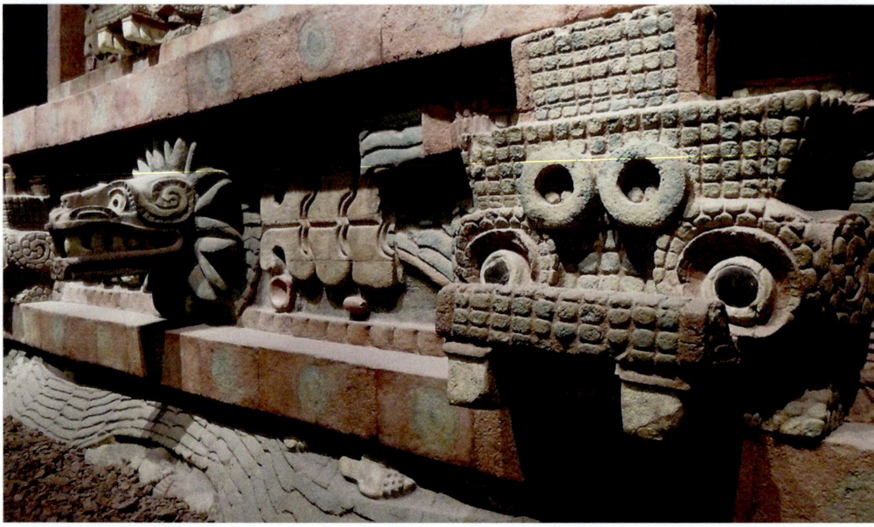

Fig. 12.8 Masks of Quetzalcoatl and Tlaloc, Temple of Quetzalcoatl at Teotihuacan (Credit Fabio Imhoff)

face of the Teotihuacan pyramid has been excavated; if the other three faces were decorated in exactly the same manner, the number of Feathered Serpent heads would equal 260. This total could have represented the 260 days of the ritual calendar. Experts suggest that a marker of some sort may have been placed in the small gap between the fangs of the serpents indicating the appropriate day of the year. At Teotihuacan, Tula and Maya cities, the feathered serpent icon was associated with the planet Venus, and especially because of its association with the beginning of the vital rainy season. Venus was also symbolically connected to warfare in all three cultures. After the fall of Teotihuacan which occurred around 550 CE, the cult of the feathered serpent god continued in the many new religious and political settlements of Central Mexico and later spread to the Yucatan peninsula as well.

Dualism is a very common quality attributed to Mesoamerican gods. The dual nature of the feathered serpent is appropriate for the double abilities of the deity; combining its supernatural traits, such as the ability to fly to the heavens, and its natural traits, the ability to creep on and beneath the ground. The dual features of the hybrid animal, combining the dominant aspects of the bird and snake, symbolized the double nature of the great Mesoamerican god. The feathers that surround the serpent's head also represent fertility; they imitate the petals of flowering crops, as well as beams of light that link the serpent god to the Sun. For the Toltec, Quetzalcoatl was also associated with the wind, one of the most powerful forces of nature that would bring rain for their crops. The god was also linked with rulership and priestly office. Some experts say that it was also a symbol of fertility, political structures and connected with agricultural deities.

During the Postclassic period, feathered serpent iconography spread significantly throughout Mesoamerica. Called Kukulkan by the Maya of the Yucatan peninsula, many temples honoring the feathered serpent were built at Chichen Itza, Uxmal and Mayapan. Among the Maya the iconography of the serpent becomes the embodiment of the sky itself; a vision serpent who could present the Maya kings with images of the Underworld. Feathered serpent images and cult devotion became dominant throughout the Maya areas. Worship of the same deity appeared to assist communications and trade transactions among people of diverse geographical areas and of differing ethnic backgrounds. The Maya's worship of Kukulkan had its origin in earlier Maya traditions but was deeply influenced by the strong devotion to Quetzalcoatl at Tula.

After the decline of Chichen Itza, the hub of the revived Kukulkan worship transferred to the nearby Postclassic city of Mayapan where they built temples adorned with serpent columns that honored the Feathered Serpent. Devotion to Quetzalcoatl continued into the fourteenth and fifteenth century in Central Mexico as well. The Feathered Serpent god was considered a highly honored deity in the Aztec Empire, one of several revered gods in their fierce pantheon, including Tlaloc (god of rain), Tezcatlipoca and Huitzilopochtli. Quetzalcoatl was associated with Venus in the Aztec world, and with Tlaloc and Xoloti as well. The god held additional influences with the wind, the Sun, merchants, arts, crafts and knowledge. Images of Quetzalcoatl appeared in the literature of Aztec codices, in their relief sculptures and in their paintings. Artists sometimes added human features when illustrating the double-natured deity. Even after the fall of the Aztec Empire, the legends surrounding Quetzalcoatl continued to play a significant role into the colonial period.

References and Further Reading

Coe, M., & Houston, S. (2011). *The Maya* (9th ed.). Thames and Hudson.

Lounsbury, F. (1982). Astronomical knowledge and its uses at Bonampak, Mexico. In A. Aveni (Ed.), *Foundations of new world cultural astronomy*. University Press of Colorado.

Miller, M. (1986). *The murals of Bonampak*. Princeton University Press.

Smith, M. (2017). Tula and Chichen Itza: Are we asking the right questions? In *Twin Tula*. https://www.academia.edu/12384659/TULA_and_CHICHEN%20ITZA

Taube, K. A. (1994). The iconography of the Toltec period Chichen Itza. In H. J. Prem (Ed.), *Maya archaeology northwest Yucatan Peninsula*. First Mahler Symposium.

Part V
Maya Postclassic Cities

Chapter 13
Postclassic Cities of the Yucatan Peninsula

Yucatan Peninsula

The Yucatan Peninsula is recognized geologically for the huge impact crater formed by an asteroid that struck the area over 66 million years ago and is now buried beneath it. Called the Chicxulub Crater, it measures about 180 km (6 mi) in diameter and reaches 20 km (12 mi) in depth. The colossal impact of this asteroid produced a worldwide blackout and freezing temperatures that were responsible for the mass distinction of the dinosaurs as well as numerous other living species.

In spite of its challenging and highly varied environment, the Yucatan Peninsula served as the most significant inhabited territory for a long portion of the over two-millennia history of the Maya civilization (Fig. 13.1). The tropical forests that permeate the southern and central peninsula are lush with flora and fauna encouraged by frequent rains, while the northern areas are much dryer with relatively low vegetation. In the northern lowlands, the year is divided by a dry season and a rainy season that lasts from May to November. Although substantial portions are still wild and untamed, the Yucatan Peninsula was densely populated throughout the Classic Period, the Terminal Classic and Postclassic Periods. Maya descendants continue to live in small villages and large cities and continue to thrive in many areas that are rich with their lengthy ancestral heritage; many still speak the ancient Mayan languages.

The northern peninsula contains many caves and underground water sources that were thought to hold sacred powers. They were also considered liminal spaces, boundaries between worlds and passages to the Underworld and they worked to use and align these powerful sites within their world. The Maya leaders recognized these unseen spiritual forces and built their cities in coordination with them. They also considered the cyclical movements of the celestial bodies and the cardinal directions. By combining these multiple aspects in appropriate and traditional ways,

M. Dolan, *Decoding Astronomy in Maya Art and Architecture*,
Springer Praxis Books, https://doi.org/10.1007/978-3-031-61092-9_13

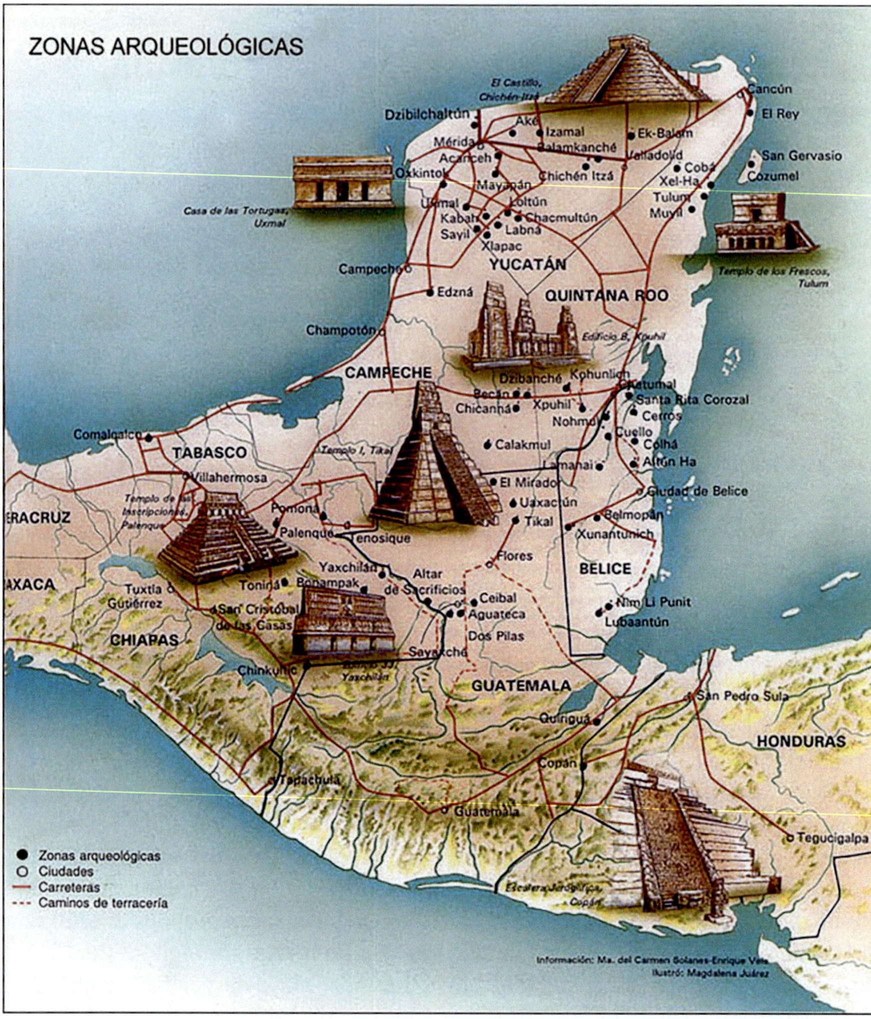

Fig. 13.1 Map of ancient Maya cities and archaeological sites of the Yucatan Peninsula (Public domain)

they designed their urban and open spaces as sacred landscapes, organized to perform religious rituals and to live in harmony with the universe.

Although most excavated sites lie in the northern portion, the central and southern areas were also highly populated at times. The dense jungles have enclosed the long-deserted ruins of many once vibrant cities under a thick blanket of foliage; ancient pyramids appear to be grassy hills that peak through the tall trees. The Campeche state, in the central part of the Yucatan Peninsula, is principally a vast, unexplored area, mostly unknown to archaeologists. A research project of the National Institute of Anthropology and History (NIAH) in Mexico has been to increase the knowledge of this vast blank area on the archaeological map of the

Yucatan Peninsula. The difficulty in accessing these areas had blocked expeditions into the impenetrable jungle where innumerable cities and much of Maya history still lay hidden. In addition to the hazards of inhospitable weather, sweltering humidity, lack of roads or even paths were the snakes and stinging insects, thorns, and poisonous plants. In spite of these countless obstructions experts took on the challenge of penetrating the punishing jungles, seeking knowledge and better understanding of the Maya civilization.

As a result, important Maya cities have been discovered recently by archaeological expeditions, such as those led by archaeologist Ivan Šprajc. His team has discovered several unknown Maya ruins: Chactún in 2013, Lagunita in 2014, and Ocomtún in 2023. In his book *Lost Maya Cities*, he describes the dense, but vibrantly alive jungles:

> An endless green thicket obstructs one's view, hinders movement, and, with trouble and danger lurking at every step, makes one feel powerless. However, the unmistakable smell of moisture, the lush vegetation, the fluttering of birds whose variety and sounds are beyond imagination, the shrieking monkeys and their boisterous acrobatics up in the trees, even the buzzing of insects—all this imposes millions of stimuli that become lodged somewhere deep in one's brain, thereby leaving a powerful impression and stirring a desire to return.

It was not until the early twentieth century that the first interest in the jungles of southern and central Yucatan took hold after being abandoned for more than a thousand years, since the fall of the Classic Maya culture. The initial desire to enter the thick forests was not exploration but instead was spurred by the desire for harvesting chicle obtained from the *zapote* or *chicozapote* trees, abundant in the central and southern lowlands. The sappy gum found in the trees was the basic ingredient of chewing gum. The *chicleros*, or collectors of chicle, hacked paths, created mule trails and set up camps to process the gum or sap taken from the trees. They would collect chicle, boil it, then form it into bricks which they would sell. During their excursions into the thick jungle, they gained thorough knowledge of the territory and became the first to venture upon ruins of totally unknown, deeply covered Maya cities. Their discoveries were spread to authorities and archaeologists and the *chicleros* were a great help in leading investigators to the hidden sites.

The better known archaeological sites of the Yucatan Peninsula include Chichen Itza, Uxmal, Mayapan and Tulum; all are fortunate to contain well preserved city centers and impressive monumental architecture. These major sites have been restored and well researched by archaeological excavations and scholarly studies, but many smaller sites still await scientific investigations.

Sacred Caves of the Maya

Much of the Mesoamerican landscape, especially the Yucatan peninsula is porous limestone where caves, rock shelters and sink holes develop naturally. The use of these abundant caves has been noted by archaeologists for decades but only recently have the activities and rituals that took place in caves been well studied and understood, often by studying spiritual traditions of contemporary societies which may hold traces of customs from the ancient past. For the Maya the word for cave had a

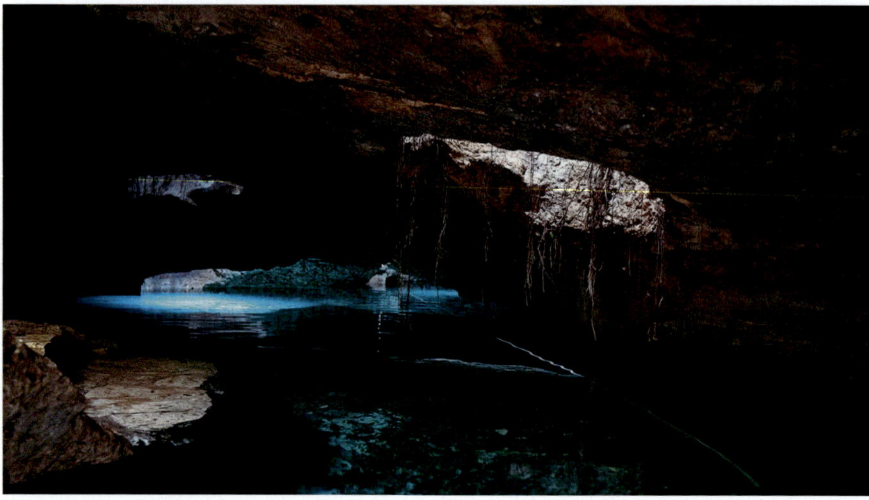

Fig. 13.2 Cave with hanging stalactites near Chichen Itza (Credit Nick Spinder)

much wider meaning; it could also refer to springs, cenote, crevasses, rock shelters or any hole into the earth. The word for cave has even additional meanings; the word was also associated with wind, rain, clouds and transformation. These impersonal forces of nature were personified and considered living sentient beings (Fig. 13.2).

Within the caves stalactites and stalagmites of all sizes form from the precipitation of mineralized water that constantly drips from the ceiling to the floor. Stalactites are icicle-shaped formations that hang from the ceiling while stalagmites are upward-growing, cone-shaped mounds of mineral deposits that capture the dripping water, slowly forming exotic icicle-shaped creations that beautify the interiors. The caves of Chichen Itza developed not only inspirational appearances with their geological changes but accumulated multi-layers of meaning for those who worshipped and offered sacrifices within them.

A basic tenet of the Maya belief system was that the universe was divided into three layers: the sky, the Earth and the Underworld and caves functioned as the entryway or portal from Earth to Xibalba, the Underworld ruled by the gods of death. Caves were often described as entrances to the watery Underworld of the Maya, a place of dread. Caves were connected to both life and death which occur at this crucial zone between the Earth and the Underworld. They were also revered as the homes of the most important gods. Caves were closely associated with Mesoamerican Creation myths, considered birthplaces where humans and the ancestors were created and first emerged on the Earth and where people go to die. When an object or person emerged from the Underworld, it came to life; if it descended into the Underworld, it died. Images were often painted and carved illustrating this action of emerging from a cave into life. Both caves and mountains were essential components of Maya beliefs and among the most sacred elements in Mesoamerican creation myths. Caves were considered avenues of communication with the ancestors.

Some Maya believed that the Sun and Moon were born in the Underworld. Certain deities, in particular Chaac and Tlaloc, were thought to live in sacred caves. Rain deities were thought to live in caves, where rain was formed; Chaac is sometimes depicted sitting in a cave house that acts as a throne for the rain god. The Maize God too is sometimes pictured emerging from the Underworld. Some specialists propose that pyramids were considered as architectural mountains and that the tombs inside were conceived as sacred caves. The entranceways into sacred temple were often likened to entrances into mountains. These doorways were often decorated to imitate the mouth of Earth Monster or *witz*. Temple 22 at Copan is the most important example of the decorative cave mouth (See Chap. 11).

Activities that occurred in caves can provide important information for understanding religious beliefs and practices in early Maya societies. Evidence attained through archaeological studies show that the caves were used for religious rituals for a very long period beginning in the Early Preclassic era (1500–1000 BCE) and lasting through the Late Terminal period. Most activities that took place in caves were for religious purposes; they were often used as places of ritual sacrifices, even human sacrifice, which as we have seen was widespread in Mesoamerica. Caves are still utilized today for traditional religious rituals.

Caves of Balankanche

In the Yucatan peninsula the most important cave complex yet discovered is the Caves of Balankanche, noted for its unusual formations, its long centuries of use, and for the huge amount of sacrificial artifacts found there. Near to Chichen Itza and directly connected by a sacbe are the highly venerated Caves of Balankanche, consistently accessed by the Maya for religious devotions from the Preclassic until the Late Postclassic collapse. The cave complex extends about 1.2 mi through the permeable strata of limestone rock. During the Terminal Classic era the water table was much lower than today which enabled the Maya to travel deeper into the extensive cave for their spiritual ceremonies. Small buildings and platforms had been built by the Maya near the entrance and inside were stairs, walls and altars. The caves had been carefully explored by archaeologists during the early nineteenth century when bones, shells, potsherds and other artifacts had been uncovered (Fig. 13.3).

But in 1959, researchers were amazed to discover a false wall that had concealed a series of passageways deeper into the cave system. The wall was opened and surprisingly revealed an undisturbed and intact Maya worship center that had been hidden for over 800 years. The newly exposed area of caves contained a fascinating variety of natural formations. Even more surprising, within the innermost section of Balankanche a thick, naturally created stalagmite-stalactite pillar fashioned into a form having the identical appearance of the World Tree or the sacred Ceibal of Maya mythology, highly honored by the Maya. The discovery of the cave tree must have seemed miraculous to the people of Chichen Itza. The World Tree was also

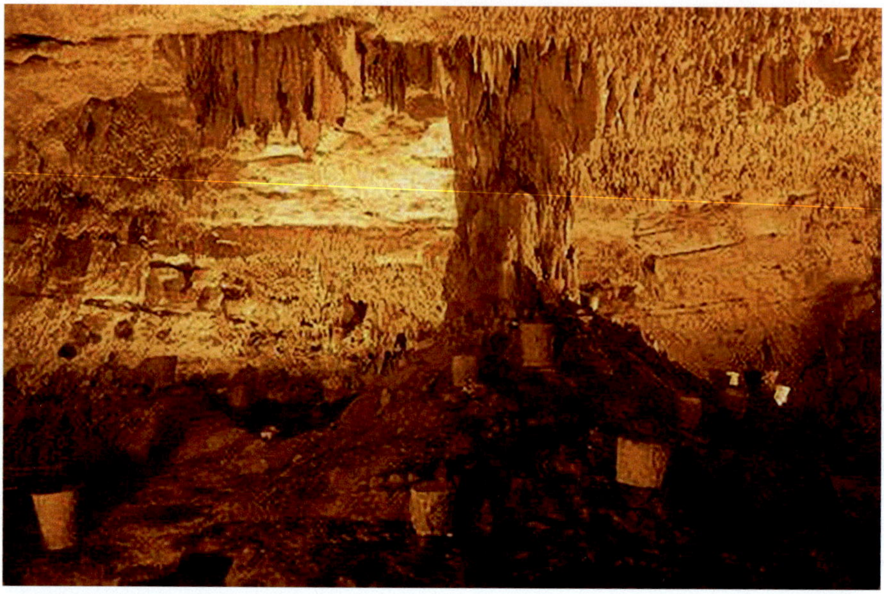

Fig. 13.3 Balankanche Cave with sacred tree and offerings near Chichen Itza (Credit expose)

associated with the world axis, around which the celestial dome turns, an archetypal symbol found in early cultures all over the world.

When initially investigated, more than 150 ritual offerings of ancient pottery, images of gods, and precious sacrificial artifacts still sat around the sacred tree exactly where they had been placed many centuries earlier. The offerings found included jewelry, carved jade, pottery, and stone artifacts. The most significant findings were 70 incense burners, many depicted the rain god, Tlaloc, who they were fusing features of Chaak. The ancient objects were photographed and catalogued so they could be studied, and then were replaced in their original positions. It is not known when or why the Maya decided to wall off the most revered area of the cave. Nor is it known which people actually participated in these sacrificial cave rituals? Archaeologists continue to discover caves which were vital components in the creation myths and spiritual lives of the Maya. Visitors today are able to explore the Balankanche Caves and witness the place of sacred Maya worship for themselves.

Maya City of Dzibilchaltun

One of the early Classic cities, Dzibilchaltun is located in the northwestern section of the Yucatan Peninsula, about 10 mi (16 km) north of Merida, the current Yucatan capital city, not far from the coastal salt-producing area on the Gulf of Mexico. The area was first settled during the Early Formative Period (1000 BCE) and reached its highest development in the years between CE 600 and 900 with a population of 20,000. Its wealth was accessed from its nearness to the coast, utilizing salt, marine life and agriculture. It is thought to be among the larger ancient cities of Mesoamerica. Dzibilchaltun was one of the few Maya city-states to remain in continuous habitation for innumerable centuries.

The surrounding area is one of the driest of the northern lowlands; clean drinking water for the city was supplied by the Xlacah Cenote situated in the southwest corner of the city's main plaza. The Cenote Xlakah was a foremost feature of the site, one of the largest and deepest in the Yucatan. Its surface water is less than 10 ft below the rim. The location of the cenote probably influenced the settlers' choice for the original foundation of Dzibilchaltun. Archaeological investigations of the cenote revealed abundant artifacts deposited in the cenote as sacrificial offerings, such as carved bones, wooden objects and vases in addition to innumerable shards of ceramic vessels and water jars. The cenote provided valuablc evidence of past socictal activities. Experts think that the city was probably the center of a religious cult at some time in its history.

The city and surrounding vicinity were densely occupied, not all sections of occupation have been excavated as yet. The ruins of Dzibilchaltun have not been unearthed and restored as extensively as some Maya Late Classic cities of the Yucatan. More than 19 square km have been mapped but the site stretches out more than 100 km into the surrounding landscape. More than 8000 structures have been uncovered in the central and extended areas, many of these are platforms created to support elite and lower class residences.

The city's name, Dzibilchaltun, has been translated from the Mayan language meaning "the site of stone writing", referring to the many standing stelae at the site. More than 25 stelae have been discovered at this point; many were decorated with royal portraits, Long Count dates and hieroglyphic text. These were details of events in the political and civil life of the city commonly seen on stelae at other Classic cities. On Stela 9 an image of a ruler holding a manikin scepter, a sign of high position, was carved along with the year 849. The stelae were all erected during a time of expansion after CE 800 when almost all of the masonry-vaulted architecture was built as well. The city retains buildings from the Terminal and Postclassic periods which were constructed over top of earlier Classic architecture, encasing them intact for the investigations of future archaeologists. Like most Maya cities, a series of raised roadways linked the different sections and connected it with nearby cities.

The city center was tightly organized with three main plazas and nearly 100 masonry buildings. The central structures included tiered pyramids and vaulted buildings that sat atop low platforms; these probably housed the elite or nobles of the city. In the remaining areas scattered architectural remains of pyramids and vaulted buildings were found. The settlement experienced a decline, a period of near desertion at the Late Formative to Early Classic and again in the Postclassic. The city reached its peak around CE 750 and then the population declined and building activity ceased for a time; a resurgence occurred around 1200 when new buildings were constructed utilizing salvaged masonry. The city was still occupied with a small population through the Late Postclassic until the time of the Spanish Conquest in 1540.

Temple of the Seven Dolls

The most renowned structure at Dzibilchaltun is the Temple of the Seven Dolls; the unusual name was acquired because seven small, crudely-made clay figurines were discovered at the base of the building when undergoing excavations in the 1950s. The pyramid, also called Structure 1-sub and the Temple of the Sun, dates to around CE 700. It is located on the eastern side of the city half a mile from the central plaza and at the eastern end of a long causeway. The temple is most recognized for its atypical architectural form and for its astronomical orientation.

The Temple of the Seven Dolls was not uncovered until the 1950s; it had been filled with rocks and then a large pyramid had been constructed atop it a century or so later, after CE 800. For many centuries the sub-temple had laid hidden inside the encasing pyramid. Remnants of the second building still partly cover the west side of the sub-building. The substructure has a quadrangular form that is roughly aligned to the cardinal points. The structure sits atop two terraces; the width of the bottom or base terrace averages 23 m (75 ft). The temple has four broad staircases with nine steep, narrow steps climbing to the second terrace and eight more steps to reach the top platform. The upper level of the temple sits upon the tiered platforms 15–20 ft tall. The almost square upper building contains a continuous vaulted corridor surrounding a small, raised central chamber (Fig. 13.4).

The building's design sets it apart from any other known Maya buildings in the northern lowlands. The temple is one of the earliest vaulted buildings in Dzibilchaltun; the vaulted ceiling forms a tower above the roof. The upper chamber

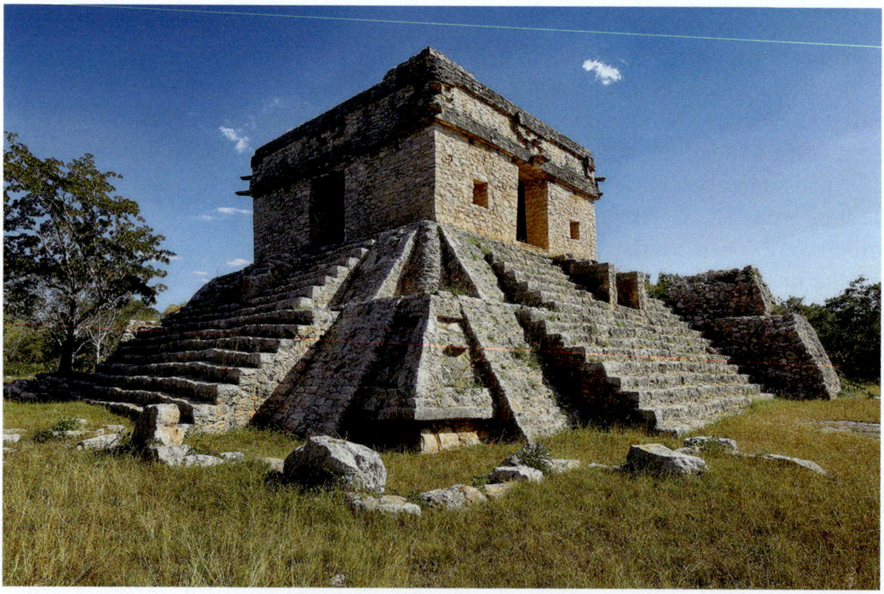

Fig. 13.4 Seven Dolls Temple at Dzibilchaltun, ca. 700 CE (Credit Camera Walk)

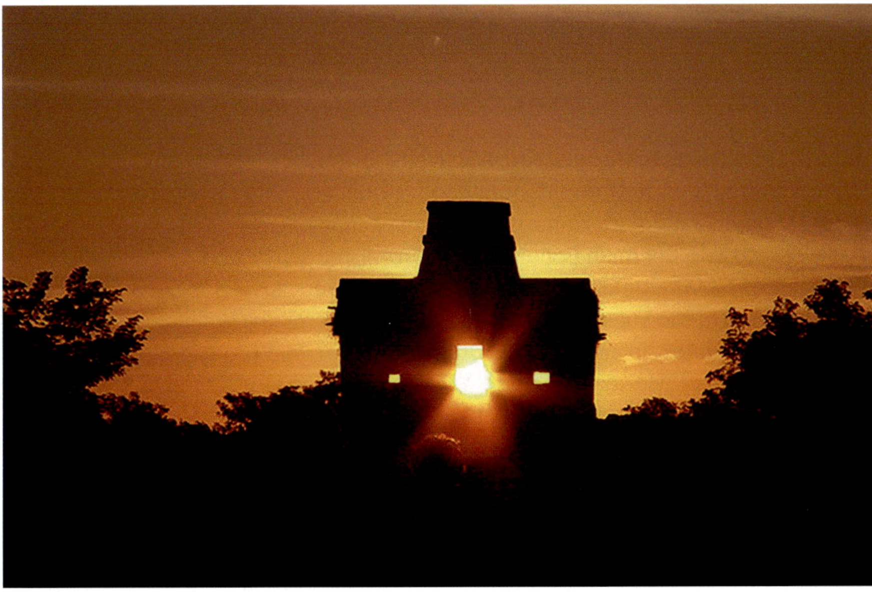

Fig. 13.5 Seven Dolls Temple sunset at spring equinox (Public domain)

of the building can be entered through four wide trapezoidal doorways; each faces a cardinal direction. The chamber's walls are thick but have deteriorated since the builders used mud instead of mortar to join the stones. Small rectangular windows were positioned on each side of the eastern and western doorways; the windows may have been utilized for celestial observations at dawn and dusk; windows are unusual on Maya buildings. The corners on the facades of the upper chamber were adorned with large but unpainted stucco masks of the Rain God Chaac with his lightning axe. There are eight stucco masks on the frieze of the temple and assorted animals (Fig. 13.5).

The temple is oriented to the spring and autumn equinoxes. As the Sun rises on the equinox, the temple door begins to glow and is brilliantly illuminated when the Sun appears entering the doorway. On the equinoxes crowds gather, as they do at Chichen Itza, to observe the Sun's hierophany as it illuminates the eastern doorway of the Seven Dolls Temple, the "Door of the Sun". Specialists believe that the vernal equinox, March 21, signaled the start of the planting season and the autumn equinox, September 20–22, signaled the start of the season for harvesting. Astronomically aligned buildings assisted in the correction of the Maya observational calendars that enabled the scheduling of agricultural and societal activities. The celestial phenomena reoccurred since the inner temple was first revealed and will continue through future time.

City of Uxmal

History of Uxmal

The initial settlement of the beautiful Maya city of Uxmal is lost in time; radiocarbon evidence from the earliest temple foundations indicates that construction in Uxmal began around the mid-sixth century CE and expanded extensively from that era. The city developed into the main population center for the Maya of the Puuc region with an estimated 15,000–20,000 people at its peak. The Puuc area is located in the northwestern portion of the Yucatan peninsula, Mexico, in a relatively flat area that extends into northern Campeche and western Quintana Roo with a prominent range of hills. Uxmal lies south of Merida, capital of Yucatan state and considered one of the largest, most famous and best preserved of the myriad Maya archaeological sites.

Because of the area's productive environmental conditions with abundant rainfall and rich soil, the Maya flourished in Uxmal for many centuries. Archaeological remains indicate that Uxmal developed into the capital of the prosperous state during the Terminal Classic period. The city reached its zenith between the eighth and tenth centuries under a ruler who took the name Lord Chac (Chan Chak K'ak'nal Ajaw) who ruled at the beginning of the tenth century. Scholars believe that he commissioned the most important and impressive buildings in the city. The structures surviving in the core area of the city are among the most remarkable architecture and artworks that the Maya produced. Most of the buildings in Uxmal are astronomically oriented 9° east of north. Space allows the inclusion of only the three most spectacular: the Pyramid of the Magician, the Nunnery Quadrangle and the Governor's Palace.

When assessing the art and architecture of the many Maya cities throughout their lengthy existence, you find that even though they shared the same written language and calendar, each Maya city-state created its own unique style. The dominant architectural style in the Puuc region first appeared at the end of the Late Classic period and reached its height in the buildings of Uxmal that are decorated with some of the finest sculpture found in the Maya world. Most surviving buildings were erected between the mid-ninth to the mid-tenth centuries. Puuc architectural articulation is more geometric than most Maya décor with complex mosaic facades but includes realistic figures, animals and symbols as well.

The limestone facades of Puuc buildings were covered with cut and shaped stonework set in a concrete core. The concrete core masonry that the Uxmal masons used could support larger and more stable interior spaces. The lower level of the facades have a smooth, plain finish between entrances but the upper level and around the doorways are richly decorated with intricate stone mosaic patterns, ornate friezes and elaborate figurative sculpture. The amazing mosaic, divided into three sections, covers the entire extent of the long façade. Masks of the Maya rain god Chac with long, curled noses were often added to the building's décor along with twisted and two-headed snakes and feathered serpents with open fangs. Although rainfall at Uxmal was consistent to fill the city's cenotes, the rain god Chac was

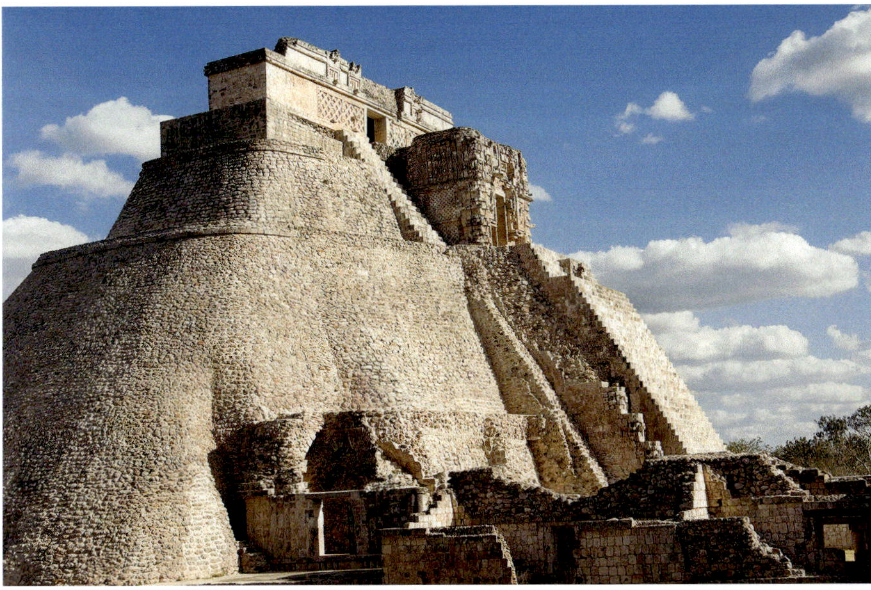

Fig. 13.6 Pyramid of the Magician, Uxmal, step pyramid with elliptical shape, built in phases from seventh to tenth century (Credit GTW)

often invoked for assistance and for assurance that the rains would continue for the Maya people and maintain their success in their constant struggle against the forces of nature. Therefore their architectural decorative programs included many Chac masks and an abundance of Chac symbolism; the sacred rain god was honored in glyphic texts and with human and blood sacrifices.

The Pyramid of the Magician

John Lloyd Stephens and Frederick Catherwood conducted two extended visits to Uxmal during their explorations and were fascinated with the well-preserved city, especially the remarkable Pyramid of the Magician. In his writings Stephens popularized the folklore tale that the huge pyramid was built magically overnight through a series of challenges presented to a dwarf by the king of Uxmal. The dwarf's mother planned the events to compete against the powerful king. The ancient magical myth lives on today. Catherwood drew many plans of the landscape and drawings of its exceptional architecture (Fig. 13.6).

The enormous Pyramid of the Magician survives as the dominant and highest structure at Uxmal; its distinctive elliptical form with rounded sides contrasts significantly from the usual square or rectangular pyramidal shape of most Maya pyramids. Two very steep and narrow staircases with an angle of 60° lead to the temple at the summit of the pyramid. The much wider eastern stairs reach a small temple that cuts the stairway in half. The western stairs are adorned with traditional Puuc sculptures of the rain god Chac which line the steps; these stairs face the Nunnery Quadrangle. Archaeologists have determined that the pyramid was built in five construction phases that took place between the seventh and the end of the tenth centuries. The Uxmal builders observed the customary Maya nesting tradition of building overtop the older, existing pyramid, greatly increasing their volume and height through the centuries. The height of this stepped pyramid measures 35 m (114 ft) with the base reaching approximately 69 by 49 m (226 by 160 ft). Archaeologists have found that the exterior of the pyramid had been finished with stucco and painted with four different colors that represented the four directions.

This overlapping technique resulted in five exterior temples; parts of Temple I, the oldest, can be seen at the base when ascending the western staircase. Temple II and Temple III are reached from an opening at the top of the eastern stairway through an internal chamber on the second level, now closed to visitors. Temple III includes a small central shrine room and an antechamber. Temple IV can be seen from the west side and it has the most decorative elements. This temple's façade is adorned with the masks of Chac; the entranceway is through his open jaws. Dating to the ninth century, Temple V, called the *House of the Soothsayer*, is located at the top of Temples II and III; it consists of three rooms. Chac masks and lattice ornaments embellish its entire façade.

The Nunnery Quadrangle

The name of the Nunnery complex was conferred by the Spanish in the sixteenth century because they thought it resembled a religious convent. Built between the early tenth and the start of the eleventh century, the four-building assembly was actually four palaces seated on platforms of different levels that surround an inner courtyard. The form of the buildings suggest that they functioned as a royal palace and included areas for meetings and administrative activities. The Nunnery's 74 separate rooms indicate it may have been used as a residence for students, priests, or military personnel (Fig. 13.7).

The facades of the four buildings are richly decorated with a variety of motifs including: lattice, columns, Chac masks, two-headed serpents, owls, Venus symbols, geometric designs, and naked and bound human figures. These symbolic sculptures represent cosmogonic beliefs related to different gods and also a strong cult of fertility. Each of the four exterior sides of the complex are adorned with ever-present Chac figures. The front walls were topped with light, airy roof combs. Surprisingly, the mosaic décor of the Nunnery also includes miniature depictions of thatched-roof huts of the ordinary people of that time (Fig. 13.8).

The northern building of the quadrangle sits on the highest platform and appears to be the most important of the four; wide steps allows access within. Two small temples flank the base of the stairs; the one on the left is thought to be dedicated to Venus. The north building contains 26 rooms aligned horizontally, decorated with monkeys, birds and human figures. While the western building holds only six

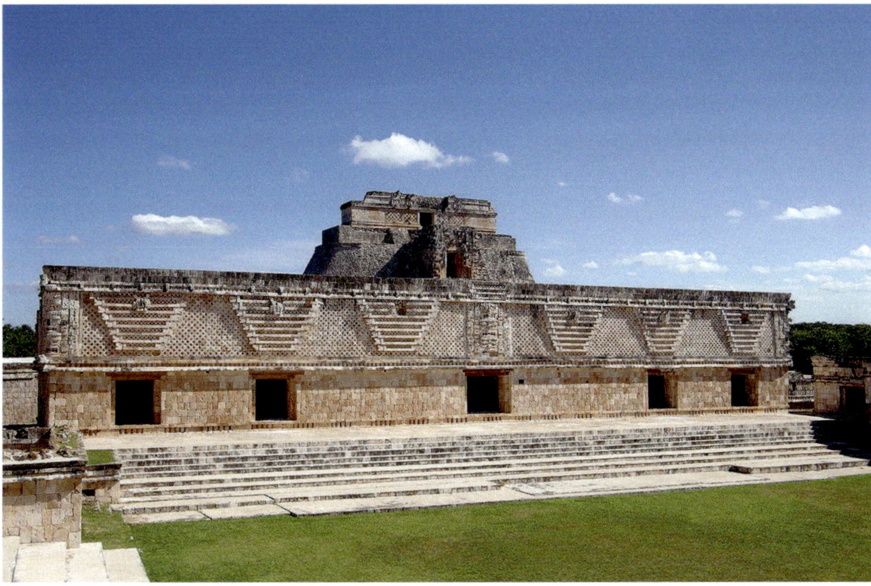

Fig. 13.7 The Nunery at Uxmal with 74 separate rooms, built in tenth century (Credit Gumbao)

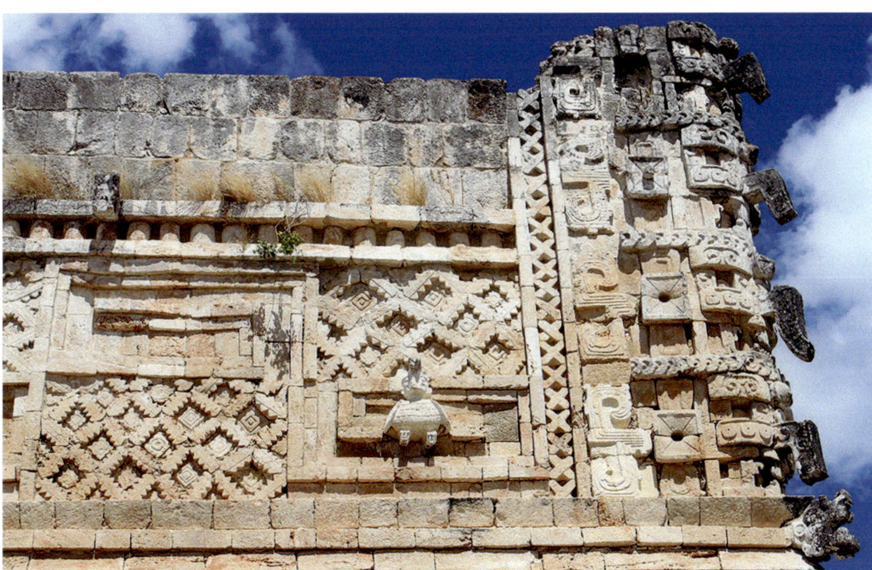

Fig. 13.8 The Nunery at Uxmal with Puuc architectural décor with Chaak masks, tenth century (Credit Daniel CZ)

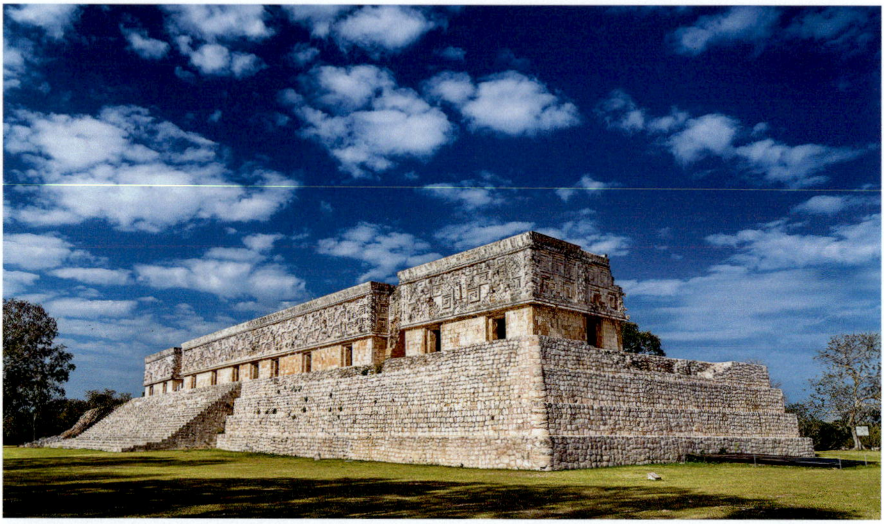

Fig. 13.9 Governor's Palace at Uxmal decorated with ornate mosaics (Credit Leonid Andronov)

interior chambers, this façade is adorned with mosaics with feathered serpents, Chaac masks and geometric designs. The eastern building has five chambers, two of the chambers have smaller interior rooms with similar décor motifs.

The Governor's Palace

The most sophisticated and admired of the many impressive buildings of Uxmal's city center is commonly known as the Governor's Palace or the House of the Governor. The Palace is thought by many specialists to be the finest building constructed in Pre-Columbian America. Mayanists indicate that it was commissioned by or in honor of the last great king of Uxmal, Chan Chak K'ak'nal Ahaw or Lord Chac and was built in the early tenth century. Scholars have determined that the architectural project would have required 1200 laborers and 33 years to construct and decorate the immense palace and its large supporting platform. As the most important building in Uxmal, the palace was decorated with more than 400 Venus symbols are still visible in the décor of the palace facade. It is carefully aligned with the maximum extreme of Venus (Fig. 13.9).

The Palace is situated in front of the Pyramid of the Magician and, like all important Maya architecture, the building stands on a huge, elongated platform that sits atop a lower platform. The rectangular palace building stretches 99 m long and faces east with the longest façade in ancient Mesoamerica. It is accessed by a wide central stairway on the east-facing side. The palace was originally built as three different structures and then later joined into one structure. The interior consists of 24 rooms with 11 doorways along the main façade with one more on the north and one on the south sides. Like the Nunnery, the palace was probably used for both residential and administrative purposes.

The decorative program adorning the palace is the culmination of the Puuc style with finely dressed stones and sophisticated façade decorations composed of stone mosaics and sculpted images. Researchers have counted more than 20,000 shaped or decorated stones that compose these remarkable mosaics. The lattice work, geometric designs, figurative sculpture, and masks of the rain god Chac combine into a single decorative ensemble. A sculpted portrait of Lord Chac, surrounded with two-headed serpents, stands over the center entranceway which is the largest of the 11 doorways and is accurately aligned with Venus positions. Considering the predominance of Venus and rain symbols, scholars suggest that this was the residence of a ruler who identified himself with the planet Venus and an incarnation of the rain deity Chac.

In the plaza in front of the Palace, a round stone monument called a *picota* and a small platform holding a double-headed jaguar throne are positioned along the central east-west axis of the structure. Although Uxmal's buildings are finished with exquisite, elaborate artwork, the city center has only a small number of poorly preserved stelae with minimal hieroglyphic text to provide historic information. It appears that most of Uxmal's stelae were intentionally damaged and toppled at the city's demise. Called the masterpiece of Maya architecture, the Governor's Palace was also astronomically oriented to honor the gods and to encode essential astronomical phenomena.

Archaeoastronomy of the Governor's Palace

The Governor's Palace has been a hot topic of investigation by archaeoastronomers in recent years. Archaeoastronomers became very interested in the Governor's Palace when they noticed that its orientation is significantly skewed relative to the other major buildings of Uxmal's city center. It is oriented 19° clockwise from the axis of the other buildings. Therefore it faces 28° south of east. In addition the Palace was honored with more than 400 Venus glyphs carved and molded onto the building's façade. The sacred status of Venus, associated with Chac one of the most important gods of the Maya world, would certainly designate the Palace as an especially revered structure.

The symbolic number eight is carved using the Maya dot-bar system above the eyes of the Chac masks at both northern corners, front and rear of the Palace. In addition, eight double-headed serpents are found above the central doorway. These numbers are thought to refer to the 8-day disappearance of Venus around its inferior conjunction. This reference could also be suggesting Venus' 8-year cycle because the Chac masks on the building's eastern façade are diagonally arranged and vertically stacked in groups of five. Five and eight are the two numbers featured in the iconography of the Palace which are thought to reflect five synodic periods of Venus consisting of 584 days and 8 calendrical years of 365 days. The Maya were aware of this Venus cycle since it was clearly written in the Dresden Codex that contains an accurate Venus table within it.

Two opposing analyses have been put forward regarding the Palace's orientation to Venus, one linking it with the southernmost rising point of the morning star. Alternately, the other theory proposes it is aligned with the northerly extremes of the evening star. The initial studies of the orientation determined that the Palace was directed toward Venus at its maximum southern extreme, although not exact.

An additional study established that the inscription on the main façade of the Throne over the central doorway refers to both Venus and to some constellations of the Maya zodiac. This analysis concluded that both the southernmost and the northernmost settings of Venus at the time the Palace was built were aligned to Venus by the Maya skywatchers but the alignments were not precise. Astronomers argued that the many years of careful research have shown that so far the astronomical orientations discovered in Mesoamerican architecture have been quite accurate. These alignments mainly mark the extreme sunrise and sunset positions of Venus. Considering their consistent accuracy in other buildings, it is unlikely that they would allow a large error in the placement of the Governor's Palace, one of their most honored and highly decorated structures.

From the central doorway the only noticeable structure is a huge pyramidal shaped monument over 25 m high, situated along the horizon in an area called Nohpat. The exact view from the Palace central doorway leads to this distant monument that marks the spot where Venus will rise at its southern extreme at the time the Palace was built around CE 800. At the same time this Palace-Venus alignment was being constructed in Uxmal, Maya masons were building another Venus aligned monument in Chichen Itza, the Caracol.

In Mesoamerican architecture, the known alignments to Venus extremes are all oriented to the positions of the evening star more commonly than to the morning star. The northerly evening extremes always occur in early May, which is the start of the rainy season, a particularly important time for maize planting. During an 8-year Venus cycle, its setting can be watched on the western skyline; this cycle will have five northern extremes that announce the rainy season and five southern extremes visible between October and December. These later dates indicate the end of the wet season and the times for harvest. Evidence shows that the northern evening setting of Venus, associated with rain, maize and fertility, would be a more likely orientation for the Palace.

The alignment of the Palace of the Governor appears to point to the northern extremes of the planet's setting but that alignment does not occur on a yearly basis and is not consistently on the exact same dates of the calendar year. Therefore the alignment cannot be suitable for keeping an accurate calendar but since the phenomena was built into the most important structure of Uxmal, it must have had a huge symbolic importance. Because the rain god and Venus as evening star shared an important relationship and the ruler often acted as a personification of the dominant deities, it is probable that the Palace was the physical manifestation of the most essential direction. The ruler who commissioned the Palace also had it adorned with numerous facial masks of the god Chac and with hundreds of Venus symbols connecting the building with his divine status.

In summary, although the intended orientation of the Governor's Palace is still under discussion among archaeoastronomers, it is aligned with the extremes of the Venus 8-year cycle. These extremes do not occur annually, so the occurrences are not on exact calendar dates. Therefore some scholars suggest that the Palace's alignment was oriented for spiritual and cosmological purposes rather than for keeping an accurate calendar. The Venus alignment was perhaps built into the grandest building in Uxmal for its symbolic impact in joining the ruler, who impersonated the rain god Chac with his costume and enjoined with the planet Venus at its extreme position on the western horizon. Through it his kingly responsibility for maintaining the ideal cosmic order was achieved and that ensured the continued success and prosperity of his subjects. The city, the individual buildings, the people and the king at Uxmal were cosmologically aligned with the stars, planets and the heavenly phenomena.

References and Further Reading

Aveni, A. F. (1980). *Skywatchers of ancient Mexico*. University of Texas.

Aveni, A. F. (1992). *Conversing with the planets: How science and myth invented the cosmos*. Times Books.

Miller, M. E. (1986). *The art of Mesoamerica from Olmec to Aztec*. Thames and Hudson.

Sharer, R. J. (1994). *The ancient Maya* (5th ed.). Stanford University Press.

Šprajc, I. (1993). The Venus-rain-maize complex in the Mesoamerican world view: Part II. In *Archaeoastronomy* (JHA) (Vol. 18, p. xxiv). Science History Publications Ltd.

Ŝprajc, I. (2015). Governor's Palace at Uxmal. In C. L. N. Ruggles (Ed.), *Handbook of archaeoastronomy and ethnoastronomy*. Springer Science+Business Media. https://doi.org/10.1007/978-1-4614-6141-8_68

Chapter 14
Maya City of Chichen Itza

History

At the same time the magnificent Classic cities of the Maya in the south were weakening and beginning their rapid collapse, the northern areas of the Yucatan Peninsula, the Puuc region, were enjoying remarkable growth. The Puuc area is located in the southern, hilly portion of the Yucatan Peninsula, and is also called the Puuc hills. The soil of the territory was fertile and widely available for productive cultivation and new settlements. At that early time period, this thinly populated region in the lowland rests of the Yucatan Peninsula was teeming with deer, rabbits and jaguars that were valued and worn by nobles and priests as a symbol of noble heritage and ceremonial status.

The territory around Chichen Itza is known as the Northern lowlands and consists of karstic limestone with many naturally formed sink holes or underground cisterns that provide the main source for water. Since the northern Yucatan Peninsula has no above-ground lakes or rivers and receives little rainfall, cenotes were essential to the survival of people who settled there. The Maya often considered the cenotes as sacred since their very lives depended upon them. The city's name "Chichen Itza" means "At the mouth of the well of the Itza" and refers to the many cenotes found around the site. Archaeologists suggest that the presence of more than a dozen cenotes in this area of the Yucatan was the prime factor for settlement; plus the readily available water for cultivation and for daily use attracted more newcomers to the area (Fig. 14.1).

The famous city of Chichen Itza is one of the most renowned yet most enigmatic cities of the ancient Maya. Considerable controversy has arisen regarding the occupational chronology of Chichen. The reconstruction of the site's history is based mainly on ethno-historical sources, such as the Chilam Balam texts and Spanish chronicles. Scholars have studied both Aztec and highland Maya histories and long-enduring legends as well, trying to piece together an accurate biography of the

M. Dolan, *Decoding Astronomy in Maya Art and Architecture*,
Springer Praxis Books, https://doi.org/10.1007/978-3-031-61092-9_14

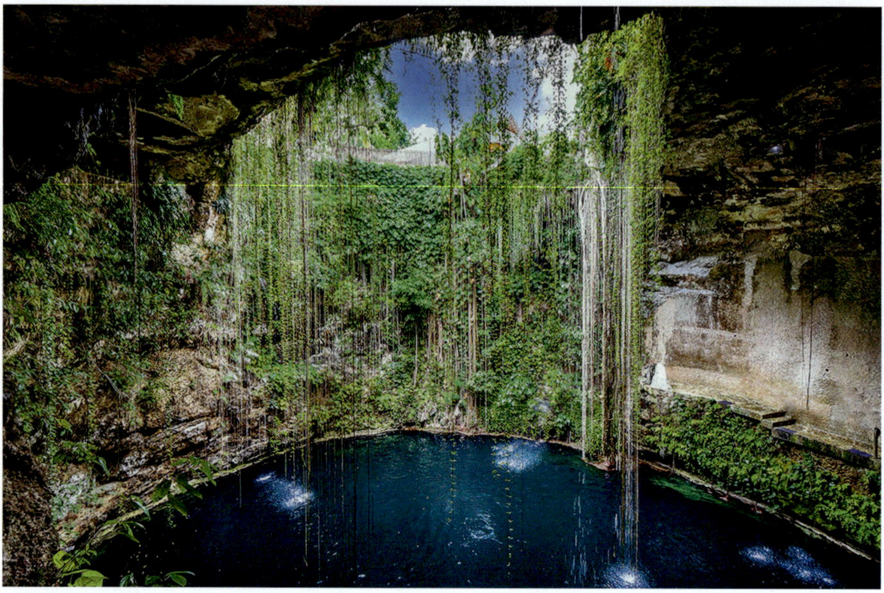

Fig. 14.1 Ik-Kil Cenote, Chichen Itza, only nearby water sources (Credit ecstk22)

leading city of its age. During the colonial period, both Maya citizens and Spaniards recorded information about the city's history and left many records. John Lloyd Stephens' vibrant description of the scattered ruins of Chichen Itza in his travelogues brought the first modern attention to the site. As in other Maya sites, Stephens' writings and Frederick Catherwood's drawings are still of great assistance to archaeologists and informative for tourists as well. Beyond its archaeological remains, important sources of information studied to learn the early Chichen history are ethno-historical texts which provide information about the chronology, their political groups and cultural relationships with other cities.

Chichen Itza grew to be one of the most successful and prosperous Maya cities, yet its time as an imperial state ruled by kings was much shorter and its ethnic mix was more diverse than most Mesoamerican cities. Through the many centuries since its fall, the remains of the city were overgrown but never totally lost within the tropical forests, like so many Maya cities. The archaeological site covers over 6 mi^2 of ruins and who knows how many more ruins may still lie buried. Even though archaeologists have spent more than a century excavating and researching the ruins, complete understanding of the various occupants through the history of Chichen Itza is still uncertain and contested. Its founding date, its basic chronology and the various populations who built and occupied the city are still debated by archaeologists.

This southeastern area of Mexico experienced the same long history of occupation as the other Maya territories. The initial settlement of the lands around Chichen Itza began in the Late Preclassic period around 600 BCE and maybe even earlier.

Pottery from this distant era and from the Early Classic has been found in the Sacred Cenote of Chichen Itza. Early Maya villages with agricultural settlements spread throughout the lowlands. This central region of northern Yucatan seemed to be mostly deserted again around 250–600 CE. Previously, the central lowlands had been dominated by the great city of Tikal, but that area had lost its leadership by the end of the ninth century and had become desolate. After this southern and central lowland decline, the northern region of the Yucatan began to rise in population and power, in particular the early phase of Uxmal, Chichen Itza and finally Mayapan.

During the early seventh century, occupation of the lands around Chichen Itza resumed and ceramic objects reappeared along with the first city terraces and a few small stone buildings were constructed. The people in the failing cities of the south needed to find fresh areas to cultivate and farm. During this period of drought and social change, newcomers moved to northern areas where they discovered that a Late Classic village already existed at Chichen Itza. Recent epigraphic and ethnohistoric studies have shown that part of the ethnic group that settled at Chichen Itza have been identified as Itza, a branch of the Putun Maya. The people of Itza had migrated to the north from the central Peten and were connected with the traditions and language of the Classic Maya area in the Central Lowlands. They seemed to have been drawn there for commercial attractions and for trading, especially attractive were the north coast salt beds.

After the political disintegration and collapse of southern Maya Classic cities, Chichen Itza became the new capital of the area. Chichen preserved key aspects of Maya culture and political organization. The city transformed these aspects by creating a new type of government that changed the form of divine kingship. It is thought that they were governed by local nobles and possibly lords of conquered or allied settlements in more joint decision-making and public activities. The extended ruling groups were incorporated into their monumental art programs as well.

Although there are no documentary records or stelae with confirming details, legends tell that an important lord from the west named Kukulkan arrived in the Chichen Itza territory in 711 CE. At that same time the Itza were thought to have 'discovered' the city of Chichen Itza. Also at that time one of the gods of the Central Highlands developed into the most revered god of Chichen Itza, named Kukulkan, the Feathered Serpent. The presence of a Feathered Serpent deity in Mesoamerican religious traditions was not new; the revered figure dates back to the era of the Olmecs, the earliest known major civilization in Mexico. The Olmecs were the dominant Mesoamerican population from about the 1500 to 500 BCE.

The newcomer named Kukulkan became worshipped as a divine entity and a cultural hero; it was thought that he came from northeastern Mexico, the land of the Toltecs. The arrival of this great figure of Kukulkan introduced many changes to the northern lowlands, such as the architectural styles and the iconography used to decorate wood and stone carvings and painted objects. The Toltecs were native to northeastern Mexico; their capital city of Tula or Tollan thrived during the Classic and Postclassic periods. The Toltecs were urban people who practiced arts and crafts with great skill and worshipped the Feathered Serpent. Close similarities have been noted between the stylistic innovations in the art and architecture of the two widely separated cities of Tula and Chichen Itza (Fig. 14.2).

Many modifications were seen in the architectural designs of Chichen Itza between 900 and 1000 CE; these newly introduced designs were seen in Central Mexico as well. Through their research, Mayanists have created the model that placed the Toltecs of Tula and the "Toltec-Maya" at Chichen Itza in the Early Postclassic period but this theory is still under dispute. Some scholars think that Chichen Itza was conquered by the Toltecs but others believe instead that the city was developed by exiled Toltec nobles. Still others believe the stylistic influences traveled the opposite direction; they developed first in Chichen Itza and were gradually conveyed to Tula. The precise relationship between the Maya who were indigenous to the area and the Toltecs who were native to the central Mexican Highlands is still not clearly understood. Even the presence of Toltecs in Chichen Itza is still contested among Toltec and Maya scholars. These great changes to the architecture of Chichen Itza are no longer called Toltec styles or influences but instead called "international styles", visible in both cities during the same period.

The diverse population of Chichen Itza slowly expanded to become the major city, spreading throughout the peninsula and gradually increasing their domination, through long-distance trade, marriage alliances, and their military might. Numerous ceramics have been unearthed dating to the Late Classic period, 600–900 CE, but only a few architectural structures with hieroglyphic inscriptions have been uncovered. Painted and inscribed ceramic vessels from that period have been excavated from the Sacred Cenote. Therefore, since Chichen Itza was constructing no major buildings at that time, it must have been of lesser importance then. The first engraved stone monument at Chichen Itza was not erected until 832. This inscription records the first group of strangers to arrive in Chichen Itza during the ninth century.

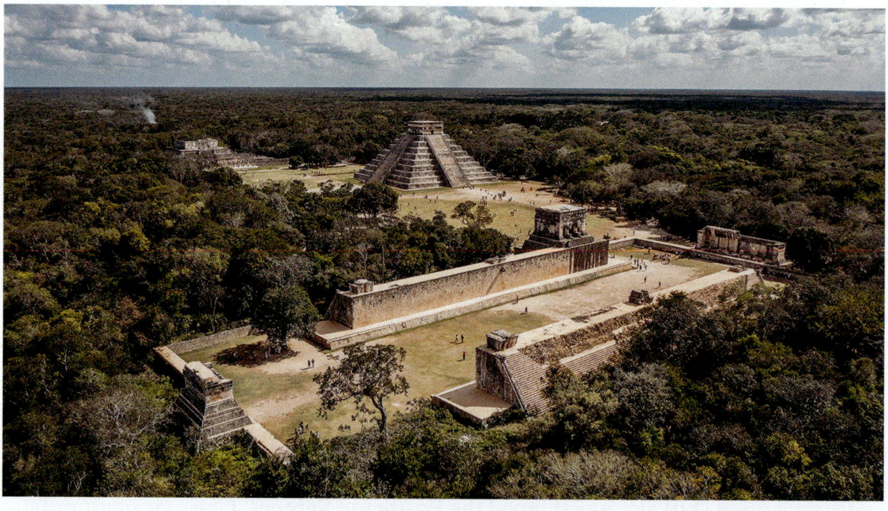

Fig. 14.2 Aerial view of Chichen Itza Castillo and Ballcourt (Credit Aleksandr Medvedkov)

By 850 Chichen Itza was established as the new capital city and had achieved political and economic dominance of the northern peninsula. The Chichen elites found means to control jade and gold trade that provided them with enormous economic advantages. From that time the city stood as the major pilgrimage and trading destination and a strategic economic power attracting merchants and travelers from various parts of the country. The elevated status of the Sacred Cenote played a crucial role in maintaining the prominence of Chichen Itza as a valued pilgrimage city. Their sturdy, seagoing canoes traversed the Yucatan northern coast, trading for cotton, cacao beans, fabric, obsidian, copper and precious gold from Central America. (On Christopher Columbus' fourth voyage he came upon a Maya trading vessel filled with trade goods.)

Together the myriad migrants formed a city that held the most diverse population in the Maya world and established the primary political and economic center. Between CE 900 and 1050, Chichen Itza ruled as a regional capital controlling the area from central Yucatan to the north coast and down the east and west coasts of the Yucatan Peninsula. Chichen Itza is said to have been the most powerful and largest city of the Maya civilization. The number of people living in Chichen Itza is only a rough estimate which varies widely; the population at its peak is thought to have been from 30,000 to 50,000. The city's occupation continued to grow into the early Postclassic until 1200, at which point Chichen began its slow decline.

Architecture of Chichen Itza

While the highly developed cities of the southern Mesoamerican areas, such as Tikal, Copan and Palenque, suffered the desertion and decay of their beautiful palaces and temples, the northern Yucatan Peninsula cities of Chichen Itza, Uxmal and Mayapan were creating some of the most admired architecture found anywhere in the vast Maya World. The architecture of Chichen Itza is unique in its large variety of styles and in the grand scale of its buildings and for their accurate astronomical alignments. The city's most impressive structures are densely clustered within the central area and dominate the extensive archaeological site. The wide-ranging styles of architecture repeat many of the usual Maya structures, such as pyramid and temples, but new structural designs appear as well, such as round buildings, clusters of piers, and standing half columns. New sculptural figures were created, to decorate the buildings such as chacmools, jaguars and feathered serpents. Chichen Itza is known not only for its outstanding architecture but for its carved lintels, stone sculpture, hieroglyphic texts and its painted murals.

A large number of Chichen Itza's stone buildings, pyramids and monuments still stand in various states of preservation; some have been brought back to their original condition through modern restorations. As we have seen, inscribed texts, dedications and exact dates are lacking on many structures, which generates disagreements among scholars. Unlike many Maya cities, especially Copan which has over 100 stelae that are inscribed with dedications and carry specific dates and details of important ceremonies, Chichen Itza has only two known stelae. The earlier buildings were not inscribed. As mentioned, the earliest hieroglyphic date discovered at Chichen Itza is equivalent to 832 CE.

Throughout its long history the diffusion of cultural groups that settled in Chichen Itza produced many different architectural styles, including styles of the Puuc and the Chenes from the Northern Maya lowlands and those of Central Mexico. After producing numerous architectural studies, scores of papers, and a myriad of differing opinions, Mayanists have reduced the two architectural styles at Chichen Itza as the initial Maya-Puuc period and the later International period, eliminating the Toltec origins theories. More accurate dating evidence indicates that certain architectural features found at Tula actually originated in Chichen Itza. In spite of the many proposed theories about the diverse population and unusual variety of artistic styles at Chichen Itza and Tula, Chichen was a totally Maya city, not Toltec. The latest studies indicate that the transition from the Puuc Maya to the International style was not a rapid change conveyed by foreign invaders but a long period of gradual development produced from the architectural and artistic choices of the local leaders and inhabitants. Studies show a substantial overlap of the two styles of art and architecture. During the years between 850 and 950, the Transitional period saw the gradual changes in the art and architecture into the fully developed International style.

No structures survive from the earliest settlements of Chichen Itza; although they may still be deeply buried beneath later buildings or they have not survived. During the Late Classic period the small settlements of the area began to develop into a true city, which included its initial examples of major architecture. The political and trading environment of the city improved considerably during the Terminal Classic period (800–900) and its population rose dramatically. The layout of the central core of Chichen Itza was undertaken during its earlier period of occupation, between 750 and 900 but only a few carry inaugural dates that furnish reliable data. Chichen's final layout took place during the tenth century when the city emerged to become a powerful regional capital.

This era of Chichen Itza history realized great architectural advancements and enabled the city's wide expansion in territory, wealth and power. Chichen Itza officials controlled the area from central Yucatan to the north coast and along the east and west coasts of the peninsula. Only a few of the ninth century buildings carry carved inaugural dates that furnish reliable data for establishing the city's history. The inscriptions on new buildings constructed between 860 and 890 tell of an important period of dedication and prosperity.

Chichen Itza has two distinct occupations with both Terminal Classic and Early Postclassic components. Each has its own characteristics and is separated by the turbulent years of the early tenth century. The first occupation dated to before ca. 900 or 1000. During this early period the southern half of the site's center contained the Las Monjas and the Caracol; other modest size Puuc buildings were also constructed. The International era is thought to date a bit later, to about ca. 1000–1250; this area has massive, blocky architecture similar to that in Central Mexico. It was during the later period that the Castillo, the Great Ballcourt, and the Group of the

Thousand Columns were constructed. At that time new artistic forms and a simpler writing system were introduced and metal objects were imported for the first time. Turquoise began to replace jade as the most preferred gemstone for jewelry, ritual costumes and grave goods.

During the first half of the tenth century at Chichen Itza, there was a pause of construction, but by the second half the city entered its second golden age. In addition to its new buildings, the elites of the city were trading for new types of vessels and innovative designs on their ceramics. The arrival of these new artistic patterns generated foreign-inspired pottery designs in their own production of ceramics. The adoption of foreign art styles along with traditional designs provide a useful dating system.

Archaeologists have divided areas of Chichen Itza into various architectural groups. Each of the groups were constructed atop a carefully levelled platform built to support them. The land was uneven and required much smoothing and flattening. The major structures of the city were built atop different supportive platforms; these architectural groups were once separated by low walls. The three most renowned groups are the Central Group, the Great North Platform, and the Osario Group. The Central Group includes the Caracol, Las Monjas (thought to be a Nunnery), Akab Dzib, the Initial Series Group and the Group of the Southwest. These were built earlier and are all south of the Great North Platform which was constructed after that group was completed. Las Monjas was the most notable and tallest building of the early group and it dominated the large central plaza. "Old Chichen" is the name of a group of structures to the south of the central plaza. The Osario Group consists of the Temple of Xtoloc and several more structures.

Surprisingly, the leveled surface of the Great Platform comprises the largest monumental construction at Chichen Itza. The height of the platform above bedrock varies because the ground surface was quite irregular. Studies have shown that the basic orientation and planning of the Great Platform was determined at an early date in the history of Chichen Itza. The Great Platform measures about 600 × 400 m (2000 × 1300 ft) and was formed during ten different building periods. The Great Platform was assembled using cut but undressed stones and was then entirely covered with a thick layer of stucco and lastly, painted bright red. This durable plaster coating was completely excavated by archaeologists, and for the most part remains in excellent condition.

The construction history of the Great Platform is complex, but it seems highly unlikely that any of the structures were built before the Late Classic period. Moreover, only the Stage IA to Stage IIA structures may date to this period, and this still must be resolved through ceramic analysis and radiocarbon dating. Conversely, the last construction episode studied probably took place between about A.D. 1050 and A.D. 1100–1200. In total, the entire sequence of the platform construction certainly represents no more than 600 years, perhaps only about 300 years in length. The leveled surface was so huge that it held El Castillo, the Temple of the Warriors, the Group of 1000 Columns, the Great Ball Court, the Venus Platform and many other structures that were constructed upon it (Fig. 14.3).

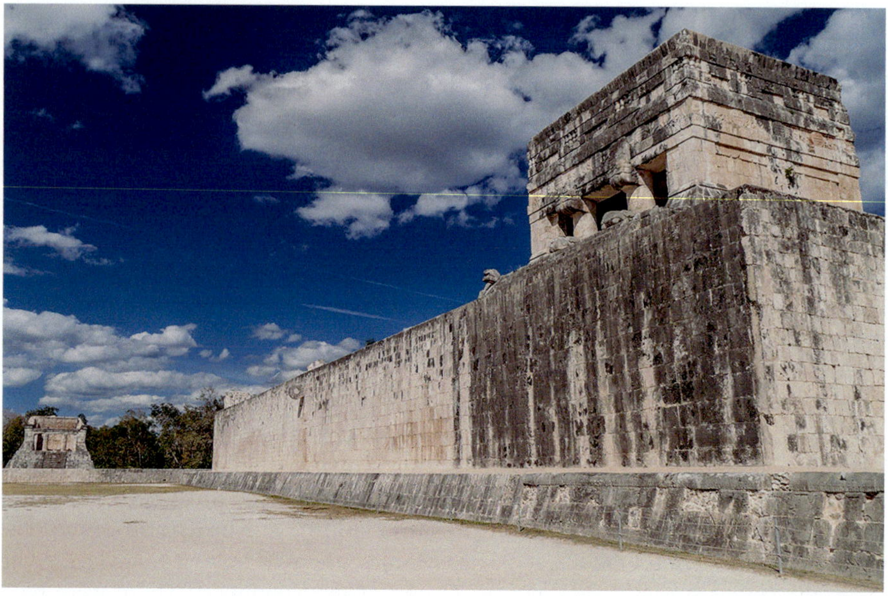

Fig. 14.3 Chichen Itza Ballcourt, long wall of the ballcourt, the largest of all Maya ballcourts, Temple of the Jaguar is overlooking it (Credit Matyas Rehak)

The earlier settlers in Chichen Itza preferred the Puuc style for their temples, buildings and monuments, the favorite stylistic innovation in the area southwest of Chichen Itza. The Puuc buildings are recognized by several distinct characteristics that were first introduced at the end of the Late Classic period. This stylistic innovation had been the favorite decor in the cities of the Puuc Hills, an area southwest of Chichen Itza. These buildings face toward the city's main plaza and they follow a basic north-south axis. In Puuc architectural methods, the masons applied an advanced technique by using a concrete core which permitted slightly larger interior rooms. This new method also produced sturdier corbel vaulting; many Puuc buildings constructed in this manner are still standing in good condition. Many Puuc-styled structures did not appear as they seem today; he upper façade of many Puuc buildings were adorned with complex stone mosaic designs and long-nosed masks made to represent the Rain God Chaac. The exteriors were often brightly painted in red, green, blue and purple. Puuc-style architecture achieved its highest level during the Terminal Classic period. Experts still debate whether there was an overlap of the two different architectural styles.

The Puuc style buildings were adorned with veneer stones placed into a concrete base; usually the lower section of the facades were left plain with a flat surface of rectangular blocks. The upper facade would be more elaborately decorated with complicated stone mosaic patterns. Some structures would have a lattice finish or a repeating geometric décor with impressive figural sculpture added as well. Many Puuc-styled structures did not appear with the unfinished stonework as they look today. The upper façade of many Puuc buildings were adorned with complex stone mosaic designs and long-nosed masks designed to represent the rain god Chaac. The Puuc-like buildings of Chichen Itza are similar to the early Puuc mosaic-style but not exactly the same, except for the Temple of the Three Lintels which is the most like the originals. To finish the structures, the exteriors were often brightly painted in red, green, blue and purple. Puuc-style architecture achieved its highest level during the Terminal Classic period.

The city center of Chichen Itza was not built at one time by "Toltec" invaders or by "Mexicanized" Maya, as one time thought; the city evolved through time and was built by the Maya settlers. The International style of art and architecture that dominates in the final stages of construction at Chichen Itza also developed gradually over time, revealing a pattern of adoption, innovation, and variation. The International period is thought to date from about CE 1000 to 1150; this style of construction produced massive, blocky architecture similar to the styles of Central Mexico. During this period the Castillo, the Great Ballcourt and the Group of the Thousand Columns were constructed. Evidence shows that the rulers, architects, and artists of the city were not influenced by foreign invaders, but instead were energetic contributors to the development of the International style. Other changes occurred around that same time which perhaps were the influenced by their widening trade areas. Chichen saw the introduction of a simpler writing system, new artistic forms and the use of metal objects. Foreign art designs in ceramic vessels were added as well and used along with the favored traditional patterns which provided a reliable dating system.

Like most Maya cities, the important buildings at Chichen Itza were connected by a network of paved causeways. Over 80 *sacbeob* crisscrossed the great city spreading out in all directions and connecting the many sections of the huge area. Their roads also extended to the far-off cities within their wide-reaching trade network. Influences from these distant cities were not only expressed in their architecture and monumental art but also in their political structure, their religious beliefs, and military actions. The development of the architecture and decorative designs by the leaders, architects and artists at Chichen Itza continued to evolve through time producing a pattern of advancement and adaptation until its final decline in the twelfth century.

El Castillo of Chichen Itza

Between about CE 900 and 1000, the Early Postclassic period, the Maya masons of Chichen Itza constructed their most extraordinary building that continues to reveal its numerous secrets. The celebrated pyramid El Castillo was erected to honor the god Kukulkan, the Feathered Serpent deity that was also honored in Teotihuacan and by the Toltec of Central Mexico. The Toltecs spread the veneration of their cult hero Quetzalcoatl between the tenth and twelfth centuries as they moved out of Central Mexico to conquer other areas. The Toltecs or their influence brought this deity to the Maya territories, where the deity was called Kukulkan. Devotion to Kukulkan by the Maya grew significantly in the early Postclassic and the god became especially venerated in Chichen Itza. Images of the feathered serpent adorned many of the prominent buildings of the city during the later stylistic period. Numerous parallels can be shown between Chichen Itza's architectural designs and those of the Toltecs' impressive capital city Tula.

The Castillo, also called the Temple of Kukulkan or Structure 2D5, survives in excellent condition; some necessary restorations were conducted in 1905 and 1921. The impressive, tiered pyramid and temple still dominates Chichen Itza's widespread ruins. Adding to its importance is its prime, central location. The northern side of El Castillo faces the Platform of Venus and a 980-ft raised *sacbe* leads from it directly to the Sacred Cenote or Well of Sacrifice, the most essential of the many cenotes in the area around Chichen Itza.

Mayanist have suggested that the location of the Castillo was established by the intersection of the axes formed when joining four nearby cenotes, including the Sacred Cenote and Xtoloc (with a north-south orientation) and Kanjuiyum and Holtun (with an east-west orientation). Each stairway of the pyramid is directed toward one of the cenotes. The western stairway at 292° 30′ points to the Sun's transit through its zenith with only 1° of deviation from perfect alignment with the entrance to the Holtun Cenote that sits almost 2.5 km away. The Holtun Cenote has a rectangular entrance that measures 237 × 140 cm with a 22 m drop to the water's surface. It is thought that the cenote was used as an observatory for marking the passage of the Sun through its zenith. This light phenomena would be observed when a beam of sunlight hits the surface of the water 22 m below. A rectangular shape would then be reflected on the cave wall in the cenote. Therefore, on days before and after the Sun's zenith, the Sun's rays enter the cavern obliquely and are reflected onto the ceiling of the cave below. When the Sun is at its zenith, the rays fall vertically into the Underworld and can return to the heavens through the opening in the ceiling.

The Temple of Kukulkan is a radial pyramid which was a common and essential form admired in the architecture of the Classic period Maya and continued into the later periods. The radial pyramid is usually the most important factor in architectural assemblages, joined with other smaller buildings. They are also associated with astronomical observations and designed for public participation in community rituals that were timed by their solar or agricultural calendar. The ritual ceremonies and processions performed on or around the pyramids were probably thought to be actions involving sympathetic magic, especially in controlling the weather. The radial pyramid would have been used for establishing the axis mundi in ceremonies concerning the equinoxes. The axis mundi was essential as the link between the nine-leveled Upperworld, the Middleworld, Earth, and the Underworld (Fig. 14.4).

The Temple of Kukulkan was carefully positioned on the *axis mundi*. Even more interesting, scholars think it is possible that an underground river system travels along these directional axes. This idea was supported by an amazing discovery in 2015. Surprisingly, archaeologists learned that El Castillo sits above a very large cavity or cenote partially filled with water; the sinkhole measures about 82 by 114 ft

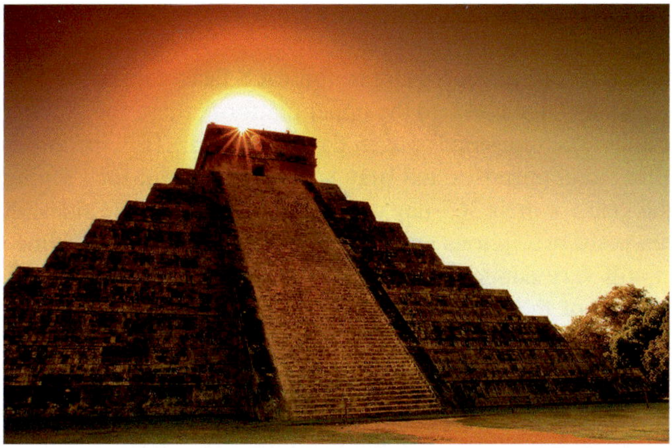

Fig. 14.4 El Castillo at Chichen Itza, radial pyramid with temple, astronomically aligned (Credit Shutterstock)

and reaches a depth of about 65 ft. Although they know it exists, scientists have never actually seen the hidden cenote.

Archaeologists believe that the Maya were aware of the hidden cenote when they selected this special site for building the pyramid, even though it was concealed beneath a thick stone covering a layer of limestone almost 16 ft thick. Certainly, the Castillo would have been more influential for the city because the Maya selected such a powerful location for the pyramid. Its prime position would increase the potency of the extraordinary Temple of Kukulkan. The presence of the pyramid seated directly over the waters of the Underworld would create a cosmological link with the Castillo situated at the center, creating and joining communication between the Lower and Upperworlds.

The nine stacked tiers of the step pyramid have a quadrangular floor plan and all four sides of the pyramid are equal, measuring 55.3 m (181 ft) at the base of each side. El Castillo rises 24 m (79 ft) to the upper platform, when including the upper temple, its total height reaches almost 30 m (98 ft). The pyramid slopes at a sharp angle of 53°. Differing from traditional Maya temples which are more often rectangular, the small sacred temple atop the Castillo is square. The temple's four doorways each face one of the four cardinal points. The most important entranceway is considered to be the north since it is much wider and separated with two columns, each decorated with the highly honored feathered serpent. Both the Castillo pyramid base and the upper temple were designed with a radial arrangement; a central focus with linear forms that radiate outward to the surroundings. This is unlike centrally organized buildings where the focus is inward to the central space. In the history of Maya architecture, radial structures were associated with solar or cyclical time counts and with public rituals in connection with their creation myth and seasonal agricultural renewals.

The Castillo includes a steep staircase on each face, capped with a balustrade; each side has 91 steps that climb to the upper platform; these four stairways plus the top platform total 365, one step for each day of the year. This specific number may have had useful ritual purposes and may have been helpful for keeping a count of the days of the year for their 365-day calendar. Steeper than the sides, the stairways rise at an even more precipitous angle of 45°, not as vertical 5as some early Maya pyramids but still very dangerous. The stairs are no longer open for visitors. Like the north doorway of the temple, the north face of the pyramid appears to be the most honored since a feathered serpent adorns both sides of the northern stairway; an enormous stone serpent head lies at the base of each side with a gaping mouth, once displaying its threatening fangs that are now missing. A large serpent head also appear at the base of the temple atop the pyramid (Fig. 14.5).

The detailed components and exact positioning of El Castillo, the 91 steps of each stairway, the four sets of stairs, the nine tiers, and the building's many solar

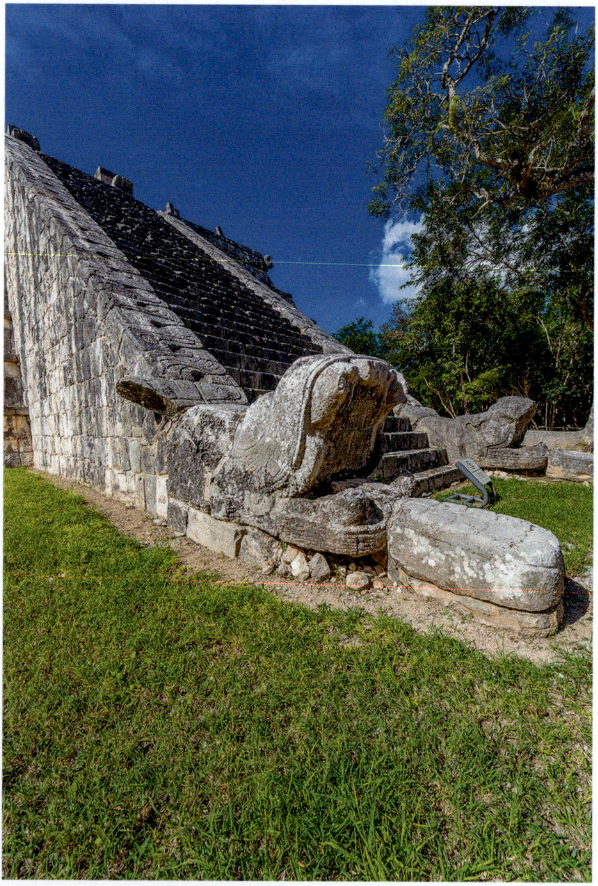

Fig. 14.5 Head of Kukulkan on El Castillo pyramid, Chichen Itza (Credit romakoma)

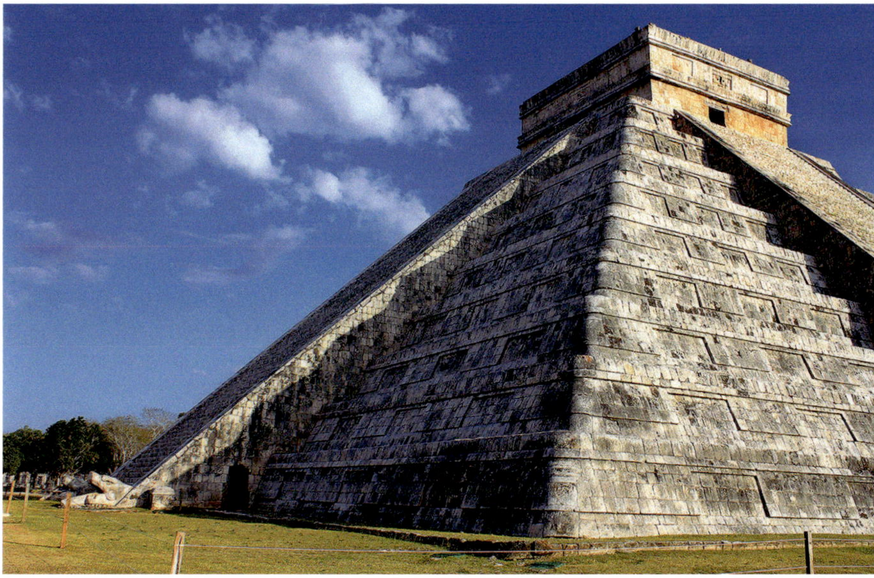

Fig. 14.6 El Castillo at the equinox, the 'descent of the serpent' (Credit Alexander Sviridov)

alignments embody their interpretation of the cosmos and demonstrate their desire to attune their most important shrine with the harmonic symmetry of the celestial sphere. The appearance of the Feathered Serpent God at the foot of the north stairway and at the foot of the temple above illustrates their dedication to the most important god Kukulkan (Fig. 14.6).

El Castillo Alignments and Light Phenomena

Additionally, the Castillo was constructed with specific positioning and also amazing light and shadow effects. Twice a year on the spring and autumn equinoxes, a remarkable shadow is projected by the setting Sun. When viewed from the west, the late afternoon Sun casts a shadow on the balustrade of the northwest stairway in the form of a serpent with seven angular shadows; the effect lasts almost 2 hours. As the Sun sets, the shadowy serpent, marked like a diamond-backed rattler, begins to recede, a solar hierophany called the "Descent of Kukulkan". The shadowy serpent descends the stairs; step by step it slowly withdraws until reaching the base and joining the fierce head of Kukulkan. This light and shadow phenomena starts about 1 hour before sunset and lasts about 45 minutes. It is doubtful that the builders of the pyramid intended this effect to record the date of the equinox; since the light effect also occurs several days before and after the exact day of the equinox. Therefore, it is not possible to determine the exact solstice date using only this solar event. The intriguing solar phenomena may have had only a symbolic function for the Maya.

But the light show is still fascinating to watch and it attracts thousands of viewers each fall and spring.

Studies of various light and shadow effects have been carried on at the Castillo and the Descent of Kukulkan on the equinoxes is not the only light phenomena. They have discovered lighting effects that occur on the solstices as well. During the first minutes of sunrise at the summer solstice, the Pyramid of Kukulkan is illuminated on the NNE and the ESE facades for 15 minutes, but the WNW and the SSO facades remains in completely dark. One half of the pyramid becomes illuminated and one half remains; this symbolism illustrates the exact moment of the solstice. Furthermore, at sunset on the winter solstice, the lighted façades are the WNW and SSO while the NNE and ESE facades remain in shadow. The light effect is due to the orientation to geographic north and the latitude of the pyramid.

In addition, the Castillo's western orientation of 291° is only 1° from the solar zenith sunset. The first scholar to suggest that the Castillo was aligned with the Sun's zenith was Susan Milbrath in 1998; the Sun's points of rising and setting are aligned as the Sun crosses directly over the pyramid. The eastern face of the pyramid is aligned toward the nadir sunrise. Perhaps solar ceremonies were enacted in the eastern chamber of the temple when the first rays of dawn on the day of the zenith nadir entered the darkened temple, but no proof of these observances has been established as yet. The days of zenith passage at that latitude occur on May 23 and July 19; these particular dates are important since they also relate closely to their agricultural cycle, the starting time for planting and for harvesting. Very few other Mesoamerican cities correspond to these dates, so there may have been additional reasons for selecting this particular alignment. This sacred building became the architectural manifestation of their calendar, symbolizing the hierophany of the eternal return of the Sun.

El Castillo Substructure

In 1930, Mexican archaeologists discovered another, smaller nine-tier pyramid within El Castillo, probably several centuries older. This sub-pyramid appears to be of similar design but has only one stairway on the north face with 63 steps. The investigations by archaeologists in the 1930s were made through a tunnel dug under the north staircase. An intact sacrificial offering was found at the base of the stairway which consisted of two wooden disks with jade, coral, shell and turquoise inlays and included other devotional objects as well. The sub-pyramid was probably constructed between 800 and 900 and stands about 10 m (32 ft) tall. This structure was built during the transitional period of Chichen Itza's history with mainly Maya stylistic elements but displaying the introduction of Central Mexico influences also. After removing accumulated rubble, the researchers discovered that this inner pyramid was also topped with a temple; this temple built atop the sub-pyramid had a chamber that was encased inside the newer temple chamber.

Further investigations a year later found a second chamber within the inner temple; these hidden rooms were intact and concealed two life-size statues in perfect condition. The first sealed room also preserved two parallel rows of human bones set into the back wall along with a life-sized Chacmool figure. The second chamber held a red jaguar in the form of a royal throne; both statues face north northwest and are still in situ. The chamber that holds the Chacmool statue was named the Hall of Offerings; the chamber with the Red Jaguar Throne was called the Chamber of Sacrifices. That early 1930s tunnel has now collapsed and is still closed; therefore any new studies of the interior structure of the Castillo will need to employ newer, more efficient scientific methods for excavations.

In 2017, subsequent investigations of the interior sub-pyramid of Castillo were carried out by a scholarly team from various Mexican universities. This study applied some of the latest imaging techniques, including Electric Resistivity Tomography (ERT). This scientific method uses a non-invasive geophysical process that does not require physical excavations and therefore does not disrupt the historical context of the archaeological site. The investigative team worked for weeks and produced a "data set that gave a total of 37,548 apparent resistivity observations". These ERT readings generated a 3D model that "illuminated" the Castillo's inner structure. The scientists were amazed to find there is not just one but two sub-pyramids within the Castillo. The dimensions of the original and earliest sub-pyramid are approximately 12 m (39 ft) in height and a width of 26 m (85 ft) at the base; it was probably built between 600 and 800 CE. The stylistic design of this oldest pyramid would be purely Maya in nature.

The investigation indicated an anomaly at the top of the initial pyramid which the archaeologists suggest is another temple but its roof has collapsed. Further study of the results determined that the innermost pyramid is not centered within the second pyramid. This original pyramid is nearer to the south-east corner and closer to the cenote that sits directly below El Castillo. Perhaps this discovery indicates that the first Mayan builders or rulers knew that the sinkhole existed and they chose that particular location to harness the power of the spiritual waters below. Further research on these two earlier pyramid/temples could provide much information concerning the early settlement of Chichen Itza and its cultural history.

El Castillo Décor

The Red Jaguar Throne and the Chacmool were two of the innovative figural forms that appeared in Chichen Itza during the later International style period. The decorative features that showed Central Mexican influence or similarities in addition to the jaguar throne and the Chacmool include mini-atlanteans, standing columns and standard bearers. These new ornate adornments were mainly used to beautify buildings devoted to rulership.

The Red Jaguar Throne in the Castillo was designed and decorated to replicate a real jaguar; it was studded with 74 turquoise spots to imitate the jaguar. The jaguar statue had fangs and teeth made of mollusk shell painted white; embedded on the jaguar's back was a turquoise disk. For the Maya, the jaguar stood as a symbol of night, of power and warfare; the color red was important since it personified a symbol of life and blood, as well as death and sacrifice. The red pigment coating the throne was made from cinnabar which was not readily available to the area around Chichen Itza; therefore objects that were painted red were especially valued. When the inner temple was functioning, the Maya ruler would have sat on the Red Jaguar Throne and looked ahead at the Chacmool lying calmly and subservient before him. The Castillo's sacred temple with its Red Jaguar Throne and Chacmool was a place of offering not only to the deities but also to rulership. Perhaps the Jaguar Throne was associated with the ritual importance of closing a temple for renewal at the end of a calendar cycle or the start of a new dynasty.

The unusual Chacmool found at Chichen Itza depicts a reclining figure carved from limestone with arms bent, knees pulled tight and his head turned at an impossible 90° angle and facing forward toward the temple entryway. Mother of pearl shell inlays were added to form the figure's fingernails, teeth and eyes. The Chacmool holds a plate on his belly that scholars believe may have held a human offering to both the gods and to the enthroned king, or perhaps the still-beating heart of an enemy warrior. The plate could have held other offerings as well, such as tobacco, feathers, tortillas or copal incense, which was considered sacred to the Maya.

The Chacmool at Chichen Itza

The inscrutable stone sculpture called a Chacmool emerged late in the history of Mesoamerican artworks, probably during the ninth century. The Chacmool found within the Castillo temple is not unique; 17 Chacmools have been uncovered in various areas of Chichen Itza, the largest number of any Mesoamerican site. Only five chacmools were excavated in situ within an architectural context; they were discovered within building entrances, near a place of ritual or a throne. Nine others were found close to important buildings but their placement does not explain their function for the Maya. Archaeologists have discovered 12 Chacmools in Tula. Eventually Chacmools were excavated in many later Maya archaeological sites, and in Aztec ruins as well.

Unfortunately, little is known about the figure's origin, its significance or how the Chacmool functioned in Maya daily life. Even the Maya's name for the Chacmool is unknown. The unusual statue is thought to have been created initially in the Yucatan at Chichen Itza and in the Toltec area of Central Mexico and then the figure spread through Guatemala, El Salvador and even Costa Rica, where one was found dating to ca. 1000. Chacmools were commonly used in the Aztec capital of Tenochtitlan in the sixteenth century, but they were unknown during the Classic period.

Most other Chacmools are designed in a style similar to the one found in the Castillo. The male figural image reclines quite uncomfortably always looking away; his raised hands appear to hold a bowl or a flat disk that sits upon the figure's stomach. There are a few slight variations in a Chacmool's basic form or iconography; some heads face right and others face left. The Chacmools may be richly dressed in a more elaborate costume, wearing a small headdress or sometimes nearly naked. They were probably all colorfully painted. The earliest Chacmools appear to be young men with warrior attributes but the figures developed different decorative characteristics in later depictions. None of the Chacmools are identical in attire or in bodily proportions but all hold the same awkward position. In Chichen Itza each one has its own individual design, but in Tula the opposite is true. There they have less innovation and show little deviation from the standard design.

The remarkable Temple of Kukulkan may yet reveal even more secrets; archaeologists suggest that the innermost pyramid may house a royal burial since this practice was frequently followed at that time. By the end of the tenth century Chichen Itza had reached its peak, but even after the decline of Chichen Itza during the thirteenth century, the Temple of Kukulkan and the Sacred Cenote remained holy places. These most-sacred sites of the Maya were visited as pilgrimage destinations long after the city was abandoned and Christianity dominated the religious climate.

The Sacred Cenote or *Cenote Sagrado*

The very existence of the Castillo is based on its connection with the Sacred Cenote, the essential element of the city's religious beliefs and ceremonies. Surrounded by sheer cliffs, the Sacred Cenote or Well of Sacrifice is nearly round, 60 m (200 ft) in diameter and 27 m (89 ft) deep. The Sacred Cenote of Chichen Itza received sacrificial offerings from pilgrims for hundreds, perhaps over a thousand years. The first investigations into the murky waters of the Well of Sacrifice were conducted from 1904 to 1910; these initial attempts searching for treasures were haphazard and damaged the cenote contents and geological setting. Its future archaeological context was destroyed which led to legal controversies; much valuable information was lost. The scientific explorations into the sacred sinkhole continued and dredging produced thousands of offerings to the rain god Chaac, mainly during droughts. The innumerable retrieved offerings accumulated the largest known quantity of carved jade objects in Mesoamerica. The precious artifacts and jewelry of jade, gold and copper sacrificed in the well proved the importance of Chichen Itza as a city of wide-spread trade and culture; also found were innumerable beads that would have formed neck pendants, bracelets, earflares, belts and clothing décor. These items may have been tossed in the well separately or perhaps worn by victims (Fig. 14.7).

Also extracted from the deep, green waters were ceramics, wooden and stone objects, textiles, obsidian tools, copal incense, and animal bones along with human

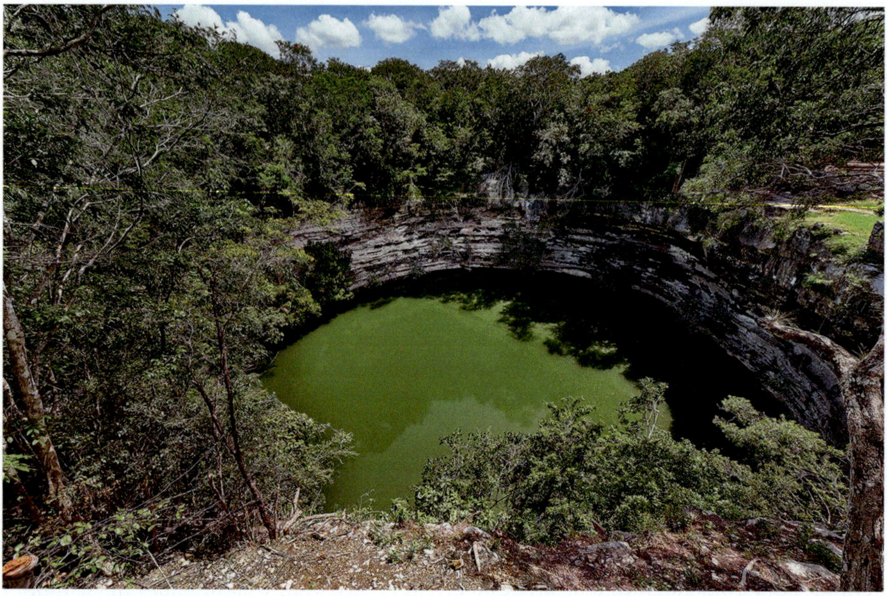

Fig. 14.7 Sacred Cenote at Chichen Itza was filled with centuries of offerings (Credit Alfredo Cerra)

bones of men, women and children. A study of human sacrificial offerings dredged from the cenote proved that the bones carried wounds typical of sacrifices. The human victims were usually killed before being thrown into the water. In the same manner, most sacrificial objects had been cut, torn or broken, typical behavior for ritually "killing" the object before offering it to the god. Not all cenotes contained offerings though, which indicates that only particular cenotes were thought to be connected with the underworld; sacrifices placed in other cenotes would serve no purpose.

Research shows that the Sacred Cenote constituted the basis of religious and ideological belief systems at Chichen Itza. Some say the city's very existence was linked to this essential natural water source. The cenote opposite, the Xtoloc has received much less attention since it was considered profane compared to the sacredness of the Well of Sacrifice, but to the Maya all water was sacred. Cenote Xtoloc was honored with a nearby temple. The paved road and important architecture leading to and around it indicated that this cenote may have also had sacred functions performed around the opening. The Templo del Xtoloc did not face the cenote but would have allowed easy access to it. There may have been a path leading directly from the temple to the Xtoloc Cenote and steps inside that would have led down to the water level. Perhaps this location was an important station on a ritual circuit or pilgrimage path through the more sacred portions of the city that displayed holy symbols, mythic references and historic or ancestral allusions.

El Caracol of Chichen Itza

Because of its unusual form and skewed orientation, the Caracol or Observatory is recognized as the most obvious structure in Chichen Itza to be associated with the astronomical interests of the Maya. According to archaeoastronomer Anthony Aveni, "The Caracol tower of Chichen Itza is surely the most famous example of a cockeyed building". This odd structure is unique among traditional Maya architecture; there is no other quite like it. Scholars describe the Caracol as probably a multipurpose building: a gnomon for casting shadows like a sundial, a watch tower for military purposes, a temple for communal ceremonies and a celestial observatory with a circular stairway for accurate star gazing.

Keeping close track of the solar and lunar years was of great importance for the Maya of Chichen Itza's agricultural cycle, for maintaining an accurate calendar and perhaps for astrological predictions. The need to align their architecture to mark certain positions in the cosmic cycle of Venus was most essential to the Maya and so they were well aware of the planet's celestial transit. Their reasons were probably numerous but since little text survives from Chichen Itza, we may never understand their obsession with the planet. Historians do understand that Venus and El Caracol were both associated with their primary deity Kukulkan and that Kukulkan, when disguised as the wind god, Ehecatl, was worshipped in round buildings throughout Mesoamerica (Fig. 14.8).

Like other major buildings of Chichen Itza, the Caracol was built atop a series of platforms, including two large rectangular platforms. Built around 800 CE, the lower and most expansive platform measures 75 × 75 m (246 × 246 ft); it was designed to support the structure and counterbalance any movements of the Earth. An additional large platform or terrace sits atop it, which raised the observatory to an even greater height. This highest platform measures 26 × 30 m (85 × 98 ft); it bears the weight of the Caracol and has features that assist in the viewing of the heavens. Archaeologists suspect that the Maya placed a drainage system within the platform to avoid the collection of water pools on the surface, although no surface penetrating radar has been applied to test that theory.

Two broad stairways rise to the highly complex cylindrical building; the base is fine but its upper portion survives in a very eroded state. The upper section was built atop a circular stone base measuring 18 m (59 ft) in diameter. The distinctive round base was formed by three concentric cylinders; the latest version of the base dates to 890 CE; the first or inner structure was built about 830–840 CE. The base section's exterior is decorated with Puuc-style decorative panels and with projecting cornices. The interior was constructed with finely cut stones. A solid central core of masonry fills the center of the building and is surrounded by two curved slots that were designed to support a moveable viewing apparatus. The inner space measures 8 m (26 ft) in diameter. The outer structure has four small doorways that are aligned to the four cardinal directions. Inside the ceiling is vaulted and a narrow spiral passageway winds up to a high tower chamber near the top of the building. Narrow openings or spyholes pierce the walls that are oriented for specific celestial views

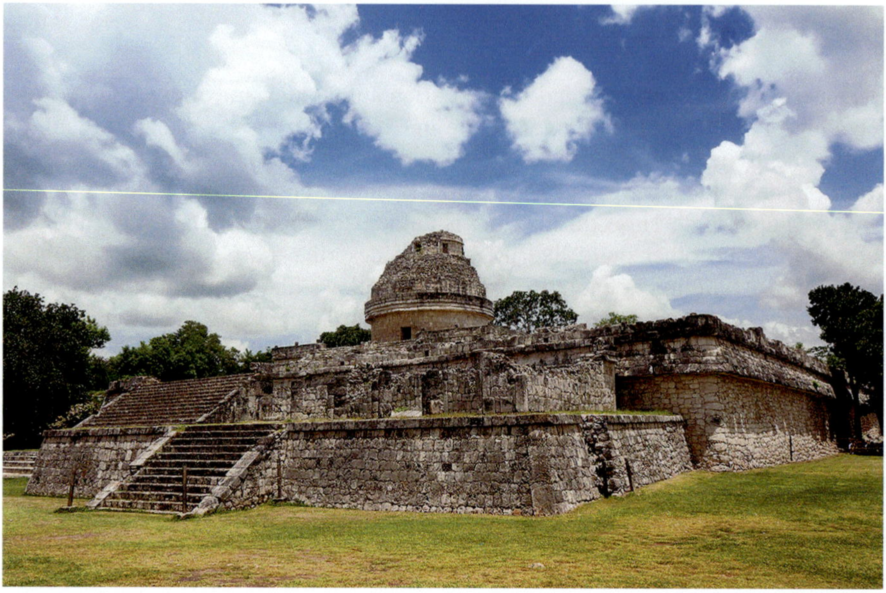

Fig. 14.8 El Caracol at Chichen Itza, the "observatory", astronomically aligned (Credit Liucaqq)

but only three spyholes survive the severe damage accrued over the past thousand years.

The principle viewing area on the upper platform stands a lofty 28 m (92 ft) above ground level. Because the observatory has been raised to such a great height and the surrounding landscape of the Yucatan lies very flat, a Maya skywatcher would have unobstructed 360° astral views from all points on the platform. The elevated position high above the city skyline would allow clear sighting of stars and planets rising on the distant eastern horizon and then setting on the western horizon; even better viewing would have been available within the Caracol's upper chamber. Discovered by archaeologists, traces of red and black paint on Chichen's Caracol's upper platform columns refer to the flight of Quetzalcoatl from Tula to the mythical city of Tlillan Tlapallan which literally means, "the land of black and red." These colors are also directional colors in Maya cosmovision; the east is associated with red and the west with black.

From the spyholes in the upper chamber there are three major viewpoints: to Venus' northern-most setting (and near to moonset at its greatest northern elongation); toward due west (the equinox sunset); and to Venus' southern extreme (and near to moonset at its greatest southern elongation). There are also alignments to the zenith passage sunset, the equinox sunsets, and sunsets on April 28 and August 18, important agricultural dates. There is also an alignment to the rising of the star Canopus. Evidently, there were many more observation points built into the upper section that has crumpled but we will never know the original appearance of the that section. Nor will we know the extent of astronomical information a Maya astronomer-priest could gather from the building's many viewing points. Astronomers have determined that of the 29 astronomical observations believed to be of great interest to the Maya skywatchers of Chichen Itza, direct sight lines to 20 points can be determined by the form and position of the Caracol.

The wide series of steps of the lower platform face 27.5° north of west, which is out of line with the other buildings in the area. The stairs are aligned to the northern extreme of Venus' setting and are only 2° off from the summer solstice sunset. Venus reaches this distant point only once in 8 years. The wide stairs up to the higher platform too, are not aligned with the central doorway. The central area of this higher stairway has a break or niche embedded that contains an altar that sits on two columns. The altar is not aligned with the upper steps but with Venus' northern extreme. One column of the altar was painted red and one was black. It was suggested that the red column represented Venus as a morning star, since the Maya word for Venus was *chak ek* that means red star or great star. The black column may have represented Venus as the evening star.

An additional orientation was found in the atypical building; a diagonal line from the northeast corner of the building to its southwest corner aligns with both the summer solstice sunrise and the winter solstice sunset. Viewing from the Caracol, the helical rising of Castor and Pollux, the two brightest stars in Gemini, can be seen at the same point where the Sun rises on the summer solstice. In Maya mythology these two stars of Gemini were known as the Turtle, an animal which was thought of as the Earth itself. These multiple astronomical readings were necessary to keep their observation calendars accurate in order to assure correct scheduling of their agriculture seasons and the associated ceremonies. The Maya astronomers had mastered the 584-day synodic cycle of Venus quite precisely. The Dresden Codex contains a Venus Table that traces the sequence of Kukulkan and some scholars think that the observations listed in the codex may have been compiled by early astronomers in this very tower. The Caracol of Chichen Itza provides some information on its possible uses but we are left with many more questions pertaining to its daily function, operation and exactly what astronomical phenomena the Maya kept close track of in addition to the synodic periods of Venus. Perhaps future astronomical studies will reveal more exciting discoveries concerning the many functions of the Caracol.

Temple of the Warriors

East of the Castillo sits the Temple of the Warriors, a massive colonnaded pyramid decorated with feathered serpent columns, Chac masks, monolithic benches and chacmools. This extensive structure is also astronomically oriented. Constructed between 900 and 1050, the temple is a colossal three-level pyramid, measuring 12 × 40.5 m (40 × 133 ft) and is thought to be the only late Classic Maya building designed to hold significant-sized gatherings. It is thought to postdate the last construction phase of the Kukulkan Temple. The Warriors Temple was almost completely built with porous limestone and was re-plastered with more than 120 layers, showing that it was used for many years after it was built. The temple was so named for its carvings of warriors and is admired for the impressive rows of 200 round and square half columns at its base that extend around the south side of the pyramid. Many of the columns are decorated with bas-relief Toltec-style warriors and tribute bearers, four of which are female. They were originally painted in vibrant colors; traces can still be seen. Mayanists believe that the columns supported some type of roof in the past; to enter into the formerly roofed space one passes through twin feathered serpent columns that end with fierce, gaping heads (Fig. 14.9).

When climbing to the Warrior Temple atop the bulky pyramid, visitors had a controlled entry; they were required to pass through the mass of guardian columns looming over them and then exit into the bright sunlight. The columns represented the military upon their return in triumph with many defeated captives with hands tied in tow. The temple atop the pyramid is reached by climbing a broad stairway that has a plain, stepped ramp on either side; each ramp holds a standard-bearer that would have carried a flag or banner. The upper platform was initially painted in tan, blue, red, green and black and would have been used for religious rites, civil ceremonies or perhaps ritual music and dance. The exterior of the temple was created in Puuc style but the interior is chiefly Central Mexican in appearance. While excavating, archaeologists discovered that, like so many other major Maya buildings, a nearly identical but smaller and older version of the temple, called the Temple of the Chacmool, lies beneath the Temple of the Warriors (Fig. 14.10).

When first uncovered, the impressive Temple of the Warriors was decorated with murals created with watercolors but the images deteriorated rapidly after being exposed to the elements. The bas-relief temple décor displays images of warriors, eagles and jaguars devouring human hearts. At the upper level a chacmool lies between the row of serpent columns that line the entranceway. The serpents on the columns bear astronomical symbols and decorative features over their eyes. On the top of each head of the feathered serpent columns sits a shallow bowl that may have been used as an oil lamp. The rear chamber of the temple is thought to have acted as a meeting place for powerful elites, indicated by the paintings of warriors on thrones in the earlier temple and by the attire of the mini-atlanteans that support the thrones.

Archaeoastronomers have determined that the Temple of the Warriors' astronomical alignment of 290° azimuth indicates an orientation towards the Pleiades setting position. In northern Yucatan, accurate sightings of the Pleiades were thought to be especially crucial during the Late Classic and Early Postclassic periods because the bright star cluster rose at dawn after about a month of invisibility, close to the time of the solar zenith. During the Late Postclassic the Pleiades' setting position also marked the horizon position of the zenith sunset. When the Postclassic period was ending, the Pleiades moved closer and closer to the position of the zenith sunset. By 1300 the bright cluster set at 292°, the exact location of the zenith sunset in northern Yucatan. Other structures at Chichen with feathered rattlesnake columns have been shown to have Pleiades orientations.

In addition to these orientations, archaeologist Ivan Sprajc published his study of the rain-maize complex in relation to images of Venus and building alignments. He

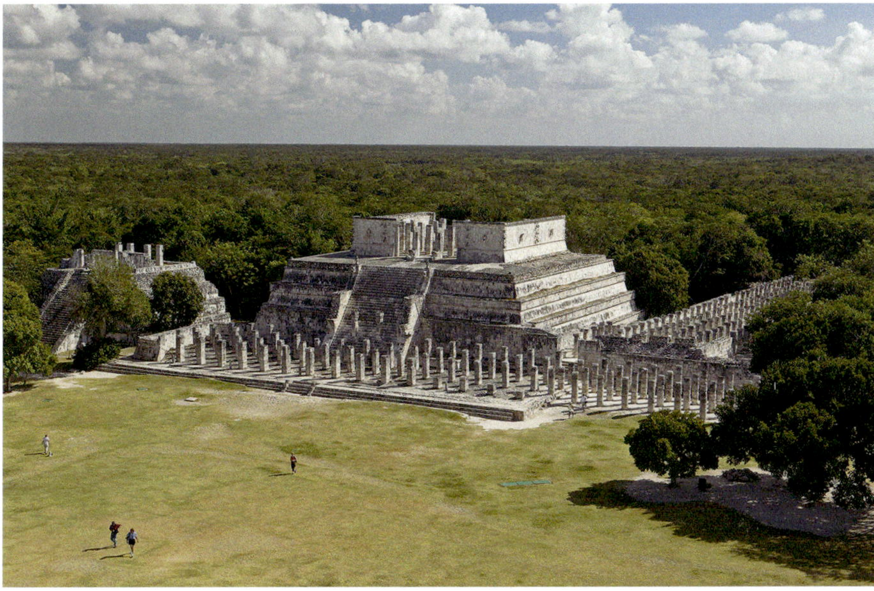

Fig. 14.9 Temple of the Warriors with rows of square and round columns and warrior decor (Credit Joseph Sohm)

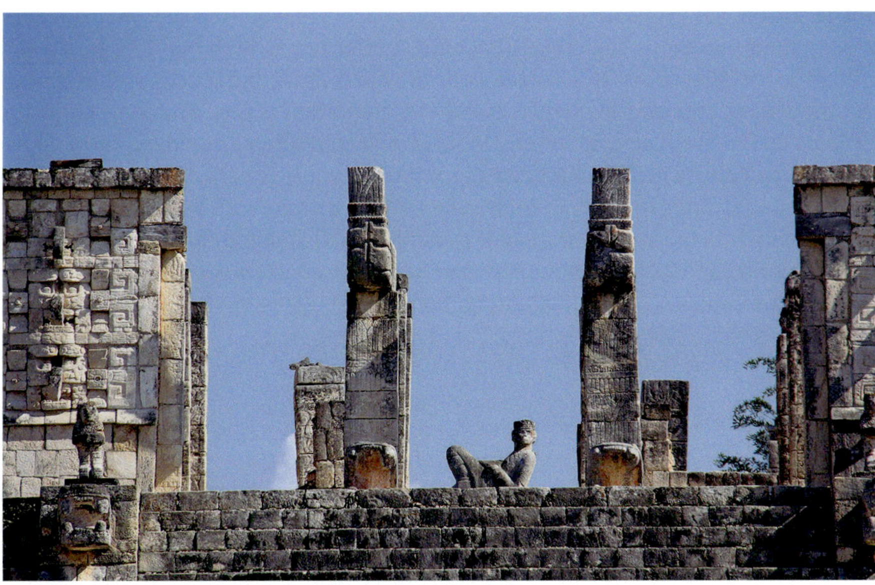

Fig. 14.10 Chacmool Temple atop Temple of the Warriors with Chacmool statue (Credit S.Moebs)

found a close association between rainy season images and the Evening Star. He discovered that Venus positions, when reaching its maximum northern and southern points on the horizon, matched the beginning and end of the Mesoamerican rainy season but at the Evening Star extremes, rather than the Morning Star positions. He affirmed that the Maya astronomers recognized this feature and observed Venus when it first appeared or disappeared as the Evening Star at the beginning of the rainy season. The orientation of the Temple of the Warriors and a few other Chichen buildings are aligned to this western Venus extreme. This important orientation is also near the position of the zenith sunset of 292°, as mentioned. Along with the Venus architectural alignments, the brilliant planet is honored with artwork on the Temple of the Warriors. At each corner of the temple are glyphs that symbolize Venus, half flowers with crosses on the petals. The claw-footed Venus monsters carved on the panels of the temple are thought to be images of Venus at inferior conjunction.

Experts suggest that the feathered serpent columns at Chichen, representing Kukulkan could relate to a symbol of Venus as the Evening Star in association with the Pleiades. When linking the longitudes of Venus with the position of the Pleiades during the Classic and Postclassic periods, there is a relationship between the conjunction of Venus with the Pleiades and the onset of the rainy season in Mesoamerica. Through its orbit of the stars, Venus shifted to the position of the Pleiades once a year between mid-March and mid-June which is the beginning of the rainy season. Watching the close conjunctions of Venus and the Pleiades would provide important forecasts for the upcoming season.

The strong, yet mysterious, association between Chichen Itza and Tula is recognized in their numerous shared architectural similarities in both building forms and in their decor. Mayanists still debate over the conditions, influences and timing of the compelling connections between the two important cities. The Temple of the Warriors and its architectural setting, including the Castillo pyramid has been duplicated by the Toltecs at Tula but there the Warrior temple is called Pyramid B. Both temples have strikingly similar colonnades, pillars carved with warriors, and chacmools but in Tula the temple atop their pyramid has disintegrated, only the gigantic, impressive atlantean supports remain standing. The consensus now seems that the two similar temples were contemporary, both were created at about the same time.

The Temple of the Jaguars

Constructed between 1000 and 1150, the Temple of the Jaguars was so named from the number of jaguar images found on the exterior of the building. The temple is located immediately to the east of the Great Ballcourt, built into the wall. Those in the temple could easily witness activities in the ball court. Its great height of 10 m (32.8 ft) provided clear sightlines to the horizon in every direction. The building consists of a pyramidal base with an Upper and a Lower Temple; both are fronted with a triple entranceway but facing opposite directions. The Upper Temple is supported by two huge, feathered serpent columns showing wide open mouths at each side of the entranceway. The walls of the temple's facade display reliefs representing the god Kukulkan, with fierce warriors wearing headdresses, military costumes and carrying atlatls. The smooth, sloping walls of the lower facade are finely dressed and the temple sides show panels that are adorned with eagles and jaguars consuming human hearts (Fig. 14.11).

To access the Upper Temple, there are long, steep staircases on each side. The interior walls and ceiling of the Upper Temple were painted with domestic and battle scenes, human figures, houses, and trees. The murals are now almost invisible, but fortunately they were copied in the nineteenth century. Facing west, the Upper Temple has an orientation of 286°, slightly off from the zenith sunset. At that time a rectangle of sunlight would pierce the inner chamber between the feathered serpent columns, brightly illuminating the door jambs and the colorful murals on the back wall. This added effect of the flash of sunlight would have enhanced the ceremonies underway during this notable date in the Maya calendar, zenith sunset.

Another celestial connection at the Upper Temple is suspected concerning the serpent columns, now identified as rattlesnakes. The Pleiades were called *tzab* or "the rattlesnake's rattle", perhaps the rattlesnake capitals on the serpent columns were symbols of the Pleiades. This would be probable since the temple is aligned with the position of an important Pleiades setting on the western horizon. This orientation is almost identical to that of the window of the Caracol, whose midline faces sunset on April 28; the right side of the window frames the Pleiades when they disappear for 1 month on that date. Also the door of the Upper Temple framed the Pleiades when they set 1 month before the solar zenith on May 25. Tracking these astronomical positions would be essential in maintaining an accurate calendar.

The Lower Temple, which opens behind the Great Ballcourt, is also fully decorated with five rows of warriors and deity figures, featuring a sun disk above. This temple holds another jaguar throne, similar to the one found in the inner temple of the Castillo, but this throne was not pristine but well-worn and lacking its paint and decorations. The outer columns and the interior walls of the Lower Temple were all adorned with elaborate bas-relief carvings. Possibly militarily focused ceremonies were performed at the temple when the rising Sun would enter the chamber and illuminate the extensive artworks.

Astronomy is also a feature of the Lower Temple which is aligned toward sunrise several weeks before the solar nadir in late November and several weeks after the solar nadir in late January; its actual orientation is 106°. This would be the dry season, a time of little agricultural productivity, when activities involved with warfare became more intense. The inclusion of this large quantity of astronomical and celestial symbolism in the architecture of Chichen Itza, along with its historical and religious symbolism adds to the depth of meaning embedded in their art and architecture.

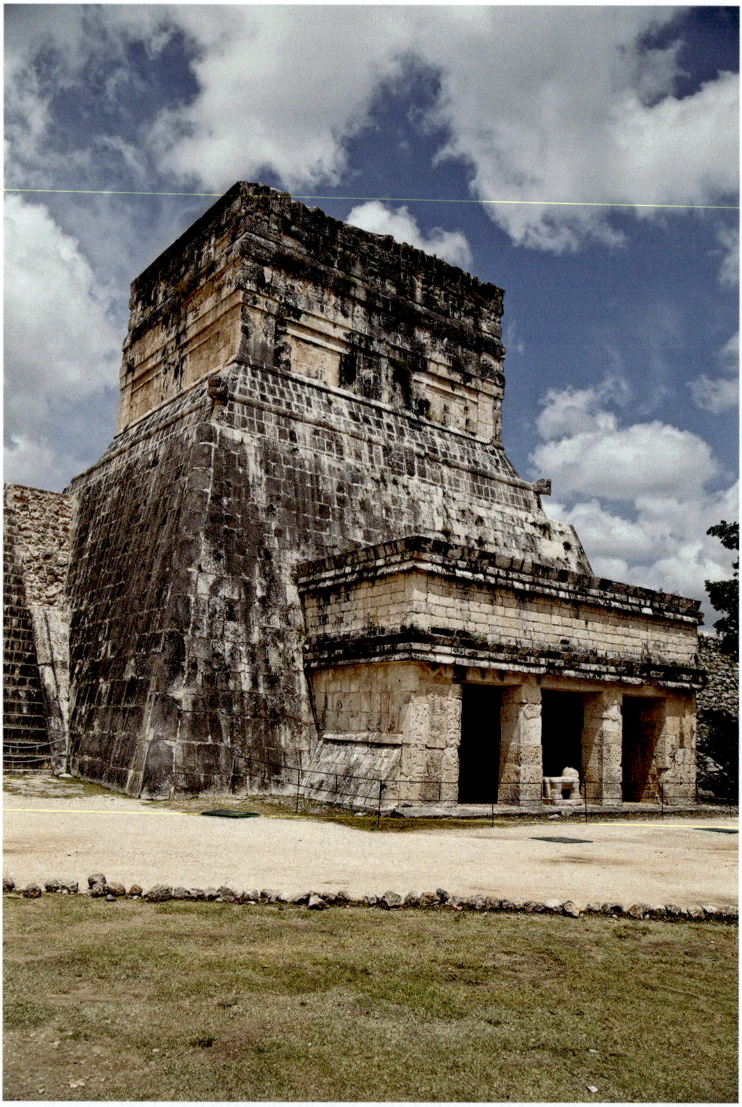

Fig. 14.11 Jaguar Temple with upper and lower temples, decorated with jaguar images (Credit Yory Frenklakh)

Chichen Itza Astronomy

Chichen Itza is one of the most important cities for determining celestial orientations and has been included in numerous archaeoastronomical studies because of the large number of its buildings that have specific alignments. These astronomical connections serve as a vital part of the city's religion, worldview and political ideology. They reflect on the success of the rulers and their approval by the gods who govern the heavens, the Earth and all life. These alignments often governed the architectural design and the urban planning in the civic and ceremonial cores of the communities. The alignments with the heavens determine the time, content and direction of many rituals, celebrations and religious processions.

These astronomically aligned buildings did not function only for that one purpose, solely as observatories, but were necessary for fundamental functions of the society as well, for religious, administrative, military and residential purposes. These special buildings do not only align to significant events in the heavens; they often include astronomical motifs as well. The decorative features and iconographic symbolism on both the exterior and interior of the structures expanded the astronomical information encoded in its geographic position. The astronomical orientation often acts as the main clue to interpreting the symbolism and implications of the relief sculptures and murals depicted on the building's walls and exterior facade.

Chichen Itza has been the subject of several advanced systematic studies in which some earlier errors in measurements have been corrected. The conclusions were based on statistically significant patterns that were carefully measured. These same astronomical patterns are seen in many Mesoamerican cities. The principal buildings at Chichen, well known to be oriented to important astronomical positions, are the Castillo and the Caracol but many other structures in the city were designed with specific celestial alignments as well. If the astronomers of Chichen simply wanted to achieve an observational calendar by watching celestial bodies, it could have been done without these monumental constructions.

The celestial points most often encoded into the architecture at Chichen Itza are those same positions selected at most Maya cities. Chiefly they utilized the Sun's key positions at the equinoxes, solstices and zenith, the synodic cycle of Venus, especially the heliacal appearance of Venus as the evening star, the maximum elongation of Venus and the setting of the Pleiades. Recent investigations of architectural orientations of cities in the Maya lowlands shows that extensive, long-lasting alignment patterns were accomplished most often by using astronomical positions at the horizon.

Customarily, when ancient skywatchers aligned Mesoamerican buildings, they would focus on horizon features, such as jagged hilltops or tall standing stones. With the lack of measuring tools, they used crossed sticks in order to pinpoint accurate rising and setting points of specific heavenly bodies. Because both the eastern and western horizons at Chichen Itza are completely flat, without hills or any distinguishing points to mark a specific locale, establishing accurate observations was usually quite difficult. As a result some of their alignments were off a degree or two from the exact position of the star or planet. By utilizing the Sun's transits along the eastern and western horizons and observing its zenith passages, the Maya amassed sufficient knowledge to create ceremonial centers that were carefully aligned to the desired heavenly phenomena. This understanding of the movements of the Sun, stars and planets enabled their accurate knowledge of their geographic position. Once an alignment was established, that directional line was often codified into a building's orientation, which explains the unorganized appearance of some regions of Chichen Itza and other Maya cities.

New studies have provided clearer interpretations and enhanced understanding of Chichen's main buildings. Many of the more significant buildings of Chichen were aligned chiefly to sunrises and sunsets on particular dates separated by calendrically important intervals. The Maya skywatchers used observational systems to monitor and make adjustments to their calendar. These periodic attunements were necessary since the Earth's revolution around the Sun does not correspond precisely with a whole number of days. It was necessary to adjust the calendar with the seasons as required for successful crop production.

The Sun's zenith was an essential marker for the Maya to maintain their calendar but they did not show much interest in the zenith sunrise or sunset. The most important measurement in tracking the zenith was the Sun's appearance directly overhead. In Chichen, it was the orientation of the Castillo and its alignment with the Holtun cenote that provided the exact moment of the Sun's zenith. Keeping close account of the Sun's zenith was essential for the adjustment of the calendar to maintain its accuracy in connection to the seasons. The astronomers at Chichen sustained their calendar's precision by observing the Sun's location at its zenith, using a marker on the horizon and a gnomon. To ensure the calendar's accuracy, they relied on three measurements. They recorded that the sunrise aligned with the north corner of the Castillo; the position of a shadow cast by a gnomon at noon, and the sunset at the western stairway that pointed to the Holtun cenote. The astronomical alignments of Chichen's foremost structures will be reviewed individually.

The Venus Platform or Temple of Venus

The Venus Platform received its name from the abundant manifestations and references to the most important planet that are found adorning the structure. The platform is carefully positioned in front of the primary entrance to the Castillo along the roadway to the Sacred Cenote indicating that the platform held a crucial role in the alignment of these two most important places of the city. Also called the Temple of Venus, the structure consists of a square, flat-topped podium with a wide stairway on each of its four sides that leads up to the flat, open surface. The steps of the stone structure are flanked with square balustrades from which large, ornate and well-preserved feathered serpents' heads project. The feathered serpent signifies Kukulkan who is closely associated with Venus (Fig. 14.12).

The platform is noted for the large Venus or star sign and is marked with the symbol of rulership and the War Serpent, a mythical and militaristic being whose origin can be traced to the ancient art of Teotihuacan. It also displays four stars in the corner of each of its sides. On its raised panels are bas-relief carvings of glyphs signifying the planet Venus in the shape of a knot of years, next to a half flower with crosses on the petals. In a recessed area an image of Venus with jaguar claws is depicted emerging from the jaws of a feathered serpent. Most importantly, the

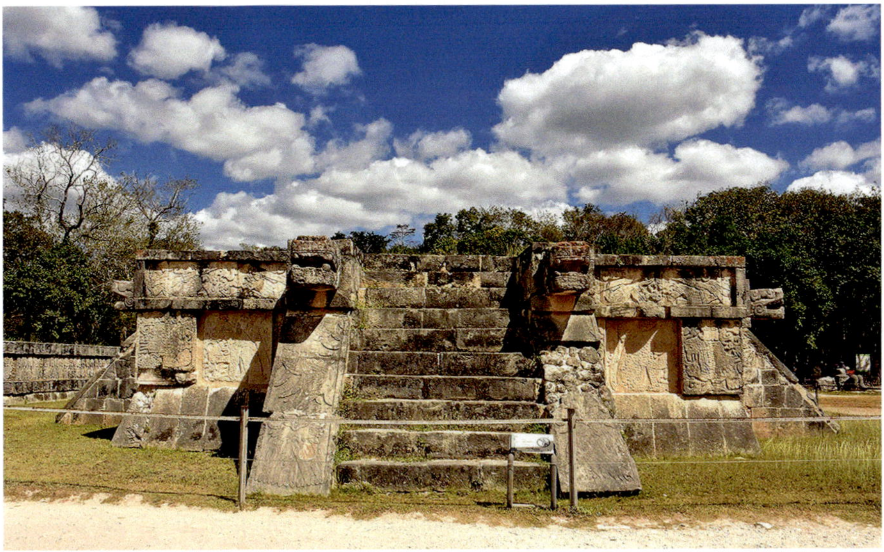

Fig. 14.12 Venus Platform, square monument with heads of Kukulkan and Venus symbols (Credit alicjaalba)

platform is inscribed with essential astronomical data equating the five Venus cycles with eight solar years.

Like other Chichen surfaces, the Venus Platform was originally painted tan, blue, red, green and black. A sacrificial offering of a decapitated male skull from an unknown date was discovered at the eastern stairway, hinting to one of the platform's possible functions. The platform was probably used as a podium for religious observances, public ceremonies, ritual dances or perhaps even human sacrifices. The platform may have functioned as a viewing deck for the tracking of Venus as well. The important artworks signifying the feathered-serpent god Kukulkan and its Venus symbolism, in addition to its prime placement within the Grand Plaza and on the sacbe to the Sacred Cenote, all indicate that the Venus Platform was a foremost edifice used for many special purposes.

A second Venus platform was erected between the Osario pyramid and the Xtoloc Cenote; it too is decorated with Venus signs and plumed serpents. The serpents are marked with cloud motifs that show they are "cloud serpents". The addition of Ek' star symbols along their bodies indicates a celestial reference, perhaps embodying Venus, as the Morning Star.

The Tzompantli at Chichen Itza

When first studied in the 1960s and 1970s, the ancient Maya were described as peace-loving and gentle people, uninterested in warfare and human sacrifices. This initial interpretation of the archaeological remains found in the amazing Maya cities has been proven quite flawed, especially after archaeologists restored and studied Chichen Itza's most gruesome structure, the Tzompantli or Platform of Skulls. This lurid expanse of skeletal decor carved on the monument displays highly graphic evidence of sacrificial death, evoking an entirely different and more realistic understanding of the Maya culture. Maya warriors were as fierce as other Mesoamerican societies; human sacrifices were just as prevalent in their military encounters as in many other early cultures, but not equal to the blood lust of the Aztecs (Fig. 14.13).

The display of trophy heads of the enemy was practiced in many medieval societies but Mesoamerican exhibitions of military deaths were much more expansive. The decapitated heads of enemies were often exhibited piled atop one another on vertical poles. The original skull racks of Mesoamerica were built from wood with rows of tall poles stacked with decapitated heads, which fulfilled the same purpose as the stone scaffolds, the public display of human heads, usually prisoners of war or other sacrificial victims. More than 30 examples of the Tzompantli have been discovered in Maya cities; their dates vary from the Epiclassic (600–900 CE) through the early Postclassic (900–1250 CE). The earliest presence of the carved skull and cross bone design dates to at least 1000 years before the creation of the wall of skulls at Chichen. The great Tzompantli of Chichen was probably built during the early to middle eleventh century; the long, rectangular platform was constructed of limestone and measures nearly 13 × 16 × 2 m (42.6 × 52.4 × 6.5 ft). The central Tzompantli was erected about the same time as the Great Ballcourt which is located nearby; it has been associated with the Ballcourt through their similar iconography.

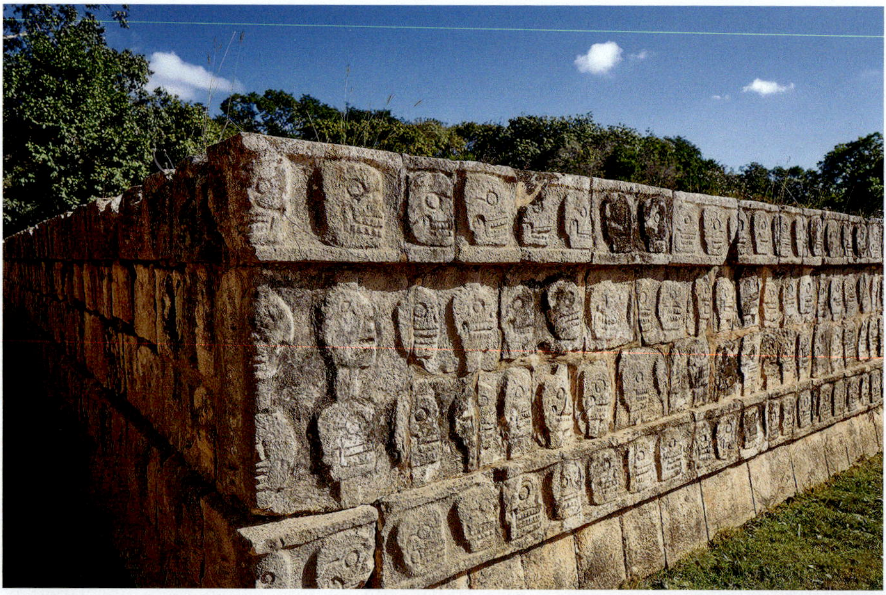

Fig. 14.13 Tzompantli at Chichen Itza, a long platform decorated with skulls (Credit lunamarina)

The skull platform forms the letter T with a wide stairway to the top; the walls are decorated with rows of decapitated heads, long bones and skeletonized warriors. The long wall facing the central plaza has four rows of continuous decapitated heads. The top and bottom rows of skulls are separated from the two central rows by raised molding. The bas relief heads are carved in profile and appear to face toward the east; the total number of skulls number 1848 surrounding the entire platform, even bending the corners. The wall behind the skulls retains bits of reddish pigment; a few heads appear to have traces of black paint. Emphasizing the platform and skulls with bright color would certainly exaggerate the effect of the artistic macabre display.

The aligned rows of skulls are all similar in size and of correct anatomical proportion but contain a variety of differing facial forms, showing the skill of the carvers. Some of the heads had been elongated, usually performed by wrapping a baby's skull until it becomes deformed; the process was an indication of elite status. It is not known whether the skulls were carved during one single episode or in stages over a long period of time.

In addition, carvings on the other wall of the Tzompantli depict two lines of ten skeletonized warriors with two large, standing figures on either side of the east-facing staircase. One warrior is deteriorated; the other warrior is decorated with feathered serpent images and flames which link the figure to Venus. The image is directed toward the Venus Platform across the plaza that was bathed in light each morning by the rising Sun. These warriors could imitate the battle between the Morning Star and the Sun which is a prominent tale in Maya mythology.

Many questions surround the enigmatic Tzompantli. Neither the civic function nor the religious or military ceremonies associated with the Tzompantli have been clearly established by Maya specialists. Nor has textual evidence of the activities that took place atop the skull rack survived to explain their significance to the people of Chichen or to those of rival cities. Nor do we know whether the Tzompantli had a connection with the afterlife. The Tzompantli was ideally placed for performances, civil announcements and public displays. Although no documentation survives, experts think that trophies of war, such as carvings, feathers, jewelry and luxury goods would also have been displayed atop the skull rack along with the enemy trophy heads.

There seems to be a close association between the ballcourt and Tzompantli, but not a mandatory connection. An enormous number of ballcourts survive in Mesoamerican, now reaching over 2000, which hardly compare to the 30 Tzompantli that have been discovered. Some suggest the ballcourt was considered as a crevice in the earth that provided access to the land of the gods and that of the ancestors. A large ring of stone, similar to those attached to the walls of the Ballcourt, was discovered among the fill within the Tzompantli. The stone ring is carved with entwined feathered serpents and seems to match the Ballcourt's rings; perhaps it used as an altar.

Carved illustrations have shown that playing ball and decapitation are closely associated in Mesoamerican cities. At Chichen, six ballcourts illustrate the beheading of a ball player; the fate of a losing player appears to be decapitation and then display of his skull on the Tzompantli; although this practice is not fully understood. The Great Ballcourt displays images of kneeling, decapitated ballplayers that depict severed arteries which become serpents and flowering vines. Across from the losing players stand the winners holding a knife in one hand and a detached head in the other. Skeletal images commonly adorn ceramic vessels and decorate the building's walls in Mesoamerica cities but none feature the vast numbers that appear at Chichen Itza; the images of death found there are of large scale and regularly appear in public spaces.

The functions and significance of the enigmatic Tzompantli are still being debated by scholars. Recent scientific studies of human remains, hieroglyphic texts, iconography and illustrations on polychrome vessels are producing revised interpretations of human sacrifices, the treatment of enemies, devotion to the ancestors and their sacrifices to the gods. The accurate explanations of the Tzompantli and its association with the ballcourt is still waiting to be written.

Chichen Itza Art

The largest amount of sculpture found in the entire Maya world can be seen at Chichen Itza, an area of great sculptural invention. Carved artworks in both bas-relief and three-dimensional form decorate the architectural facades, standing columns, stone lintels, benches and interior walls. Unfortunately, great quantities of Maya artworks were created using perishable materials, such as carved wood and painted pottery and woven textiles that have not withstood the all-pervasive tropical climate. An enormous number of jade art pieces were offered to the gods by deposit in the Sacred Cenote, mostly very thin plaques with bas-relief carving on one face only, probably drilled with sections of cane using jade, sand, garnet or quartz grit. What also remains in great numbers are the stone and mineral carvings and painted ceramic vessels, even considering this large amount of artifacts provides a limited perception of the actual appearance of a Maya city or village. Surely the temples, palaces, and civic buildings were embellished with brightly painted wall murals, woven curtains and rugs, multicolored wall hangings, and flamboyant flags. Maya burials preserve many artworks and ceramics but tease scholars with scant traces of codices written on bark paper of which there must have been thousands in existence.

After 900 CE the production of artworks in the south had ended, but art creation in the northern cities continued to be plentiful and very creative with new designs and artistic techniques. The Terminal Classic and the Early Postclassic sculpture in the northern Maya lowlands, especially Chichen Itza, developed its own particular features, quite distinct from the southern designs where the climate and raw materials were different. As an example, carvings of long processions became popular scenes on Chichen buildings, not seen elsewhere.

The carving and erecting of tall stelae filled with timely information, so popular during the Classic era, was almost eliminated; carvings of text and images were engraved instead on interior and exterior walls, doorways and lintels. But later, textual inscriptions declined in Chichen as well and representational art became the main mode of communication. The emphasis on rulers changed from the individual powerful king or single leader common in the south to pairs or groups of important figures of equal status in the north. A leader's status came from his strength and reputation as a warrior-king rather than his lineage or ancestors. The previously popular ruler portraits disappeared but the names of rulers were recorded; the noble images were replaced by carved thrones that could hold any qualified candidate.

The use of symbolic artworks was a manifestation of political power and relationships among the elites and religious leaders of the city. Scribes and artist held prominent positions by maintaining the hieroglyphic and iconographic records which showed that they held special knowledge unknown to others. At Chichen, accurate identity of images of the gods is problematic since it is difficult to distinguish between historical people, political offices and sacred gods.

Repetition of the plumed serpent appearing in building after building emphasized its role and reinforced the power of the cult image, spreading it throughout the peninsula. Images of the Feathered Serpent god, Kukulkan are abundant at Chichen Itza probably the most prevalent and most honored subject in Chichen artworks. The devotion to Kukulkan in the Yucatan began during the decline of the Classic period around 900; the mythical deity became the focus of a religious cult around which the community began developing political organizations and noble lineages. The artistic images combined the traits of both Maya and Central Mexican styles. Kukulkan was not a new deity but had been esteemed in various parts of Mesoamerica. It was first seen in the artworks of the Olmec, pictured as a crested rattlesnake, and also appeared in the images surviving in Teotihuacan. The venerated images of the Feathered Serpent played a vital role in the cultural development of the Mesoamerican cultures.

The Feathered Serpent God joined the plumage of a creature able to soar to the skies with the snake who could freely descent into the dreaded Underworld and just as easily return to Earth at will. The multi-functional Feathered Serpent possessed power in both the terrestrial and celestial worlds; it had the ability to pass through the clouds to generate rainfall to enable the growth of the most precious crop, maize. Its dual nature imitated the supernatural abilities of the gods. Association with the powerful forces of the Feathered Serpent enabled certain chosen individuals to establish dominance over others. Kukulkan was the most revered deity and his image was displayed in the architecture, sculpture, murals and ceramic vessels. The Feathered Serpent came to dominate the art and architecture of Chichen Itza, the protruding heads and threatening jaws of the fierce snakes added drama and power to its most important structures.

Another dominant subject of artworks at Chichen was Venus symbolism and Venus-related figures. Some Venus figures hold darts, some are intertwined with serpents and others appear with skull faces or masks which are seen on several piers at the Temple of the Warriors. The latest phase of art production at Chichen Itza ends around 1100 and shows no datable inscriptions. After that date almost nothing new was built, even ceramic vessel production ceased. The artworks of Chichen Itza began their long era of decay.

Chichen Itza Final Decline

The decline of Chichen Itza appears to have occurred in two waves, both associated with droughts which became an ongoing problem for the area. The first sign of decreasing wealth and political difficulties occurred between 850 and 925 CE. The scarcity of inscribed dates and chronological details makes both the development and the weakening of Chichen Itza difficult to state with certainty. Radiocarbon dates derived from the archaeological site are few in number. As a result the conclusions on its decline are drawn mainly through the study of its art and architecture. The lack of information on the history of both Tula and Chichen Itza allows for many competing theories to be presented in the literature of the archaeological site.

The city's decline began to drop again from 1000 to 1100 CE, when the drought became even more severe and the population levels dropped considerably throughout the northern Lowlands. By the start of the eleventh century, the production of monumental architecture and carved sculpture had ended; very few ceramic vessels were created. The decline in population, political disintegration and the final fall of Chichen Itza followed about 100 years after the collapse of the other Puuc Maya cities. It was one of the few large cities to survive the collapse of the early tenth century. The reason for this may have been the city's prominent position as a pilgrimage destination and its importance as a cult center for the worship of Kukulkan-Quetzalcoatl, the Feathered Serpent. From the eleventh century to the start of the sixteenth century some Mayanists call the "dark age" of the northern Maya lowlands, but the average rainfall returned to normal.

Chichen Itza is considered the most important of the hundreds of archaeological sites in the Mexican state of Yucatan. The possibility of Toltec inhabitants and their influence, the presence of foreign invaders, and the origin of architectural designs in the city remain hotly debated topics among archaeologists and historians. Most scholars now agree that Chichen Itza was a Maya site throughout its history, but the precise origin of the Itza people is still undetermined. Even though the city was long deserted by the time the conquistadors arrived in the Yucatan, the Maya people continually made pilgrimages to the honored places in the ancient city, especially the Sacred Cenote and El Castillo. The fascinating city of the Yucatan Peninsula now attracts more than two million people each year, with as many as 48,000 visitors per day.

References and Further Reading

Boot, E. (2000). Architecture and identity in the Maya lowlands: The Temple of K'uk'ulkan at Chichen Itza, Yucatan, Mexico. In P. R. Colas et al. (Eds.), *Sacred and profane: Architecture and identity in the Maya lowlands*. Verlag Anton Saurwein.

Braswell, G. (2015). *The ancient Maya of Mexico: Reinterpreting the past of the northern Maya lowlands*. Routledge.

Braswell, G. E., & May, N. P. (2015). In the shadow of the pyramid: Excavations of the great platform of Chichen Itza. In G. E. Braswell (Ed.), *The ancient Maya of Mexico*. Routledge.

Carlson, J. B. (2013). Chacmool, who was that enigmatic recumbent figure from the Epiclassic Mesoamerica? Reposing the question. *The Smoking Mirror, 21*(4), 1–8.

Garcia, I. A. M., & Cano, D. W. (2014). El Castillo en Chichén Itza; Un monumento al tiempo. *Arqueologia Mexicana, 21*, 80–85.

Marley, C. (2015). *Templo de Kukulkan*. https://www.academia.edu/9230874/Templo_de_Kukulk%25C3%25A1n

Šprajc, I., & Sánchez, P. F. (2016). Astronomy and architecture in the Maya lowlands. *Journal of Systems Architecture, 2*(2), 189–215.

Teiero-Andrade, A., et al. (2018). Illuminating' the interior of Kukulkan's Pyramid, Chichen Itza, Mexico by means of a non-conventional ERT geophysical survey. *Journal of Archaeological Science, 90*, 1–11.

Volta, B. O., et al. (2017). The archaeology of Chichen Itza, its history, what we like to argue about and what we think we know. In *Landscapes of the Itza: Archaeology and art history at Chichen Itza and neighboring sites*. University Press of Florida.

Chapter 15
Later Cities of the Yucatan Peninsula

Maya City of Mayapan

Around 1100 CE the city-state of Mayapan, located about 100 km (60 mi) west of Chichen reached its ascendency and served as the political and cultural capital of the Maya in the Yucatan Peninsula; it was the last and least known of the principal Maya cities. The earliest evidence of occupation dates as early as the Late Formative period. Archaeological remains in the region confirm that Mayapan was occupied long before the great city was actually established. Samples of pottery typical of the area appear as early as 1050/1100, when monumental construction in the small, early city first began. Mayapan was settled during the Classic period but was redesigned after it became a political and religious center around 1263 CE. The archaeological record of the earlier city was lost after many modifications and rearrangements of its buildings.

During the Late Postclassic period until the 1440s, the city had ruled for over a century and it reached the peak of its power; the population of Mayapan is estimated to have peaked at 15,000–17,000 people. According to ancient legends Mayapan was founded by the famous, or perhaps mythical, ruler, semi-god Kukulkan in CE 1250 or 1263, but new studies indicate the actual foundation date have been much earlier, even back to the eleventh century. It may even overlap with the fall of Chichen Itza. Whether mythical or real, Kukulkan was honored at Mayapan, just as he was at Chichen Itza and Tula. His cult was celebrated and conveyed by images and iconography of the feathered serpent in the most important architecture and in public art, such as the columns and balustrades of the Kukulkan temples found at both locations. The cult's significance within Mayapan's social, political and religious systems is still a topic of research for scholars. Whether the major cult figure was considered a god or a living mortal, Kukulkan was connected to the formation of many important Mesoamerican cities in both the Yucatan and in Central Mexico, including Tula, Chichen Itza and Mayapan.

M. Dolan, *Decoding Astronomy in Maya Art and Architecture*,
Springer Praxis Books, https://doi.org/10.1007/978-3-031-61092-9_15

Earlier assessments of Mayapan's position and power found in scholarly literature were mostly negative, but these early appraisals have been proven to be inaccurate. Refuting those early opinions, recent studies have shown that Mayapan was a thriving import/export center that traded highly desired luxury goods, such as cotton, salt, maize, honey and cloth. They also traded in obsidian and metal objects that they produced in their own workshops. After Chichen Itza began losing its political supremacy toward the end of the tenth century, Mayapan rose to fill the power vacuum. The rulers of Mayapan continued the prosperous trading and cultural exchanges established by Chichen merchants and leaders. They secured and maintained its political and economic relationships. It held the prime location on the crossroads of international contact. Their extensive trading networks spread as far as Guatemala, the Gulf Coast and Central Mexico. Mayapan succeeded Chichen as the political, economic and ceremonial hub of the Yucatan. The city's successful operations intensified so that for the next 200 years it reigned as the most powerful urban center in the waning Maya world. Mayapan continued as a vibrant international center during the final centuries of the Maya civilization, a politically turbulent time, when all other important city-states of the Maya Classic period in the south had failed. Mayapan also had an enormous influence on the smaller city-states around them. Perhaps their great success was even the cause of Chichen's decline (Fig. 15.1).

The first serious investigations of Mayapan were undertaken by John Lloyd Stephens and Frederick Catherwood in 1841 when they measured and mapped the

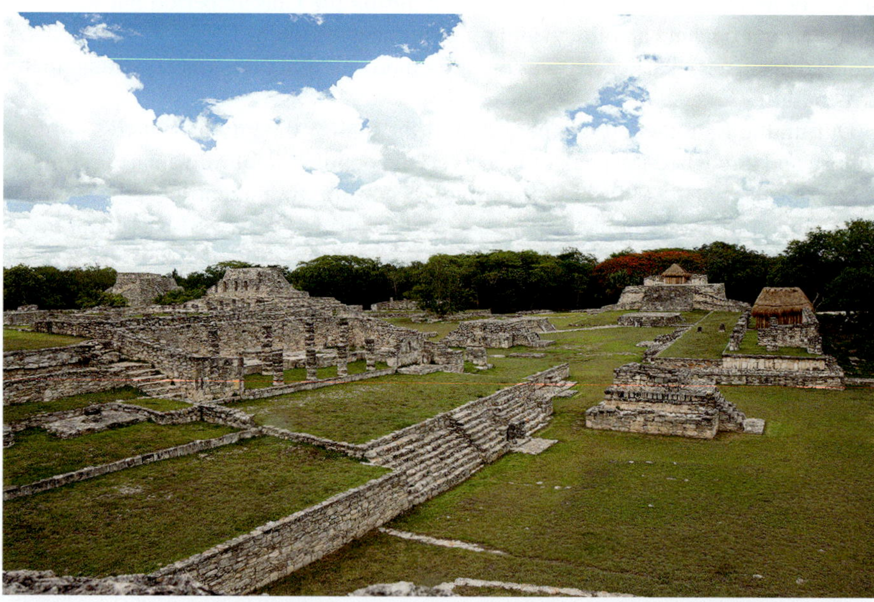

Fig. 15.1 Ruins of Mayapan, central plaza and platforms (Credit Claudia Luna Mtz)

buildings and drew illustrations of the most important central structures. The site has now been professionally surveyed and mapped; it was excavated by archaeological teams and research studies of its art, architecture and history still continue. The ancient ruins include more than 4000 structures, all within the city walls, with more than 2000 modest residences, dwelling groups, clusters of house lots share boundary walls and are separated by footpaths. Even more houses sit outside the walls, extending in every direction and enclosing almost 4 mi^2 but these areas are less densely occupied. The residential structures were quite crowded, all tightly packed within the city walls which measured over 9 km (5.6 mi) and had 12 gates, 7 of which have vaulted entrances.

Most residences were simple structures with only a front and back room, built using only wattle and daub material and then covered with a roof made of palms. Elite residences in the city number about 50 and are much more elaborate; they are part of an assemblage: a colonnaded hall or residence structure, a small one-room colonnaded oratory and a personal altar or shrine; these are rarely found in modest homes. The prestigious homes often display stone columns, masonry walls and beam-and-mortar roofs and are located throughout the site. The streets were not organized in grids and houses were situated haphazardly with lanes winding between the houses and walls. The residences were often arranged in small patio groups with courtyards.

Like other areas of the Yucatan, Mayapan's karstic environment includes of a series of dry caves, cenotes and exposed rock overhangs. The site includes perhaps as many as 40 cenotes and many underground limestone canals necessary for their existence and to enable agriculture in an area with no lakes or rivers. These entrances to the Underworld were long knows as ideal places for specific ceremonial rites and ancestor worship. The location of the essential water abundant cenotes determined the placement of residences as well as the positioning of both civic and religious architecture. Evidence of offerings found in certain cenotes demonstrates that, like the Sacred Cenote at Chichen Itza, particular cenotes at Mayapan were the pivotal positions for religious ceremonies and sacrificial offerings.

The cenotes were often paired with public architecture. For example, the Hall of Chaac Masks does not face the Central Plaza but instead faces south toward the Cenote Ch'en Mul and is linked with the cenote cult. The cenote has a cave-like entrance where pottery fragments have been found as ceremonial offerings. The reverence for cenotes as essential water sources survives until today; some cenotes in Mayapan are considered "alive" needing specific ceremonies for entry. Excavations at Mayapan have discovered no elaborate tombs and most burials had no luxury goods or offerings; burials were found beneath most homes. The goods uncovered in various houses suggest varying levels of social status, some give indications of workshops.

Architecture of Mayapan

The ceremonial center of the city is located in Square Q (or Quadrant Q) found in the western half of the walled enclosure. Studies of the city disclose a long occupation with an almost continual building program, demonstrated by the 13 sequential renovations of the stucco floor in the central plaza. The city center is tightly crowded with various architecture, including temples, palaces, pyramids, columned buildings oratories, shrines, sanctuaries, altars and platforms. The crowded plan and close proximity is thought to be a reaction to the warfare among the towns of northern Yucatan. The grandest religious and political buildings are located in this dominant ritual center; the Observatory, the Palace of Kukulkan and the Temple of the Painted Niches will be discussed in greater detail.

At Mayapan some of the architectural forms that were most important in earlier societies were reconstructed there to support the survival and revival of Classic and Terminal Classic traditions and essential activities. The architecture of Mayapan was designed to reflect the greatness and continuity of architecture from earlier times, but as the buildings were constructed minor details were changed that reflect revisions that supported Mayapan belief system. These alterations found in the different forms, dimensions and decor between the architecture of Chichen Itza and that of Mayapan were often considered a result of the skill of builders, but there may have been other purposes motivating the changes. Scholars continually work to better understand the role of architecture in the cosmically timed community rituals that relate to ancestor worship and the primary role of the gods in assisting the connection between the community and nature. The rulers and priests of Mayapan maintained their relationship with the Terminal Classic Maya traditions and with the architecture of that era to add authenticity to their positions and to insure their political power.

The Postclassic Maya cities were originally thought to produce buildings and artworks of lesser architectural and artistic quality when compared to the most prominent Classic cities. Their architectural programs have been described as derivative and uninspired with inferior, shoddy workmanship but recent discoveries have discounted those previous theories. One of the reasons for the poor evaluation of Mayapan's buildings was that the city's architects appeared to favor stucco for finishing their facades and décor. When the stucco was exposed to the elements over time, it resulted in the erosion of details of their sculptures and architectural decorations. The carved stone sculptures of the Classic cities have survived the centuries in relatively good condition but the stucco sculptures of Mayapan have not. New findings of sophisticated murals sealed beneath the stucco show well preserved paintings that indicate the very high quality of their artworks; some are even considered masterworks of Maya art.

Rather than decadent, Mayapan was actually one of the most artistically and architecturally varied urban centers in all of the Mesoamerican world. Their leaders, architects and artists borrowed ideas and popular concepts from Classic Maya cities as well as from later Terminal Classic regions. Mayapan's widespread interactions relevant to their extensive trade relations are reflected in the variety of artistic forms discovered there. The architecture and artworks of Mayapan's central core reference both Maya and central Mexican worlds. Its principal structures are close copies of those at earlier Maya cities and many of its mural paintings and sculptures are comparable to those from important Central Mexican cities.

Several of Chichen Itza's principal buildings too were repeated at Mayapan. They preserved building forms that were part of other powerful Maya cities but they were presented in new and creative ways. The structures were designed to conform to the features of Mayapan's own natural environment. The buildings at Chichen were widely spaced with broad plaza and not closely associated, so that each building appears even larger. Standing separately Chichen's architecture creates a more intimidating and dramatic impression when either near or far away. The spatial relations among the buildings at Mayapan create a very different impression; they are crowded and physically much closer to one another and can be viewed together as a whole unit. Mayapan's architecture was concentrated in its large ritual center and the presence of walls extending throughout the city appears to strengthen the sense of closeness. Another difference between the two major cities is the presence of ball courts; Chichen Itza has the huge number of 17, while Mayapan, surprisingly has none.

Its city center was adapted to make these structural designs accommodate and conform to the local landscape. Similar to earlier Maya cities, the central plaza, Quadrant Q, was the site of public events that were highly ritualistic. Mayapan's central Quadrant Q was planned according to cosmological ideals; the communal gathering placed held the architectural forms common to other powerful cities throughout the Mesoamerican world The combination of art, architecture, landscape created the ideal location for enactments of their sacred cosmologies and shared religious experiences. The rites were performed both day and night in the open air supported by the cyclical motions of the heavenly phenomena. Their carefully designed architectural spaces, facade sculpture and painting programs evoked and honored their sacred cosmology. Their elaborately costumed ceremonies were conducted to maintain world order and confirm the leaders' power and right to rule. The public presentation were deemed necessary to sanctify and dedicate new temples and monuments; they celebrated royal births, marriages, and accessions to the throne; they commemorated military victory and sacrificial ceremonies.

One of the most common structures at Mayapan were the colonnaded halls on raised platforms; these halls were long rooms with open colonnades on the exterior and a second row of columns on the interior. Near the end of the room is usually a bench; the rear walls display an altar in the middle. Mayapan has 26 colonnaded halls; 22 of them are surrounding Quadrant Q, but their function is still disputed. They could have been used as housing for pilgrims, secular or religious lodging, or administrative buildings. There are additional types of structures found at Mayapan not usually seen in Maya cities; these are oratories and small shrines. The exact function of the colonnaded halls, oratories and shrines are still questioned.

The oratory is a small building perhaps unique to Mayapan; they are described as a one-roomed construction having a doorway divided by two columns. The oratories usually have a bench and an altar placed against the center of the back wall. It is thought that the oratory was used as a private retreat before certain rituals, perhaps even funerals. Oratories have been found in community temples, and in the houses of lords, priests and the leading members of the city. Likewise, the shrines are cell-like enclosures having an altar or a statue. Shrines have been discovered within temples, colonnades and houses but also seated atop small platforms. Some of the altars found in colonnaded halls are enclosed in small shrines. Some shrines were placed on staircases or on low platforms placed before temples; some are even more elaborately designed structures similar to a temple.

The artistic forms and decorative programs at Mayapan show that they were reviving architectural styles popular at Chichen Itza many centuries earlier. Their decorative program includes an eclectic variety of features from Puuc stonework, Toltec-Maya serpent-column temples and Central Mexican and Mixteca-Puebla style paintings. The artisans of Mayapan displayed interest in a variety of cultures and applied ideas, iconography and styles from as far as Oaxaca and from the Aztecs. Artistic creations at Mayapan show remarkable innovations in three-dimensional forms especially in its variety of sculptures in both clay and stone. Some large and small ceramic sculptures survive intact and depict popular deities, many associated with agricultural fertility. Artists returned to the early Maya tradition of creating large standing stela, mostly missing at Chichen. The surfaces of the stela were carved as if it were a house or shrine with the text forming the roof; images of the gods were illustrated within.

The iconography of the serpent balustrade carvings and depictions of the rain god Chaac and the Monkey-man god are seen throughout the ritual center. They appear in murals on temple walls, stone carving or carved stone covered in plaster. The cardinal directions were acknowledged as well; the east was linked with life, males and heat, while the west was linked with death, females and cold. Ring-shaped stones located at ground level have been discovered around Mayapan which experts suggest were used to tie men or animals waiting for the sacrificial altar. They also produced large stone turtles that register the count of *k'atuns*, the 20 year periods celebrated on the round altars of the southern lowlands. Remains of chacmools have turned up in archaeological excavations as well. Although the Yucatan had economic declines, the revival and renewal of traditional, favored building forms and artworks at Mayapan in the twelfth and thirteenth centuries is remarkable, a revival of ancient traditions that survived some very difficult eras.

El Castillo

The most massive structure at Mayapan is the Castillo (the Castle or Temple of Kukulkan) overlooking the monumental central plaza, Quadrant Q, the ritual core of Mayapan. The pyramid/temple was designed as a copy of Chichen Itza's Castillo which dates to 1050 CE. Although it was certainly a temple built to honor Kukulkan, it will be referred to as the Castillo to avoid confusion with the earlier Temple of

Kukulkan at Chichen Itza. The popularity of the cult of Kukulkan continued from the Terminal and Postclassic Periods and remained strong at Mayapan as well. Examples of symbolic iconography referring to the Kukulkan cult was included on the architecture there, especially on those building forms associated with the god.

Called Structure Q162 by archaeologists, the Castillo, a radial pyramid, was built about 1300, halfway through the city's lifespan. The Castillo reigns as the most important of Mayapan's temples but ten additional temples can be found in the city's central precinct. There are several radial pyramids at Mayapan; they imitate those at Chichen Itza. Similar to the Chichen Itza Temple of Kukulkan, this temple is a four-sided radial pyramid with a long, steep staircase on each side; each face is oriented to one of the cardinal directions. Repeating Chichen's orientation, the front of the temple faces north and held the serpent balustrades that depict the god Kukulkan carved along the staircase. Interestingly, Mayapan and Chichen Itza are located on almost the exact same latitude (Fig. 15.2).

The most noticeable differences between the Castillos in the two major cities are their dimensions and placement within the central plaza in regard to other buildings. Unlike Chichen Itza's Castillo which stands totally alone, Mayapan's has six later structures attached to its Castillo; two are colonnades that were thought to have particular functions in connection with special ceremonies at the temple. Both arcades originally had stone columns decorated with figures similar to atlanteans. Unfortunately they were created with stucco that has a short lifespan; so only their feet survive. The pyramid at Mayapan is of a lesser height and volume and sits close

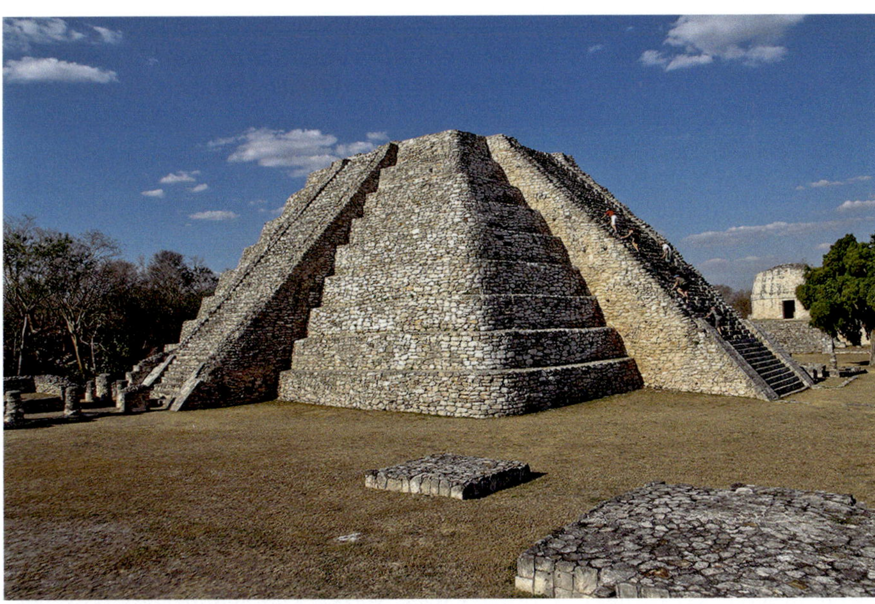

Fig. 15.2 El Castillo at Mayapan, Temple of Kukulkan, a radial pyramid, ca. 1300 CE (Credit Gonzalo Buzzoni)

by several other principal structures, including their version of the Caracol. Modeling Chichen's Temple of Kukulkan, the pyramid was constructed with the same nine tiers, referring to the levels of the Underworld, but it reaches only about 15 m (50 ft) in height, compared to 24 m (79 ft) at Chichen, with its upper temple it stands almost 98 ft.

In comparison, the Temple of Kukulkan at Chichen Itza has 91 steps on each of its four sides, plus the top step, adding to a total of 365, which refers to the solar year or *Haab*. It is thought that the Castillo may have had 65 steps on each face originally, but some were covered over in subsequent remodeling. The number of the four staircases would have totaled 260, rather than 365, but this is another most significant number in the Mesoamerican calendar. This perhaps indicates a greater interest in the *Tzolkin*, or ritual calendar, rather than the 365-day agricultural calendar. Like Chichen Itza one of the largest but unnoticed and significant constructions at Mayapan is the plaza floor that covers Quadrant Q. The present Castillo was built at the same time that Plaza Floor 6 was laid down, one of the 13 successive floors created within the Central Plaza.

At some earlier time, a section of the Castillo collapsed revealing an older building and like Chichen, it exposed a much smaller pyramid buried inside. The inner pyramid was given no Maya name but is rather formally referred to as Q162a. The earlier version of the pyramid is thought to have been the city's central highlight when the central plaza and the other main ceremonial buildings were planned, since it was constructed when the central plaza was first paved. Archaeologists excavated only the nine terraces and the stairway on the east face of Q162a. The inner structure was not a radial pyramid and it is not known if it has a north staircase, but if it does, the solar display of light and shadow may have also been apparent on the north stairs of the earlier pyramid since it has the same astronomical orientations. The decorative program found on the Q162a pyramid included serpent imagery along with vultures and skeletal imagery.

When considering the physical design of primordial time and place, both Chichen Itza's Temple of Kukulkan and Mayapan's Castillo functioned as the axis mundi or center of the world for their own city and community center. The authoritative ideology and the iconography that was so essential for the success of Chichen was repeated at Mayapan to preserve their traditions, although the transference of these features was not exact. Some structural forms were maintained between the two cities and the two temples but others were changed. One of the differences in planning the city's architectural center was the result of Mayapan's landscape. Some areas of the central core had exposed rocks which were incorporated into the base and structure of the buildings. The location of the many essential cenotes at Mayapan dictated the design of the layout and placement of structures.

The Castillo is located in the southern section of Quadrant Q and the most important cenote at Mayapan, Cenote Ch'en Mul, sits at the temple's base. Ch'en Mul has a cave system that passes beneath the temple and radiates from it and also connects it with Structure Q151, the Hall of the Chaak Masks. This significantly positioned cenote seemed to have determined the placement of the pyramid temple. Another cenote Itzmal Ch'en has a ceremonial temple built close by and was known to be a location of religious activity. This ideally placed religious setting over the caves and water pool would function as a portal to the Underworld. This combination of temple with cenote below was found at temple/pyramids in Mayapan, Chichen Itza and Teotihuacan, and probably at other Mesoamerican sites as yet unknown. Like other sacred architecture, the Castillo symbolized the central world axis that linked the Upperworld, the Middleworld and the Underworld; the temple was created using the ideal location and incorporated sacred numbers.

Several more radial pyramids were constructed at Mayapan; they were created to encode and preserve the associations to the fundamental traditions that reach back to the Maya's distant past. For the Maya radial pyramids were associated with time, celestial bodies and sacred geography. Specialists connect this concept as a vertical four-point diagram. These special points sit where the Sun rises on the horizon, where the Sun reaches its high point or zenith, where the Sun sets, and where it reaches the bottom of its cycle, an imaginary point never seen. This represents the span of 1 day, a two-dimensional diagram that is equivalent to the enactment of a solar cycle. The similar structure of the radial pyramids at Mayapan and at Chichen imply that Mayapan's architects and rulers were attempting to continue aspects of Chichen's identity. They wanted to demonstrate Mayapan's position as the heir to the authority and political standing connected with Chichen Itza.

Although not as well known, the north staircase of the Mayapan Castillo produces a play of light and shadow similar to the "Descent of the Serpent" phenomena seen at the Chichen Temple of Kukulkan. Mayapán's Castillo north balustrade is lit in much in the same way as Chichen's but it occurs around the winter solstice rather than the equinox. The Castillo's orientation is 58° east of north and the event of light and shadow occurs at sunset around the winter solstice. Since the Sun appears to move more slowly in its yearly cycle when close to the solstices, the sunset event at Mayapan is visible for a month on either side of the winter solstice. While at Chichen Itza, the sunset light and shadow snake-like pattern occurs for only about a week on either side of the equinoxes since the Sun moves rapidly along the horizon at this time of year. Like Chichen, balustrades ending in large, feathered serpent heads were once present on the Castillo but are no longer in-situ. The Castillo has other solar orientations as well. It was not closely aligned to a significant horizon position of Venus or to the zenith position of the Sun, as the Temple of Kukulkan is, but its northern and northwest walls are lit by the rising Sun on the summer solstice, and its southern wall is lit by the rising Sun on the winter solstice.

The switch from the Chichen pyramid's alignment with the equinoxes to Mayapan's alignment with the solstices shows a transfer in the importance of the ritual calendar between the Early Postclassic and later Postclassic period. The difference provides some evidence for the reasons Mayapan architects decided to recreate the radial pyramid. The connection with Chichen continued in many ways; the main pyramid was altered for the social and religious needs of the new capital city. These alterations seemed to be guided by the movement of the Sun during the solstices at Mayapan. This change of alignment and new time associations may also explain why the pyramid at Mayapan is smaller than the Temple of Kukulkan.

At the base and attached to the Castillo is a small building or arcade named Structure Q161, also called the Room of the Frescoes (or Sala de los Simbolos Solares) for the images that beautify the room. Within this small room and under two coats of stucco, archaeologist discovered a precious collection of painted murals, especially important since very few examples of Maya painting have survived the tropical climate. These are the final paintings of Mayapan made during the last decade of occupation. Some of the paintings depict colorful sun disks; the murals were painted in the style popular in Central Mexico. Eight different figures appear on the solar disks and there is a grouping of eight solar disks that astronomers suggest represent eight solar years in the Venus almanac, equal to eight Venus cycles. Scholars found another connection between these murals and the cycle of Venus. The light of the rising Sun struck these paintings on particular dates that divided the year into a 2/3 ratio linking it to the summer solstice. These particular dates also connect the solar year with the synodic cycle of Venus. They also indicated the dates that Venus could be viewed behind the tower of the Temple Redondo (Q152) when watched from Q161.

After the fall of Chichen Itza, Mayapan initiated and developed its own style of architecture and decorative illustrations; in many instances, they revived and reworked ancient forms. Mayanists suggest that the Castillo at Mayapan was designed to reflect the greatness of earlier times and to enhance its cultural continuity. The Temple of Kukulkan at Chichen and the Castillo at Mayapan were designed to correspond in some but not all ways. They were both constructed atop earlier substructures, as common in Mesoamerican architecture. The designers of Q162 positioned their replica in a similar manner to the Temple of Kukulkan, atop a sacred cenote. They used the same number of levels or terraces and staircases, placing the main staircase on the north face. They would have been similar to Chichen's images of Kukulkan. Serpent columns were used on the upper temple; similar serpent iconography was added as well. Both pyramid/temples were principal edifices to conduct ceremonies connected to ancestor worship and perform cosmically timed rituals honoring the gods and maintaining their close relationship with society and nature.

The elites and architects of Mayapan chose what they considered the most important architectural forms to duplicate. Like the Castillo, the Caracol at Mayapan copied and preserved sacred and political elements from Chichen's urban setting but also relied on formal archetypes including the radial pyramid and round buildings to transfer the ancient, mutual ideologies. These forms became associated and identified with the culture of Mayapan.

Templo Redondo

The most unusual building at Mayapan is the circular observatory called Templo Redondo or Structure Q152 to archaeologists; it was modeled on the Caracol at Chichen Itza which had been built around 1050 CE. The first round temples at

Mayapan appear at the beginning of a conscious revival of Chichen's structures. Surprisingly, there are three round buildings at Mayapan which is most unusual for Maya cities. The earliest round temple at Mayapan may date to about 1280. The Mayapan structures were built centuries later than Chichen Itza's Caracol. The three circular buildings at Mayapan, Q126, Q214 and the Castle Redondo, all have an interior chamber formed by a wall bisecting the circular plan. For the Maya, round structures were also connected with the feathered serpent, Quetzalcoatl-Kukulkan. They were also linked with Venus worship in Mesoamerica. Scholars have noted that round buildings were related to the wind god Ehecatl, a god considered to be an aspect of Kukulkan, they were both associated with Venus. In Aztec mythology it has been suggested that were associated with the appearance of the wind god because the wind could travel around the building's curved sides without disturbing it.

These two major structures at Mayapan, copied from Chichen Itza, the Castillo and the Temple Redondo, functioned to integrate traditional ceremonial, astronomical and calendric principles into their architectural plan. The intention of the city rulers was to reflect the success and excellence of earlier eras when the Itza dominated the city-states of northern Yucatan. Like the Castillo at Chichen Itza and at Mayapan, the Round Temple is associated with water. Cenote Ch'en Mul has an underground passage that leads toward Structure Q152, the Round Temple. Early survey drawings show a small chamber with several pools of water that seem to lie directly beneath the Round Temple and probably determined its position. Many ceramic artefacts, mural paintings and stucco sculptures have been discovered at the site that help date and explain their interest in the Maya history. Borrowed from the past, the familiar forms and décor of these two most important buildings provided a measure of cultural continuity to more readily understand the role and beliefs fulfilled through the cosmically timed ceremonies (Fig. 15.3).

Two drawings of the Caracol were made in the nineteenth century, before a bolt of lightning caused the building to collapse in 1869. Although it appeared as just a mountain of stones when archaeologists began working on the Round Temple in 1997, they were able to establish that the building had four doors but had no inner staircase for which the Chichen Iza Caracol was named. Temple Redondo sits north of the Castillo (Structure 151) and was built sometime later. This structure was not designed as a radial temple but like the Castillo at Chichen Itza, it does have nine levels and probably was used for astronomical observations.

The circular tower was built on an extensive rectangular base which was divided into two semi-circular areas. The temple can be accessed by a wide staircase on the west side. This lower platform of the Caracol was created during the early Puuc period, while the upper platform and the tower were associated with Plaza Floor 7 which dates to 1323 CE. The platform base is 20 m long by 18 m wide; the platform sits 3.5 m above ground level. The Round Temple measures 10.2 m (33 ft) in diameter and 7.5 m (24.6 ft) high on average. It is quite similar to the Caracol at Chichen Itza with the same four doorways, but they are not oriented to the exact direction of those at Chichen but have close alignments with the cardinal directions. The Round Temple has no inner structure but archaeologists showed that a colonnaded hall on

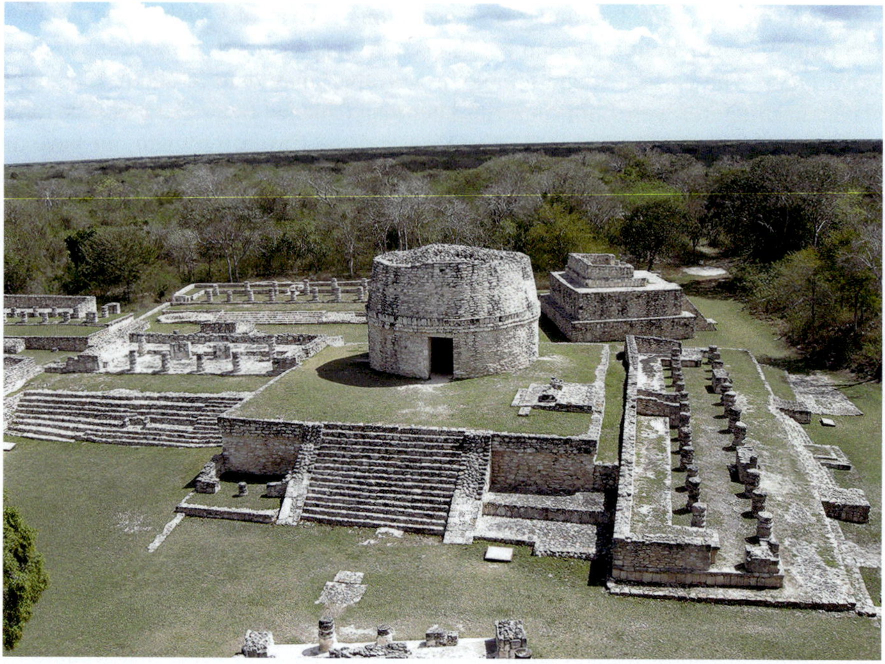

Fig. 15.3 Temple Redondo at Mayapan, round building similar to the Caracol at Chichen Itza, ca. 1280 (Credit LAFS)

a north-south axis was torn down to build Q152 and its base platform. The temple walls are 1.15 m thick and inside was constructed with a circular vaulted form; early researchers described paintings on the vaulted chamber's walls. The cylindrical wall extends 4.5 m in diameter with four niches inside, spaced evenly around the central core. Two of the niches have fragmentary murals with colorful figures outlined in black paint and costumed with feathers, rosettes and step frets; originally the rooms were entirely covered in stucco and painted.

Like the Castillo at Mayapan, the Round Temple has two additional small structures attached; these are a small altar at the basement and a sanctuary on the north side of its staircase. Traces of paint left from its murals were discovered inside the temple. It also has stucco reliefs of decapitated warriors. Because bits of bones and human skulls were found there, the site was probably used for sacrificial ceremonies.

Although the Round Temple has no astronomically aligned windows to make observations and it is not quite high enough to see above the trees and the height of buildings, it has been called an observatory. There is no evidence of having an upper chamber on the tower or specific Venus orientations, as seen at Chichen, but it does have points aligned with sunrise and sunset events that may have possibly served as calendrical markers. The Caracol at Mayapan replicates many features of Chichen's Caracol which enable similar astronomical observations. The western stairway of the Round Temple has the same orientation as the midline of Window 1 of the

Caracol, which was used as an astronomical observatory between about 800 and 1100/1200 CE. The two alignments face the sunset in late April and mid-August, marking intervals of 105 and 260 days, important dates for agricultural purposes. Recent restorations of Mayapan's Round Temple have enabled the determination of other alignments; evidence suggests a definite repetition of certain Caracol orientations, including an equinox orientation. An additional astronomical event involves halo-like light effects that can be observed during the winter solstice.

A distinct spatial distance exists between Chichen's Caracol and the Temple of Kukulkan, but the Caracol and the Castillo of Mayapan were constructed much closer together. These two huge buildings at Mayapan may have been designed to function together; principally, in tracking the spring equinox as a method for calculating the important "burn period", the time when their agricultural fields were set afire to prepare them for planting. The north stairway of Mayapan's Castillo shares an identical alignment with the doorway of the Caracol. At Mayapan archaeoastronomers have discovered that when the equinox Sun is viewed from the west doorway of the Caracol, it sets directly over the mid-point of the Castillo.

Both the Caracol and the Round Temple were carefully oriented to chart the cycle of Venus. The painted platform in the east could have served as a shrine to Quetzalcoatl/Venus as the Morning Star and the western platform to its role as the Evening Star. The orientation of each Caracol shows that it is also important for its Venus/Quetzalcoatl-Ehecatl association. The northern and southern extremes of Venus settings can be determined through the windows of the Caracol at Chichen but only its zenith can be determined at Mayapan's Round Temple. But building was useful in determining the 584-day Venus year, which is discussed in great detail in the Maya Dresden Codex. By reusing the impressive structures, such as the radial pyramid and the round temple, they made it possible to provide Mayapan with the ancient structural forms and their cosmological significance. The transformed building positions and adjusted landscape at Mayapan changed the ceremonial experiences held at each building, especially relating to celestial alignments.

Mayapan's Final Decline

As Chichen Itza was being abandoned in the thirteenth century, its neighboring city Mayapan was rising to power. Mayapan developed into the principal city among a group of allied city-states that was comprised of their trade partners that spread in many directions, from northern Yucatan to Honduras, Belize, Cozumel and Mexico. Mayapan reigned as the most powerful northern city until it, too, was eventually abandoned around 1450, only about 70 years before the Spanish arrived. Compiling Mayapan's history has been quite complicated because of the many political, economic and religious factors involved and documenting texts are few. During the supremacy of Mayapan, Maya cities were no longer ruled by the earlier traditions of a divine king or queen, the *K'uhul Ajaw*. Instead Mayapan, like Chichen Itza, developed its own political system of shared rule between noble households. The wealthier ruling factions were often hostile, which caused social unrest among the lower classes. Some evidence implies that the ruling factions were often at war with one another, showing hints of violence and conflict. Burnt roof beams and broken architectural pieces found at the base of the Castillo indicate intentional destruction of the temple; examples of destruction appear around many other Mayapan temples and altars. Eventually the internal conflicts among city leaders led to a total revolt and Mayapan was sacked, burned and abandoned.

The complete story of Mayapan's collapse is still being studied and contested but its end was similar to many other Maya cities. The area was stressed with overpopulation, diminishing agricultural production, periods of drought and perhaps disease. The Yucatan Peninsula declined from successful trade partners to warring city-states. Mayapan was completely abandoned by the mid-fifteenth century. There is little evidence that the city was occupied after that time but the sacred temple-pyramid of the Castillo may have continued to attract visitors as an honored pilgrimage destination.

Isla de Cozumel

The Maya settled on the small Island of Cozumel in the early first millennium CE, perhaps even earlier, and were still inhabiting the island at the time of the Conquest. Its Mayan name is "Cuzamil" which means "the island of swallows". Cozumel is located just off the northeast coast of the Yucatan Peninsula, part of Quintana Roo, Mexico. Like most of the Yucatan Peninsula, Cozumel landscape is relatively flat with a karst limestone base and riddled with deep sinkholes or cenotes. The island's population reached 10,000 at its height but in 1520 a Spanish ship brought a smallpox infestation and caused the number of Maya on the island to drop to less than 400. The archaeological ruins surviving on the island date mainly to the Postclassic period.

After the decline of the Classic period city-states, Cozumel became an important center of regional power. Reaching its peak during the Postclassic period, Cozumel played a recognized role in the history of late Maya settlement designs, trade and social interactions. Although the island was a major trading center, there were no grand monumental or defensive structures and no high defensive walls that had become more evident in the Postclassic period. Unfortunately, there are not sufficient archaeological remains nor historical documentation to provide a definite understanding of the political and social foundations of Maya history at Cozumel.

Cozumel is not like the major sites of the Yucatan Peninsula, such as Uxmal and Chichen Itza, admired for their spectacular architecture. The island is not noted for soaring pyramids and monumental, highly decorated buildings but for its cosmological, geological and religious landscapes. Archaeologists have documented eight early Maya settlements that spread throughout the entire island, including one at the present town of San Miguel in the northeast region. The towns were located at the most strategic areas, the four directional extremes of the island with the ceremonial and administrative town in the center, now called San Gervasio. In addition to the eight large settlements, archaeologists uncovered a considerable number of isolated monuments scattered within the island's forests and along the eastern seacoast. They have also found many circular and rectangular platforms and sites that show evidence of religious ritual activities. A raised causeway system connected the many separate settlements and monuments found throughout the island. In addition a network of ancient low walls separated different areas of the landscape; clearly, the overall topography was well organized from the beginning of the island's settlement.

Archaeological surveys have also uncovered four elite residential groups in San Gervasio which indicates that the city was the political and economic center of Cozumel. The growth of San Gervasio's power and reputation seems to coincide with the decline of Chichen Itza during the Late Postclassic period. Several building groups excavated at San Gervasio imitate the alignments commonly used in Chichen Itza; all of the major buildings of the city were constructed during the Late Postclassic period.

The Cult of the Moon Goddess at Cozumel

Experts have provided evidence that the Island of Cozumel was highly revered as the center of a major cult devoted to the Moon Goddess called Ixchel. The island became the most renowned and sacred pilgrimage center of the Maya world dedicated to the worship of the Goddess Ixchel. The main temple of the capitol city San Gervasio is thought to be the cult home, shrine and oracle site devoted to the Moon Goddess. Early Spanish records state that pilgrims often traveled to the island's religious facilities to fulfil their religious vows and to present offerings to the goddess (Fig. 15.4).

Cozumel's unique setting at the furthest eastern extent of the Maya world appeared to be the ideal location for a pilgrimage center, just as Santiago de Compostela on the western edge of Europe became one of the most sacred pilgrimage sites in the medieval Christian world. Documentation in early Spanish chronicles tell that Cozumel was a particularly important pilgrimage site for the Maya in the Postclassic era.

Because the Cozumel channel could sometimes be hazardous to cross, the island, like many pilgrimage centers such as Mont-St-Michel in France, was often difficult to reach. This also brings to mind the difficult pilgrimage route across northern Spain to Santiago de Compostela, where the pilgrims are required to make visits to many churches along their way, or the long treks of devout Muslims in their sacred journey to Mecca. Experts think that the many temples along the coastline of the island may have designated the symbolic perimeter which enclosed the sacred space for enacting the cult ceremonies. The ritual use of causeways and the increased construction of isolated coastal shrines were both reflections of Postclassic Maya religious orientations regarding the importance of the eastern Caribbean Sea.

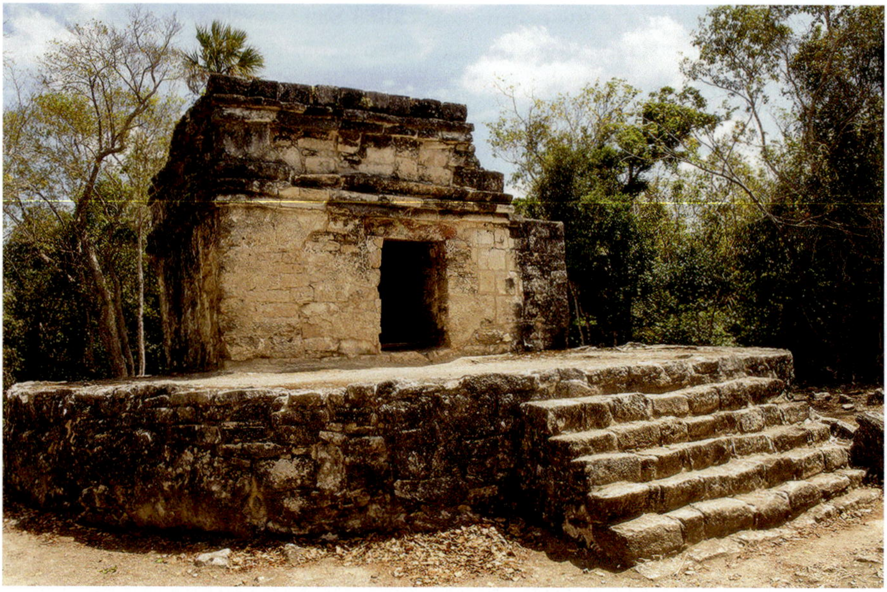

Fig. 15.4 Remains of small temple at San Gervasio, Cozumel, center for worship of the Moon, astronomically aligned (Crredit B Campbell)

The Moon held a special position in Maya thought. During the Classic period specialists in iconography have shown that the Moon could be represented by both male and female personifications. The full Moon is often considered masculine, possibly because it appears to be a second Sun; it is also correlated with the jaguar since they are both nocturnal. To the Maya the waxing and waning Moons were thought to express feminine traits. Moon images of the Maya also tend to merge with the iconography of Maize God representations, emphasizing the Moon's significance in the cultivation of maize. When full, the Moon is in opposition to the Sun, because it rises when the Sun sets and then it sets just as the Sun rises, seemingly a second Sun and referencing the masculine principle.

During the Postclassic period the image of the Moon changed for the Maya; the masculine lunar personification disappeared and two goddess images appeared. The Maya Moon goddess took on numerous associations, especially with corn and plant growth; the Moon was worshipped by Mesoamericans as a fertility goddess associated with weaving, medicine and childbirth. The Moon was also closely linked with water and with sending precious rains when needed. Since the Moon moved quickly or jumped about the sky at different hours, the Maya connected the Moon with the hare. The phases of the Moon through its monthly cycle had different meanings; the waxing of the Moon was connected to the youthful appearance of the goddess and the waning period was connected to her older form. The young goddess was the patron of pregnancies and childbirth, while the old goddess was the patron of weavers, mid-wives and seers. The cenote was considered the dwelling place of the Moon when it was in conjunction, not visible.

The goddess Ixchel has been well documented by Maya scholars; she represents only one of the many aspect of the Moon, noted for healing, wellness, fertility and childbirth. Unfortunately, iconographic examples of her personification as an idol or in paintings or relief sculptures are now quite scarce due to the Spanish goals of destruction of all pagan idols. Ixchel encompassed two different expressions of the goddess and two different manifestations. As a young woman she was called the 'mother of the gods' and as an older woman, Ixchel was called 'our grandmother'. Mayanists know the waxing deity also as Goddess O, who appears in Postclassic sculpture.

The temple honoring the Goddess Ixchel is thought to have been in San Gervasio; the exact place of the main sanctuary has not been confirmed as yet. Some experts believe that the pyramid Kana Nah was the chief temple of the goddess. Not only a place of worship and a pilgrimage destination, the Ixchel temple was also the site of a famous oracle. Spanish accounts tell that there were oracle activities before an image of Ixchel; the strange idol that the Spanish wrote of was a large, hollow statue of fired clay that hung on the temple wall. The oracle was managed by a Maya priest; there was a secret doorway behind the statue through which the holy priest answered. Those who came to the oracle seeking answers offered many gifts from the pilgrim's harvests in return for the priest's services. Perhaps the participants were looking for personal requests or supernatural visions to influence one's fate. According to Spanish records, the oracle at Cozumel had an important part in the foundations of new settlements and was consulted when they were threatened with warfare. They compared the Maya sacred site to Jerusalem and Rome, as it was the most sacred destination of the Maya world at that time.

Archaeologists have uncovered hints of a ceremonial path around the island landscape. The activities along the path may have included prayers and rituals with incense and blood offerings. They found evidence of stopping points and shrines along a pilgrimage trail that would be visited and receive prayer and sacrifices before reaching the final, most sacred destination, the Ixchel sacred shrine and oracle. Not only pilgrims visited the oracular sanctuary and sacred spaces at Cozumel; traders also would have most probably visited the revered locations while trading on the island. Specialists recommend a satellite-based tracing of the ancient Maya roads of the island to indicate the possible devotional stopping points which would be a huge asset in better understanding the history and pilgrimage activities on the island.

There appears to have been a centralized management of the pilgrimages and of their timing. Many of Cozumel's buildings were astronomically aligned; ceremonies dedicated to the Moon Goddess would be timed to coordinate the special phenomena of the heavens with the buildings. Pilgrimages to the island that were timed to the movement of celestial bodies would be planned ahead by astronomer priests who understood the heavenly cycles. Building alignments to lunar events, such as the rise of the full Moon would be especially effective against the starry sky. Interactions with the rising Sun or Moon that inspired a hierophany may have been timed to the pilgrimages. The appearance of a solar or lunar hierophany at the end of the pilgrim's devotional path would have been an extremely powerful and unforgettable experience.

The celestial event or hierophany that occurred would also promote the leader as a semi-god. History has shown that astronomy was often practiced and celebrated within religious complexes in connection with legitimizing and amplifying the ruler's political power. The use of astronomical hierophanies in the management of religious power was not new to Cozumel but a technique regularly utilized by rulers. The management of religious power too is often associated with the repetition and 'control' of the celestial cycles to elevate the position of the priests. Experts suggest that perhaps power and pilgrimage at Cozumel was a state affair, organized and managed at the highest level to promote the rule of the leaders and reaffirm their positions. When the rulers conducted rituals in connection with movements of the Sun and Moon, they often arranged it to associate themselves with a hierophany.

Astronomy of Cozumel

Cozumel's landscape at the eastern edge of the Maya world created a dramatic setting in relation to the heavens. Its privileged position surrounded by water embodied the Maya concept of sacred geography of locations amidst water. To the Maya the east is symbolically the place of the gods and of rebirth; the west was the place of death. Cozumel's flat expanses and its setting relative to the horizon, looking to either east or west, assured perfect conditions for astronomical viewing and measurements. Observations of the rising and setting Sun on the solstices, the zenith and other significant dates connected with Maya agriculture and ceremonies could be easily observed. The gradual emergence of the full Moon from the sea would provide a dramatic and emotional experience for those who considered it a celestial god. The spectacular heavenly events, along with the monthly phases of the Moon, were celebrated at Cozumel in association with the important lunar cult devoted to the Goddess Ixchel. Archaeologists question also whether the sites of embarkation from the mainland may have been oriented to the cycle of the Moon. Moonrise over the island viewed from the mainland is still a most impressive sight. All of these factors made Cozumel a very special destination.

Through time the practice of aligning major buildings with the heavens became highly specialized and closely associated with religious ceremonies. The astronomer priests sought perfect harmony with the cosmos on particular dates in their calendric system. These alignments imbued the buildings with supernatural significance to draw the approval of the gods. The meaningful alignments along with mural painting, decorated stelae, relief sculpture and sacred rituals in the buildings all convey additional strength to the spiritual power felt by those partaking in the religious rituals. The island setting of Cozumel's architecture added even more to the mystical effects.

Investigations by archaeoastronomers have verified the orientations of 271 buildings at 87 archaeological sites on the island. Most of the civic and ceremonial structures were aligned toward the position of particular celestial bodies on the horizon. Many of the east-west orientations in the Maya lowlands were directed at points within the solar span, between −23.5° and +23.5°. Some of the building at San Gervasio are oriented to these solar extremes north and south. The most common alignments were to sunrises and sunsets on specific dates, set through the use of observational calendars which enabled them to schedule the agricultural tasks and the required ceremonies. The studies of the orientations show that they used the same rules throughout widespread areas and during long time spans.

Cozumel differs from many Maya sites since it has alignments with the standstills or extremes that are not of the Sun but of the Moon. This is not surprising since abundant historical and iconographic evidence reveals the dominant status of the cult of the Moon Goddess. The major standstills of the Moon always occur about the time that the Sun reaches the solstice. The full Moon reaches its northern extreme near to the Sun's December solstice; the full Moon reaches its southern extreme when the Sun is near the June solstice; this makes the full Moon cycle especially important and observed at Cozumel.

San Gervasio contains three major architectural complexes; each had five groups of buildings. The complexes were all interconnected by 'white roads'. On Cozumel archaeoastronomers have found unusual alignments directed to lunar standstills. Many buildings on Cozumel are oriented to major lunar standstill events. The largest complex of monumental structures is Group VI of San Gervasio; it was also placed on the highest area of the island. Its largest building is oriented to the lunar and solstice extremes. The unusual number of solstice orientations on the island shows that full-moon standstills were especially noted.

The city also retains the remains of a pyramid called Kana Nah, also 'High House' or Structure C22-41-a. Though not especially tall or impressive, Mayanist suspect that this sacred space was the temple of the oracle of Goddess Ixchel. The pyramid is only 5 m (16 ft) tall; it has a square design and four tiers with a main staircase bordered by a railing or balustrade on the east. The two-room sanctuary at the top of the pyramid still contains traces of red and blue paint on the walls. The upper sanctuary has an entranceway facing west and an altar which extends in a north-south direction but does not touch the northern or southern wall. The wall constructed over a part of the altar has a central opening. The pyramid faces northeast; measurements have determined that it is not aligned with the Sun but with the major lunar standstill, when the Moon sets at its extreme northern position on the horizon. The Moon reaches these extreme positions only once every 18.6 years (Fig. 15.5).

Another building on Cozumel, Structure C5-1-a at La Palma faces east; a *sacbe* from the building led directly to the sea. It was oriented to sunrises on February 5 and November 6, dates divided by 91 days. Structure C15-1-a of El Cedral was aligned to sunsets on May 7 and August 6; this marking is also divided by 91 days. The building at San Gervasio, called Structure C22-34-d, has alignments with

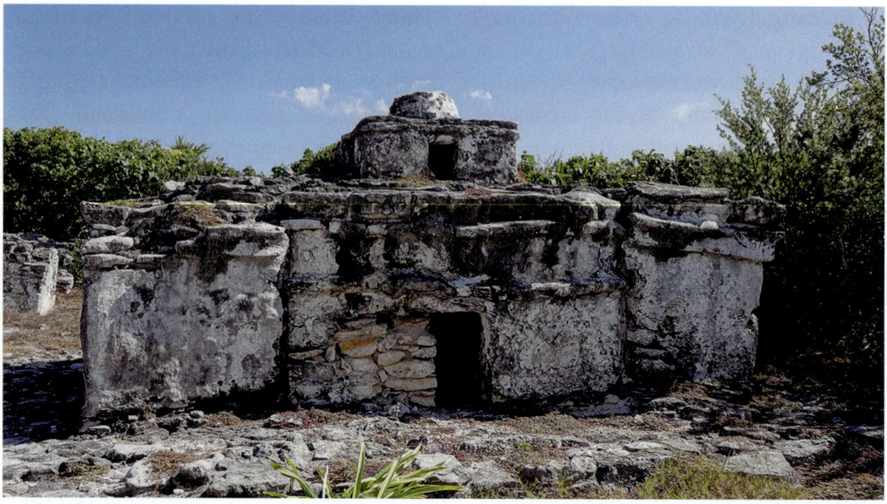

Fig. 15.5 Remains of small temple at El Cedra, astronomically aligned (Credit Aleksandr Kutiavis)

sunrises on February 12 and October 30. This building is rather small but was considered important because a buried collection was uncovered beneath an altar, under the floor. The treasure included jade and greenstone celts, a basalt metate and two basalt manos, used for grinding maize. Adding to the alignments, one of the ceremonial roads, Sacbe 2, was also oriented with specific lunar positions. Amazing and unexpected discoveries resulted from the recent archaeological work done on the island. Much more investigation needs to be undertaken at Cozumel to better understand the lives, architecture and activities of its Maya inhabitants.

Maya City of Tulum

The best preserved architectural remains of the Late Postclassic period are those at Tulum, located on the east coast of the Yucatan Peninsula. Specialists think that the site was probably called Zama, which means City of Dawn, since Tulum looks directly at the rising Sun. The dramatic ruins of Tulum stand atop the sea cliffs 12 m (39 ft) high, overlooking the Caribbean Sea, just south of the channel between Isla de Cozumel and the mainland. It was one of the last important cities built by the Maya. The Tulum archaeological site is relatively small compared with many other Maya sites in the area but is one of the most interesting. The most prominent archaeological site of the area, Tulum is still a landmark to mariners sailing along the coast. The first explorers to write a detailed summary of Tulum were John Lloyd

Stephens and Frederick Catherwood who described the city in their book on travel adventures published in 1843, *Incidents of Travel in Yucatan*. As they did at most sites they visited, they made accurate maps and drawings. Tulum's buildings were not scientifically mapped and investigated by archaeologists until the early twentieth century.

Little is known about the first settlers in the coastal area, but archaeological evidence found ceramics that indicate an early date around 300 BCE for a few settlers. Some inscriptions with dates in the sixth century have been uncovered at Tulum, but most monumental architecture was constructed in the Late Postclassic, beginning around the start of the thirteenth century when Tulum reached its peak. Its success continued through the fifteenth century. One motivation for the city's founding was to establish a prime trading center on the east coast. The city established extensive trading routes; they traveled both along the Yucatan coast and deep inland in sturdy merchant canoes. A wide break in the sea cliffs near the site would have provided a landing beach and port access for trading canoes arriving at the city. Abundant evidence of non-local artifacts has been discovered, such as ceramic vessels and incense burners from various cities throughout the Yucatan and the highly valued obsidian and jade from northern Guatemala, located nearly 700 km (430 mi) from Tulum. Occupied until the Spanish Conquest, it was probably the Maya center sighted in an early account of Spanish exploratory voyages and compared to Seville in Spain.

In the Mayan language Tulum means "wall"; this is significant since it was one of the very few walled city in the Maya world. The great stone walls around the city protected it from invasion by warring enemies. Approach to the city from the east is well protected by sea cliffs averaging 12 m (39 ft) in height. The fortified site is bounded on its landward sides by a lengthy stone wall that was once equipped with a walkway and parapet. The wall averages 8 m (26 ft) thick and 3–5 m (10–16 ft) high and 400 m (1300 ft) long on the side parallel to the sea. The wall on the landward side is broken by five narrow gateways, one on the western side and two each on the north and south. Small watchtowers mark the landward southwest and northwest corners. Construction of the impressive wall must have required an enormous amount of time and energy, which indicates the importance of defense to the city leaders. A small cenote near the northern wall provided a source of drinking water within the walls.

Tulum's architecture shows parallels with Chichen Itza and Mayapan, although on a much smaller scale. In general Tulum's masonry is very crude and often covered with thick coatings of plaster. Within the walled city stands a series of masonry platforms and buildings, colonnaded palaces and raised temples. Tulum's architecture appears typical of Maya sites on the coast of the Yucatan. The artisans at Tulum did not approach the masonry skill apparent at Chichen Itza; their buildings were usually covered with thick layers of plaster. The three major structures are named El Castillo, the Temple of the Frescoes and the Temple of the Descending God (Fig. 15.6).

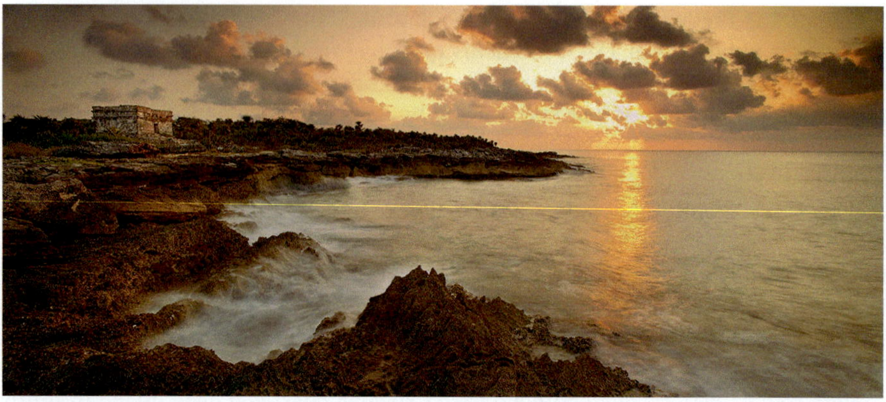

Fig. 15.6 Sunrise at the late Maya city of Tulum (Credit Red Square Visual Arts)

El Castillo and Astronomy

East of the central plaza is the El Castillo (the Castle) pyramid which reaches 7.5 m (25 ft) high; it has a small two-room temple that is reached on the western side by a staircase with a wide balustrade. On the west side of the temple building is a single doorway supported with two circular columns. The columns were later remodeled into feathered serpent columns similar to those commonly seen at Chichen Itza. The lintels in the upper rooms have serpent motifs carved into them as well. The pyramid seems to have been built in stages. The Castillo was constructed atop an earlier building that was colonnaded with a beam and mortar roof; the wings of the earlier building extend north and south from the platform. A small shrine was thought to act as a beacon for incoming canoes. The Castillo is aligned in a similar manner as the Castillo at Chichen Itza. It is aimed directly to the setting Sun on the day of its zenith passage. The location of the temple pyramid on the coastline of the Caribbean was though to symbolize the sacred mountain of the east. The Castillo was not only a temple but considered a symbolic form of Flower Mountain (Fig. 15.7).

Structure 1 at Tulum appears to have influenced the orientations of many structures around it but Structures 21 and 25 have different orientations that are linked to solstices and major lunar standstills. On the south side of Structure 25 is a stairway leading to Room A that has two small openings in its east and west walls. These openings appear to have provided certain astronomical observations. The alignment along the two apertures is parallel to other east-west orientation of the building.

The fact that the alignment of the two openings had particular importance is reinforced by its relationship with other architectural elements. Just west of the Structure 1 is Structure 29, a low platform that has a small staircase with a midpoint placed exactly along the directional line signified by the two openings in Structure 25. If extended further west, the same alignment passes precisely over a small staircase in the west part of the defensive wall of the city. It appears to be important that this sight line corresponds exactly with the two apertures in the east and west wall

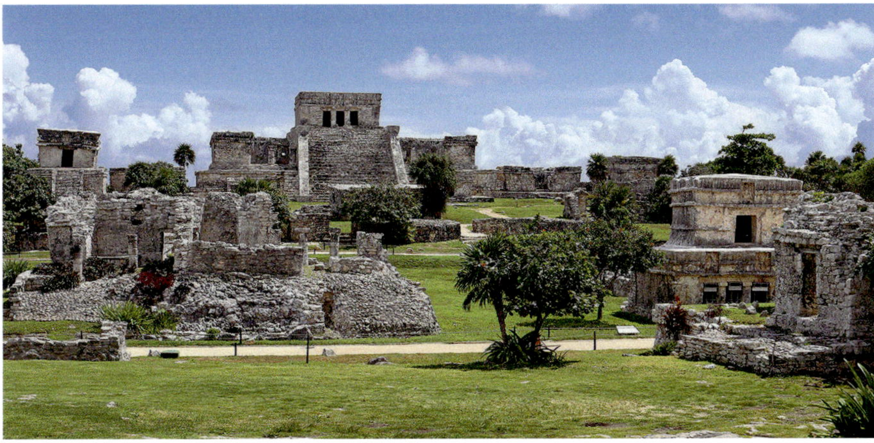

Fig. 15.7 Ruins of Tulum, E-Group complex faces east overlooking the sea (Credit Nik Waller Productions)

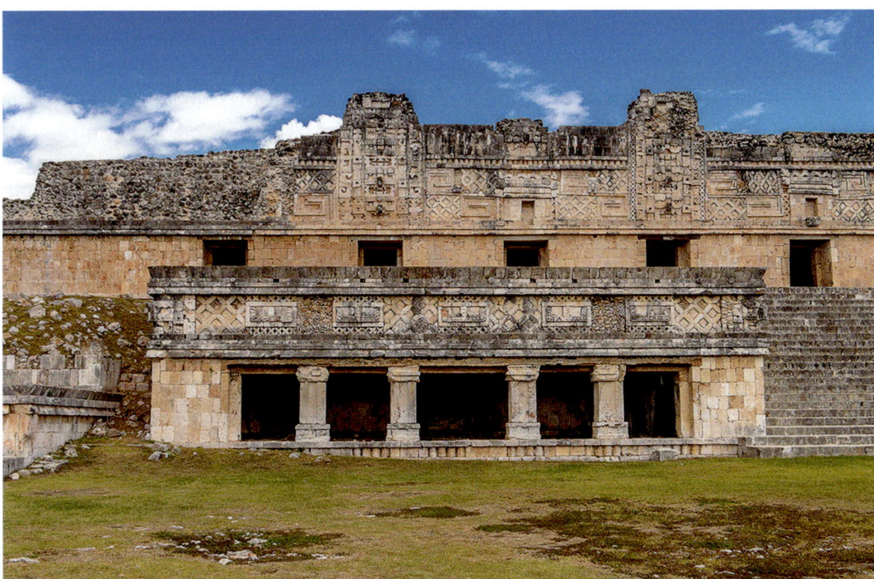

Fig. 15.8 E-Group buildings reverse view on the eastern edge of Tulum (Credit Alexander Sviridov)

of Room A. Even though the orientation shown by the two openings in Structure 25 links precisely to northern major lunar standstills on the western horizon, it is possible that the stairway in the defensive wall designated the spot for observing southern major lunar standstills on the eastern horizon, close to the direction marked by Structure 25 (Fig. 15.8).

Just south of Structure 25, Structure 21 has an alignment that corresponds to the June solstice sunsets. It is possible that Structure 25 was oriented to the southern major lunar standstills on the eastern horizon; it could have been aligned to the major northern standstills on the western horizon. If the lunar extreme position were viewed during the full Moon, Structure 21 could be oriented to the December solstice sunrises. Aligned buildings could be functional to the east or to the west depending on the surrounding buildings and landscape that may block sightlines. That Structure 21 was aligned to the eastern horizon is entirely possible, designating sunrises at the winter solstice. It is also possible that both Structures 21 and 25 were aligned to the west, designating summer solstice sunsets and marking the northern major standstills of the dark or nearly new Moon. Scholars suggest the first option appears more likely. Because Tulum is located on the northeast coast of the Yucatan, it is suspected that Structures 35 and 45 at Tulum were also oriented to lunar positions. The two buildings appear to be oriented to the minor lunar standstill.

Temple of the Diving God

Just north of the Castillo sits a small, elevated structure, called the Temple of the Diving God or Structure 5, one of the most beautiful buildings at Tulum. Dating to about CE 1400, it was built atop a promontory overlooking the sea. The temple was accessed by a narrow staircase on its western side. The temple was built over an earlier structure that served as its foundation. The temple sits upon a circular platform and consists of a single room with a door to the west with a single doorway on its west side. The temple room contains a sacred altar that was still being used for religious activities when early archaeologists first carried out their research. On the interior eastern wall of the room is a splendid mural painting that is still partially visible; above the doorway is a depiction of the building's namesake, a stucco image of the Diving God, thought to represent Venus. The god is shown with wings, a headdress and he holds an object in his hands. From numerous depiction in murals and other artworks at the site, Tulum appears to have been an important site for the worship of the Diving or Descending god (Fig. 15.9).

Buildings at Tulum have also been coordinated with the movements of the celestial bodies. The Diving God temple western face is astronomically aligned to the setting Sun on the solar zenith which announces the start of the planting season in May. The temple's east face has a window that focusses a ray of light at sunrise directly into the interior on the solar zenith, brightly illuminating the area directly below the Diving God image. The window facing the sea could also be used to view maritime transit that was synchronized with the seasonal cycle. There is evidence that trading rose considerably during the dry season (November–April).

The Temple of the Diving God also holds an image of a sky band showing a scene with two goddesses seated on sky-band thrones. The overhead sky band has Venus glyphs alternating with solar rays. The goddesses are dressed in costumes with shell decorations and wear earrings with rays that seem to represent lunar light.

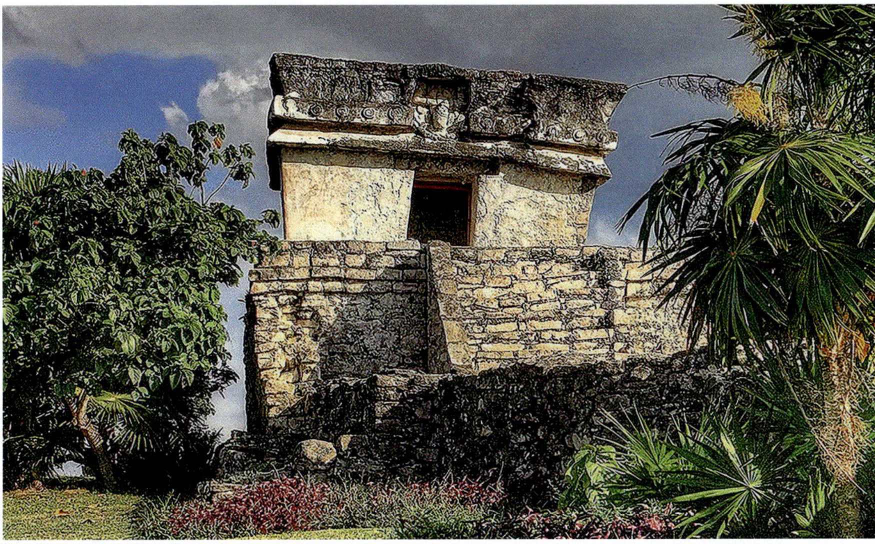

Fig. 15.9 Temple of the Diving God at Tulum with stucco image of a diving god, ca. 1400 (Credit Vladimir Korostiyshevskiy)

The goddess to the right has a line around her mouth meaning that she is older than the other. A myth or story may be referred to by the youthful and older variations of the Moon Goddess. Tulum was a location on a pilgrimage route connected with female cults especially associated with childbirth. An important shrine for pregnant women was located at the Island of Cozumel. Scholars suggest that Tulum was dedicated to a cult of the lunar goddesses, in particular the aged aspect of the Moon, a patroness of midwives. Participants could watch the waning Moon descending into the waters of the Caribbean.

Temple of the Frescoes

The Temple of the Frescoes or Structure 16 was another of Tulum's astronomically aligned structures and one of the more spectacular buildings. The temple had a lower gallery and a small second story gallery. The Diving God also adorned the façade of this temple. Above the entrance on the west wall, the niched figurines of the Diving God are formed of stucco. A mural that gives the temple its name can still be viewed on the east wall which is similar to the style that originated in highland Mexico. Because of its fragile condition, visitors are no longer allowed to enter the temple. The temple was used as an observatory for keeping track of the movements of the Sun (Fig. 15.10).

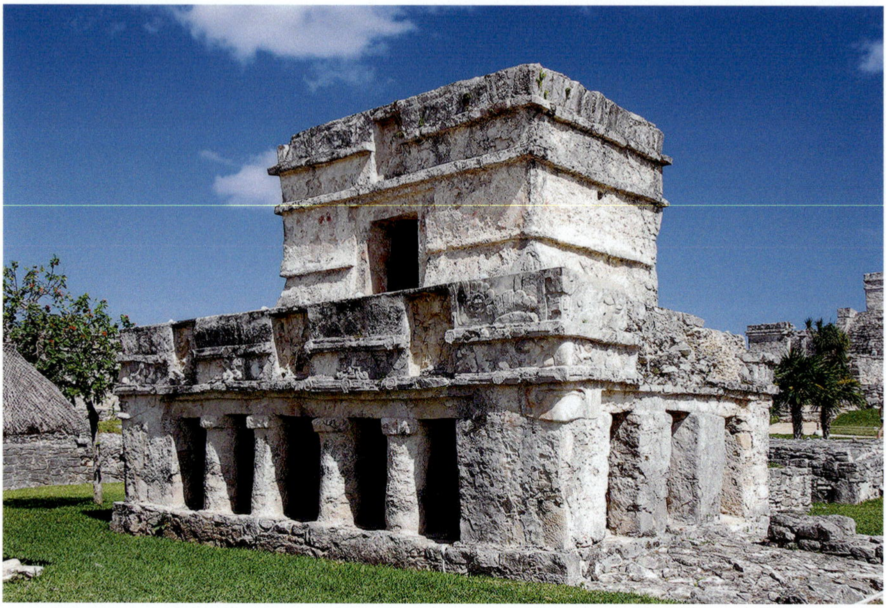

Fig. 15.10 Temple of the Frescoes at Tulum, two tiered building with murals (Credit ecstk22)

The aged moon goddess appears in the Temple of the Frescoes as well. Known as Goddess O, she appears in the lower band of Mural 2 and wears a composite spindle and knotted-serpent headdress. The goddess holds a serpent staff in her hands and has a maize symbol on her back, which links her with agriculture. She is likely near the end of her life since her face is completely covered with lines. The fresco shows that she is about to be consumed by the reptilian jaws of the Earth Monster. The monster also swallows the sea beneath her which is filled with creatures of the sea such as fish and stingrays. She seems to be between the world of the living and the dead. Another goddess is drawn above her, but not shown as quite so aged as the goddess floating on the sea. The image is thought to depict the waning Moon descending into the sea and dropping into the maw of the Earth Monster.

The presence of so many beautifully painted murals reveals the prosperous trade status of the east coast cities during this late period of Maya history. The expensive décor and the expert skill shown by the artists show that the Tulum rulers were interested in creating a site of high status. The numerous blossoms decorating the facades at Tulum were not just ornamentation but label the structures as floral temples. Like Cozumel, Tulum was considered a place of pilgrimage as well as a highly functioning trade center. The daily rebirth of the Sun dramatically rising out of the eastern sea inspired a great deal of ritual and symbolism. The Maya occupied Tulum for about 70 years after the Spanish began the settlement of Mexican cities. By the end of the sixteenth century though, Tulum was abandoned. It is still rated as one of the best-preserved Maya cities on the Yucatan Peninsula.

References and Further Reading

Iwaniszewski, S. (2016). Time and the Moon in Maya culture: The case of Cozumel. In *The role of archaeoastronomy in the Maya World: The case study of the Island of Cozumel*. United Nations Educational, Scientific and Cultural Organization.

Magli, G. (2016). Astronomy and power: The case of the Island of Cozumel. In *The role of archaeoastronomy in the Maya World: The case study of the Island of Cozumel*. United Nations Educational, Scientific and Cultural Organization.

Milbrath, S., & Lope, C. P. (2003). Revisiting Mayapan Mexico's last Maya capital. *Mesoamerica, 14*, 1–46. Cambridge University Press.

Milbrath, S., et al. (2010). Religious imagery in Mayapan's murals. *The PARI Journal, 10*(3), 1–10.

Patel, S. (2016). Pilgrimage to the Island of Cozumel. In *The role of archaeoastronomy in the Maya World: The case study of the Island of Cozumel*. United Nations Educational, Scientific and Cultural Organization.

Šprajc, I. (2016). Lunar alignments in Mesoamerican architecture. *Anthropological Notebooks, 22*(3), 61–85.

Taube, K. (2009). Tulum, Santa Rita and floral symbolism in the international style of Late Postclassic Mesoamerica. In G. Vail & C. Hernández (Eds.), *Astronomer, scribes, and priests*. Dumbarton Oakes.

Trejo, J. G. (2016). Calendric-astronomical alignment of architectural structures in Mesoamerica: An ancestral cultural practice. In *The role of archaeoastronomy in the Maya World: The case study of the Island of Cozumel*. United Nations Educational, Scientific and Cultural Organization.

Chapter 16
LiDAR Technology

The introduction of the newly developed LiDAR technology into studies of ancient Maya archaeological sites in 2018 has revolutionized the Maya scholars' perceptions of the vast Maya civilization. Former estimates of dates, locations, numbers and abilities of the earliest Maya populations must all be rewritten. New LiDAR surveys in Guatemala and the Yucatan Peninsula have provided much more information than previously thought possible. Impenetrable jungle areas can now be visualized within the forests as if experiencing X-ray vision without spending days or weeks slashing through dense foliage in sweltering heat. LiDAR stands for Light Detection and Ranging; it is a remote sensing method that uses light in the form of a pulsed laser to measure variable distances to the Earth. Streams of light flashes, combined with other data recorded by the airborne system, generate precise, three-dimensional data concerning the shape of Earth and its surface characteristics cleared of overgrowth.

The LiDAR instrument consists of a laser, a scanner and a specialized GPS receiver. The most commonly used platforms for obtaining LiDAR data over large areas are helicopters and planes. A flying plane pulses light waves that bounce off objects below and then return to the sensor. When an airborne laser is pointed at a targeted area on the ground, the beam of light is reflected by the surface it strikes. This reflected light is registered to measure its range. Sonar uses sound to locate structures, while LiDAR sensors mark the amount of time each pulse takes to return to the sensor. The gathered data is used to create a 3-D map of the landscape surveyed.

Using LiDAR systems, scientists and mapping professionals can accurately examine both natural and manmade environments with precision. LiDAR is able to generate more accurate shoreline maps and supply data for geographic information systems. Initially, LiDAR systems have been used to create high-resolution maps, but recently they have become extremely important in surveying, geodesy,

M. Dolan, *Decoding Astronomy in Maya Art and Architecture*,
Springer Praxis Books, https://doi.org/10.1007/978-3-031-61092-9_16

archaeology, geography, geology, and seismology. The technique can make digital 3-D representations of large areas of the Earth's surface and can even reach the bottom of the ocean. Large amounts of information can be gained without harming the rainforests or the areas under survey.

The dense vegetation of the rainforests of Guatemala and the Yucatan Peninsula was hiding hundreds of Maya towns and cities. These areas are exploding with new sites, more than ever imagined. Below the dense tree canopies, innumerable structures have been detected, exposing an early civilization that was much more complex than anyone could have imagined. Ancient cities, roadways and pyramids, long buried deep in vegetation and partially eroded, have now been reproduced in three dimensions. With its in-depth capabilities LiDAR is able to pick up specific patterns in the landscape; it reveals the distinctive features of each structure with amazing clarity (Fig. 16.1).

These recent studies have opened a treasure trove of unidentified archaeological remains, soaring pyramids, complex systems of raised roads, reservoirs, and large urban centers. The LiDAR studies have mapped new cities and found new site boundaries. Thanks to the new technological resources, architectural groups have been identified and amplified. Maps have demonstrated the presence of dense concentrations of new and previously unknown sites. The massive platforms and pyramid constructions, including numerous E-Groups, causeway networks, ballcourts, and reservoirs discovered would have required vast amounts of labor and resources.

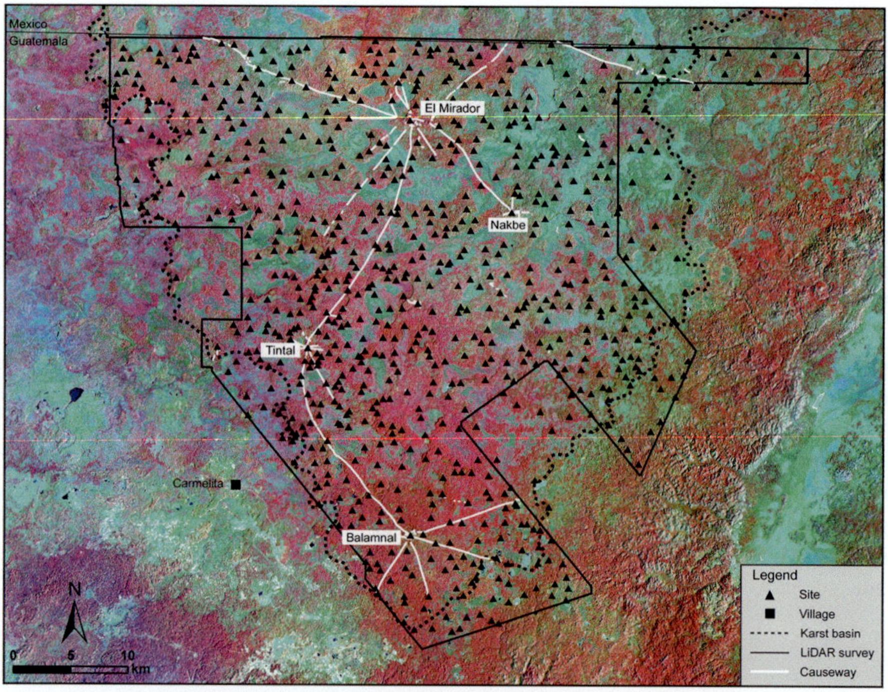

Fig. 16.1 LiDAR image of Mirador Basin (Public domain)

Even at the largest and most widely explored sites, such as Tikal in Guatemala, the three-dimensional maps produced by the aerial surveys generated many surprises. LiDAR mapping showed that the Classic Maya city Tikal was at least four times larger than previously thought and it was partly surrounded by a huge ditch and rampart reaching for many miles. Even more, two large hills, always thought to be natural, were actually shown to be monumental pyramids. The larger of the two hills is now suspected to be an important ceremonial building and may contain the tomb of a powerful king.

Since the full extent of its enormous size and extensive population was totally unknown, the way of life of the early Maya population is only partially understood. These great new numbers of sites and structures found are considerably more than scholars every imagined. The first phase of LiDAR information gathering has generated the largest amount of data ever gathered for archaeological research. New findings must be pieced together and integrated with knowledge already established. Archaeologists hope to eventually map the entire Maya Biosphere Reserve area from Guatemala to the Gulf of Mexico which will undoubtedly reveal even more unexpected results.

New Findings with LiDAR Technology

A groundbreaking investigative study of previously unknown Maya territory was undertaken and published in 2022 by a large team of archaeologists headed by Richard Hansen. The area studied was one of the largest regional sections of the Maya lowlands in Guatemala. This vast region, mostly buried in dense tropical forest, is called the Mirador-Calakmul Karst Basin (MCKB). The forested region was only marginally understood as most sections had never been penetrated in contemporary times. Just a few of the larger cities had been visited or studied. Archaeologists suspected that parts of the Basin might contain a few early Maya cities but the vast extent of the settlements had been totally lost. Beneath the dense forest canopy, the new LiDAR technology revealed evidence of hundreds of ancient cities that were totally unknown.

Using the new tool of LiDAR, the archaeologists analyzed new maps produced through advanced LiDAR technology; their studies have concluded with amazing results. More than 775 ancient Maya settlements were discovered in the initial land surveys and 189 more towns in the surrounding areas. The unimaginable number of unknown archaeological settlements that the study detected totaled 964; as yet 645 are unnamed. Archaeologists determine that a grouping or cluster of buildings is considered a settlement but a group of settlements, connected by causeways or other construction formats, is called a site or a polity. An urban setting with at least 4000–5000 residents is classified as a city; a town would need about 1000–3000 people; a village would have less than 1000 inhabitants. The researchers condensed the large number of settlements that seemed to have political or social relationships with other settlements into larger formations of at least 417 cities, towns and villages (Fig. 16.2).

Far-reaching causeways connected the numerous sites discovered which created a web of social, political and economic interactions among the multitude of cities. The causeways were necessary because they rose above the seasonal swamps and low vegetation of the areas. Some raised causeways were as wide as 40 m (131 ft). The roads were composed of a mixture of mud and quarry stone among several layers of limestone cement. The Maya coated their raised roads with a thick layer of white plaster that would enhance visibility at night when the plaster would reflect the moonlight. This vast scope and scale of the Maya road system was previously unknown. The LiDAR surveys along with satellite imagery and GIS software has added immeasurable information to the study of the interactions among the cities. The extensive road systems made efficient interactions between cities possible, one of the most important achievements of the Middle and Late Preclassic Maya.

Many of the newly found settlements date to the Middle and Late Preclassic periods; some of the largest Maya cities were functioning during this era. The research team divided the settlements into six different tiers based on their size, their architectural density and architectural groupings. The raised roadways and more sophisticated structures of many cities suggest that they had centralized administrations along with social and economic advancements. These city-states were capable of managing water resources for agriculture and for creating 177 km (110 mi) of elevated causeways which shows obvious organizational abilities and interest in uniting the major sites into a network of complex and interacting cities. Richard Hansen says, “The basin is a contained, enclosed, integrated cultural and natural system, unique in the world”.

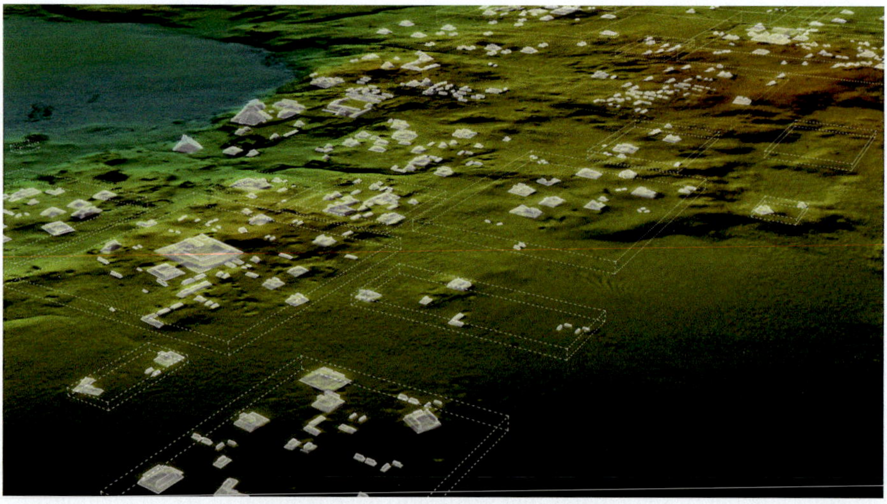

Fig. 16.2 LiDAR image of newly discovered Maya city (Public domain)

Considerable new information from archaeological research, LiDAR studies, and the multidisciplinary approach of scholars has provided a better understanding of the cultural process of Maya community development. These studies use multiple sources, along with LiDAR data to explain how the societies established their infrastructure and demonstrate their economic and social organizations. The recent LiDAR studies have included 85.8% of the huge Guatemalan Basin area and have revealed a surprising density of ancient Maya sites. Within the newly found cities of the large basin area, scientists have been able to map huge numbers of monumental architecture, which includes E-Groups, triadic complexes, ballcourts, causeways, and reservoirs. The laser imaging focused in on some remote regions of northern Guatemala that they thought totally unoccupied to find 20-ft-high walls, watchtowers, moats and indications that ancient Maya societies fought large-scale warfare there for many years. The discovery of structures built especially for warfare and defense shows that conflicts were a larger part of the Maya existence than formerly believed.

Gathering the numerous resources and controlling the huge number of workers necessary to accomplish the monumental construction would have required a centralized organization. E-Groups were quite common throughout the MCKB area but they vary in size and numbers. The LiDAR mapping project has identified 42 E-Group complexes located in at least 30 centers. The causeways and many facilities within the settlements suggest social and economic unity and demonstrate the beginnings of state-like formations. These advancements indicate high levels of administration and political power that existed during the Middle and Late Preclassic (c.1000 BCE–150 CE). The populations of these cities dropped during the Late Classic period (600–900 CE).

To manage the labor necessary to construct the massive platforms, towering pyramids, palaces, civic buildings and causeways uncovered in these large lost cities, the leaders would have needed the power and labor of thousands of workers along with specialists. They all had to work together to produce these highly functioning cites. The many architectural buildings suggest that the monuments were used by many people but there is no strong evidence of the presence of powerful elites. The vast span of towns and cities recently discovered indicates a social organization that worked together for amazing results. This architecture was created when Maya societies had less social inequality than existed in later periods when royal families controlled all aspects of the cities.

The studies indicate that the long-hidden ruins of the sprawling Maya civilization was much more complex and interconnected than specialists had suspected. The new maps have lead researchers to conduct on-site excavations of more than a dozen of the most promising settlements; most of them would have been completely undetectable without the LiDAR surveys. Some sites are so completely camouflaged that you could walk atop them and you would not even notice them.

After the development of LiDAR, archaeologists conducted sweeping aerial surveys of many uninvestigated areas of Guatemala and the Yucatan Peninsula searching for evidence of early Maya settlements. Although LiDAR has revealed an enormous amount of new information about the history and accomplishments of the

Maya civilization, it cannot see below the ground to deeper levels or give any indication of dates of occupation. Unfortunately, archaeologists must still hack through the tropical jungles and dig into the earth for thorough study of the materials and architectural structures. To gain a true understanding of a site or a building, an expert must also study the structure's artistic adornment and hieroglyphic texts.

LiDAR mapping can now help fill in the many missing gaps that exist throughout the long centuries of Maya history. So much of the earliest Maya history is still unknown; large areas that were thought to be uninhabited were actually teeming with cities and immense populations. These ancient Maya sites had been concealed for well over a thousand years; all are now waiting to yield exciting new information. This new research technique opens pathways to decades of further discoveries. The introduction of LiDAR technology to Maya studies has revealed an entirely new understanding of the breadth and depth of the vast Maya habitation. Scholars now realize that the Maya civilization was as extensive and successful as the most important civilizations of antiquity. Previous ideas on the ancient civilizations of Mesoamerica will now have to be reassessed and Maya history rewritten.

Reference and Further Reading

Hansen, R. D., et al. (2023). LiDAR analyses in the contiguous Mirador-Calakmul Karst Basin, Guatemala: An introduction to new perspectives on regional early Maya socioeconomic and political organization. *Ancient Mesoamerica, 34*(3), 587–626. Cambridge University Press.

Endnote

This book presents just a fraction of the enormous extent of Maya history, knowledge and capabilities. The Maya were able to carve enormous cities from the dense jungles of Mesoamerica and align their cities and their architecture in attune with the movements of the heavens. From the small pyramids at Cerros in Belize to the monumental pyramids at Calakmul, they built with accurate orientations and proper proportions. The Maya decorated their pyramids, temples, buildings and monuments with beautiful artworks, mosaics and paintings. Living within the rainforests, their workmanship was constantly exposed to rot and decay; repairs and repainting were continuing problems.

The Maya worked to understand the cycles of the universe to the best of their ability through their observations, calculations and judgements. Astronomy and the calendar were closely linked in Maya thought. Their calendric system and astronomical observations, made only with the naked eye, were astoundingly accurate. They studied the heavens continually to keep their accounts up to date, since positions of the stars changed over the centuries. The most exact of these measurements is the length of their tropical year which exceeds the length of the 365-day *Haab'* by 1 day every 1508 years. Their charts recording the Venus cycles were extremely accurate, off only 1 hour in 400 years. The long count calendar tracks specific astronomical events back to CE 143; the sacred calendar round of 260 days was used to track eclipses and the cycles of Venus. The unfortunate loss of most of their codices denies scientists so much of their knowledge and experiences. The role that astronomy played in artistic representations, ceremonies, calendars, architectural constructions and urban planning was enormous.

Current research is consistently discovering new findings of how the Maya included astronomy in their daily lives. These studies show the enormous impact astronomy held over Maya culture throughout the 3000 years of Maya history. The advances made through new readings and deeper understanding of the Maya hieroglyphics depict a culture steeped in literature and poetry. The archaeological findings will continue to disclose more evidence of the length and depth of Maya

M. Dolan, *Decoding Astronomy in Maya Art and Architecture*, Springer Praxis Books, https://doi.org/10.1007/978-3-031-61092-9

developments and understanding in astronomy, calendrics and mathematics. Archaeologists recommend that the traditions and folklore of living Maya population be consulted and studied since ancestral teachings, customs and ceremonies may carry on traditions from the ancient past. Shamanic rituals are timeless and may provide a glimpse into the supernatural world that was so dominant in the ancient world of the Maya. But the Maya and their culture continue today, as over seven million people still speak one or more of 30 Mayan-descended languages. We are only beginning to understand the crucial role that astronomy played in the Maya world; many pieces of the puzzle are still missing.

Glossary

Ajaw "Lord, King", the primary title of Maya rulers, by the Middle and Late Classic periods its meaning had become extended to the general nobility.

Alignment An arrangement in which three or more objects are in a straight line, in astronomy one of the points is a rising or setting celestial body on the horizon.

Altar Usually a large, flat stone or monument suitable for accepting offerings or burning sacrifices to the gods.

Anti-zenith The anti-zenith, or nadir, occurs on the two dates when the sun is directly beneath the observer on the opposite side of orbiting the Earth (such as the Moon or an artificial satellite) is furthest from the Earth, the opposite of perigee.

Apotheosis Elevation to divine status often depicted in works of art as the deity among the stars. Elevation to a preeminent or transcendent position.

Archaeoastronomy The scientific study of the beliefs and practices concerning astronomy that existed in ancient and prehistoric civilization and the way people's knowledge of the skies were used, also called cultural astronomy.

Ascending and Descending Nodes The two points where an orbit intersects a plane of reference to which it is inclined. A non-inclined orbit has no nodes.

Asterism Groupings of stars, fewer than constellation, which may not correspond to the International Astronomical Union constellations. The Big Dipper, for example, is one of the most famous asterisms, and is part of the large constellation Ursa Major.

Axis Mundi World axis: a line or stem through the earth's center connecting its surface to the underworld and the heavens and around which the universe revolves.

Azimuth A direction in the horizontal plane, usually specified in degrees or as compass bearing. Star azimuths are commonly measured from true north, increasing in a clockwise direction: east is 90, south is 180, west 270 and north.

Bakabs, or Pauahtun Bacab is the generic Yucatec Maya name for the four Prehispanic aged deities of the interior of the earth and its water deposits.

M. Dolan, *Decoding Astronomy in Maya Art and Architecture*, Springer Praxis Books, https://doi.org/10.1007/978-3-031-61092-9

Bas-relief or Low Relief Sculpture in which the projection from the surrounding surface is slight; the raised artwork is low such as on coins.

Bicephalic Monster A two-headed monster which appears in Maya artworks; also known as the Celestial Monster, the Witz and the Cosmic Monster. The supernatural creature also has a crocodilian or a sky band body. The front head of the Monster usually has a Venus symbol or crossed bands for eyes. The back head is skeletal and upside down.

Bilateral Symmetry The symmetry in which similar parts are arranged on opposite sides of a median axis so that only one plane can divide the individual into essentially identical halves.

Bloodletting The Maya let blood regularly for ritual purposes throughout Mesoamerica. Bloodletting was the most important sacrificial offering; the event was made to the gods regularly by kings, nobles, priests and queens. Military captives were taken alive and their blood was offered to the gods during public ceremonies at temples and shrines.

Calendar Round A combination of the Tzolk'in and Haab' calendars. Any specific combination between these two calendars repeats after 18,980 days, the 52-year 'Mesoamerican century'.

Celestial Equator The great circle in the heavens where the Earth's equator, if its plane extended outward into space, would intersect the celestial sphere.

Celestial Horizon The circle on the celestial sphere which is 90° from the zenith. The celestial horizon is the approximate boundary between the Earth and the sky.

Celestial Pole The celestial poles are imaginary lines that trace the Earth's rotational axis in space.

Celestial Sphere An imaginary sphere surrounding the Earth. The celestial bodies appear to carry out their motions on the celestial sphere.

Chaac Masks Hook-nosed Chaac masks, dedicated to the rain god, commonly appear on the corners of Maya buildings, often one atop another.

Chaac The Maya god or rain and lightning is one of the most continually worshiped gods throughout Mesoamerica and often holds a lightning rod.

Chacmool A recumbent male statue with disk or plate on stomach possibly for sacrificial offerings during rituals.

Chultun Underground chamber carved out of bedrock by the ancient Maya. Presumably used for storage and often subsequently used for burials.

Circumpolar Stars Those stars that never set or go below the horizon for observers at specific latitudes. If we face the north celestial pole, the stars turn counterclockwise around it.

Classic Period The period from CE 200 to 900 is considered the height of the Maya civilization, characterized by great accomplishments such as: distinctive writing, calendric systems, polychrome ceramics, corbelled vault architecture, major public architecture with pyramids, palaces and ball courts. Maya Classic cultures included scholars who maintained astronomical observations, established sophisticated mathematics, and passed down written texts.

Codex A screen-fold book made from strips of pounded bark paper or deer hide, usually thinly painted on both sides with white lime gesso; only four survive today.

Conjunction An alignment of two celestial bodies and the observer, so that the two appear to be together in the sky, appear to be at the same ecliptic longitude.

Corbelled Vault Architecture Stone slabs placed in layers with each layer slightly projecting over the previous one up to cover the room.

Cosmic Monster or Celestial Monster A dragon-type monster with a crocodile head and deer ears. The body has legs with deer hooves and water scrolls at the joints.

Cosmology The scientific study of the origin, evolution and eventual fate of the Universe.

Cultural Astronomy The various ways in which cultures perceive celestial objections and cyclical motions and integrate them into their view of the world.

Declination The angular distance of a celestial body north or south of the celestial equator. The declination is analogous to the terrestrial coordinate system. It is joined with right ascension.

Early Classic The period between approximately CE 200 and 400.

Eclipses The alignment of three celestial bodies so that one body passes between the other two and its shadow darkens the other. They were considered by the Maya as 'biting the Sun' or 'biting the Moon' Solar eclipses were considered much more dangerous than lunar eclipses.

Ecliptic The apparent path of the Sun through the stars that it passes through the course of a year. The constellations through which the ecliptic passes define the traditional zodiac. The Maya saw the seasonal movement of constellations along the ecliptic as the creation story.

E-Group Complex Unique architectural complexes found among a number of ancient Maya settlements. They are central components to the settlement organization of Maya sites and, like many other civic and ceremonial buildings, could have served for astronomical observations.

Emblem Glyph The royal title of the ancient Maya, reading 'Divine X Lord', where X is the name of the local polity.

Equinox Time of year when day and night are of equal length, when the Sun is on or nearly on the celestial equator, around March 20 and September 23.

Ethnoastronomy The study of the knowledge, interpretations and practices of contemporary cultures regarding celestial objects of phenomena.

Foliated Cross A maize tree representing the central axis of the world, its branches are ears of corn with a living human head, referring to the myth of creation where human flesh was created from maize.

Geomancy A method of divination that interprets markings on the ground or the patterns formed by tossed handfuls of soil, rocks, or sand.

Geomorphology The study of the physical features of the surface of the earth and their relation to its geological structures.

GIS Systems A geographic information system creates, manages, analyzes and maps all types of data. GIS connects data to a map, integrating location data with all types of descriptive information.

Gnomon An upright stone or wooden post from whose shadow time is estimated.

God L One of the ancient gods who appears in scenes of Xibalba; he is frail and bent with age.

Ground-Penetrating Radar A method that looks at the deflection of radar pulses to find features below ground level.

Haab' The Maya 'vague solar year' of exactly 365 days, composed of 18 'months' of 20 days each.

Heliacal Rising/Setting The rising of a star or group of stars just before sunrise or its setting just after sunset. This happens on the first day that the star is visible in the dawn light after its conjunction with the sun.

Hieroglyphic Stairway A staircase on a pyramid or structure that is decorated with Maya glyphs that tell the history of the ruler and the city.

Hierophany A meeting of a person with a deity after death or a symbolic alignment with celestial light, a manifestation of the divine.

High Relief Carving Sculpture in which the forms project at least half or more from the background and may be in parts completely disengaged from the ground; near to sculpture in the round.

Iconography Symbols that identify the person or element depicted in works of art.

Inferior Conjunction A meeting of an inferior planet with the Sun in which the planet is aligned between the Earth and the Sun.

K'atun A period of 7200 days, or 20 *tuuns*.

***K'atun* Ending** The end of a k'atun when all units of a Long Count below k'atun were at 0. *k'atun* endings are named after the *Tzolk'in* day on which the k'atun ended.

Kinich Ahua During the Classic and Postclassic periods, the Maya Sun God was called Kinich Ahua, which means sun-faced or sun-eyed lord.

Late Classic The time period from CE 700 to 900.

LiDAR Often deployed from low-flying aircraft, LiDAR instruments fire rapid pulses of laser light at a surface, and then measure the amount of time it takes for each pulse to bounce back. The pulses are measured, and they are then used to created 3D images of areas that are hidden under dense vegetation.

Limb The edge of the apparent disc of the Sun or the Moon.

Lunar Standstill The varying range between the northern and southern limits of the Moon's declination over the course of one-half a sidereal month (about 2 weeks) or 13.66 days. One major (maximum) or one minor (minor) lunar standstill occurs every 18.6 years due to the precessional cycle of the lunar nodes at that rate.

Lunistices The most northerly and southerly positions of the moon in its monthly cycle. They are the lunar equivalent of solstices.

Magnitude Star brightness is measured in magnitudes 1–6, as classified by Hipparchus in the second century BCE and by Ptolemy in the second century CE. The smaller the magnitude the brighter the star. With low levels of light pollution, humans can see stars with the naked eye to a limit of about 6 magnitudes. At 5.5 magnitudes 2860 stars are visible.

Maize God The Foliated Maize god appears with a stylized maize ear sprouting from the top of his head and continues in this form until the Late Postclassic period. In the codices his is referred to as God E.

Meridian The plane containing the northernmost and the southernmost points of the horizon, the north celestial pole, and the zenith overhead: or that part of a great circle on the celestial sphere through the last three points.

Metonic Cycle The cycle of 19 years or 235 months that brings the Moon back in step with the Sun, so that new and full moons repeat on the same dates.

Middle Classic The time period from CE 400 to 700.

Middle World In Maya cosmology the Middle World is the Earth, the world of men and animals.

Milky Way The Maya considered the Milky Way as a white road and the Road to Xibalba.

Nadir Passage The nadir is the direction pointing directly below the zenith Sun; the Sun's passage over that position cannot be seen. The Maya carefully tracked the Sun's zenith passage but not its nadir.

Neolithic Era The period in which agriculture was first practiced, pottery was made, and fine tools mainly in stone.

Node A point where two great circles on the celestial sphere intersect. The two circles of much importance are the apparent paths of the Sun and Moon through the stars and where the lunar nodes cross. The lunar nodes are not fixed but move slowly round the sky in a sense counter to the Moon's monthly motion, completing the circuit in about 18.6 years.

Palenque Triad Three gods most fully described in the inscriptions and imagery at Palenque where they are asserted to be ancestors of that kingdom's dynasty.

Paleolithic Era The period of human existence when stone tools first played an important technological part in human evolution, approximately 700,000 years ago to about 10,000 years ago, Lower to about 200,000 years ago, Middle to about 40,000 years ago, Upper until about 10,000 years ago.

Perigee The point at which a body orbiting the Earth (such as the Moon or an artificial satellite) is closest to the Earth, opposite of apogee.

Period Ending A station in the Maya Long Count calendar that is a multiple or a fraction of a *k'atun*. Half, quarter, and even occasionally eighth *k'atun* markers are known to have been commemorated; 3 and 13 *tuuns* were also commemorated as Period Endings.

Planisphere A circular map of the night sky showing the stars and constellations. Planispheres are made for observing the night sky at different latitudes.

Precession of the Equinoxes The slow, periodic conical motion of the rotational axis of the Earth or another rotating body. The effect causes the drift of the measured ecliptic longitudes of the stars to increase by about 1.5° per century. This changes the rising and setting of stars on the horizon over long periods of time.

Proper Motion The apparent change in position of a star usually expressed in seconds of arc per year that results from the projection on the celestial sphere of its motion with respect to the solar system.

Right Ascension In the equatorial coordinate system, the celestial equivalent of terrestrial longitude. It divides the celestial equator into 24 hours, each of 60 minutes.

Serpent Bar A bar carried in the hands of the ruler, usually held against their chests. Its original function in the Late Preclassic was to represent the sky.

Sidereal Month The time needed for the Moon to return to the same place against the background of the stars, 27.321661 days.

Sky-Band The sky was often depicted as a band marked with diagonal land vertical elements. During the Classic period the sky band is usually segmented; within the spaces are glyphs stating Sun, Moon, stars, darkness and other celestial phenomena.

Solstice The times of the year at which the Sun is at its maximum northerly or southerly declination. The days are then at their longest and shortest respectively, usually June 21 and December 21.

Stela/Stelae A standing stone or slab of modest size with one face only sculpted in low or bas relief. They celebrated the reigns or rites of passage of each city's ruling elite.

Superior Conjunction A conjunction of a planet with the sun in which the sun is aligned between the earth and the planet.

Synodic Month The average period of the Moon's orbit with respect to the line joining the Sun and Earth: 29 days 12 hours 44 minutes and 2.9 seconds. This is the period of the lunar phases, because the Moon's appearance depends on the position of the Moon with respect to the Sun as seen from Earth.

Tropic of Cancer The parallel of latitude that is approximately 23 ½ degrees north of the equator and that is the northernmost latitude reached by the overhead Sun.

Tropic of Capricorn The parallel of latitude that is approximately 23 ½ degrees south of the equator and that is the southernmost latitude reached by the overhead Sun.

Tuun The Long Count 'year' of 360 days. Formed of 18 *winals.*

Tzolk'in The 260 day ritual calendar of ancient Mesoamerica, composed of 13 numbers associated with 20 day names.

Tzompantli A skull rack or wooden scaffold used to hang human skulls pierced horizontally by cross beams; some are made of stone as well. The most ominous example was constructed at Chichen Itza.

Underworld or Xibalba A fearful, dreaded place that was not just for sinners but for those who escaped violent death, rich, poor, good or evil. Only those who experienced a violent death went to one of the heavens. The Maya believed that there were 9 levels of the Underworld and 13 levels of the Upperworld.

Vision Serpent An image of a two-headed snake that is conjured during bloodletting ceremonies; it rises from the smoke of the burning offering. Gods, ancestors emerge from its jaws. It is thought to be a metaphor for birth and rebirth of divinities or humans.

Winal The Maya 'month' of 20 single days.

***Witz* or Cosmic Monster** See Bicephalic Monster.

World Tree/Cosmic Tree The World Tree or axis mundi connects the three parts of the world; the Underworld with its roots and the Upperworld with its branches; the Earth with the sky. It is marked with God C to denote it is holy or a divine thing. The World Tree is the path of communication between natural and supernatural worlds.

Xibalba See Underworld.

Zenith Passage The point in the sky that is directly overhead from the perspective of a particular location on the Earth and the sun's transit of that point. This only occurs in the tropical zones and is often signified in Maya architectural structures, important for calendric calculations.

Zodiac The zone of the sky that extends approximately 8° north or south (in celestial latitude) of the ecliptic, the apparent path of the Sun across the celestial sphere over the course of the year as observed from Earth. The Sun, Moon, and visible planets appear to travel across a band of 12 zodiacal constellation within this belt as the Earth orbits the Sun.

Index

M. Dolan, *Decoding Astronomy in Maya Art and Architecture*, Springer Praxis Books, https://doi.org/10.1007/978-3-031-61092-9

X

Y

Z

MIX
Papier aus verantwortungsvollen Quellen
Paper from responsible sources
FSC® C105338

If you have any concerns about our products,
you can contact us on
ProductSafety@springernature.com

In case Publisher is established outside the EU,
the EU authorized representative is:
Springer Nature Customer Service Center GmbH
Europaplatz 3, 69115 Heidelberg, Germany

Printed by Libri Plureos GmbH
in Hamburg, Germany